The Traveling Anatomist

The Traveling Anatomist

NICOLAUS STENO AND
THE INTERSECTION OF DISCIPLINES
IN EARLY MODERN SCIENCE

Nuno Castel-Branco

The University of Chicago Press   CHICAGO AND LONDON

The University of Chicago Press, Chicago 60637
The University of Chicago Press, Ltd., London
© 2025 by The University of Chicago
All rights reserved. No part of this book may be used or reproduced in any manner whatsoever without written permission, except in the case of brief quotations in critical articles and reviews. For more information, contact the University of Chicago Press, 1427 E. 60th St., Chicago, IL 60637.
Published 2025

34 33 32 31 30 29 28 27 26 25   1 2 3 4 5

ISBN-13: 978-0-226-84227-1 (cloth)
ISBN-13: 978-0-226-84229-5 (paper)
ISBN-13: 978-0-226-84228-8 (e-book)
DOI: https://doi.org/10.7208/chicago/9780226842288.001.0001

Published with support from the
Lila Wallace—Reader's Digest Publications Subsidy at Villa I Tatti.

Library of Congress Cataloging-in-Publication Data

Names: Castel-Branco, Nuno (Historian of science), author.
Title: The traveling anatomist : Nicolaus Steno and the intersection of disciplines in early modern science / Nuno Castel-Branco.
Other titles: Nicolaus Steno and the intersection of disciplines in early modern science
Description: Chicago ; London : The University of Chicago, 2025. | Includes bibliographical references and index.
Identifiers: LCCN 2025003180 | ISBN 9780226842271 (cloth) | ISBN 9780226842295 (paperback) | ISBN 9780226842288 (e-book)
Subjects: LCSH: Steno, Nicolaus, 1638–1686. | Steno, Nicolaus, 1638–1686—Travel—Europe. | Anatomists—Denmark—Biography. | Geologists—Denmark—Biography. | Scientists—Denmark—Biography. | Science—History—17th century. | Anatomy—History—17th century. | Geology—History—17th century. | Europe—Intellectual life—17th century.
Classification: LCC QM16.874 C37 2025 | DDC 509.2 [B]—dc23/eng/20250319
LC record available at https://lccn.loc.gov/2025003180

Authorized Representative for EU General Product Safety Regulation (GPSR) queries: **Easy Access System Europe**—Mustamäe tee 50, 10621 Tallinn, Estonia, gpsr.requests@easproject.com
Any other queries: https://press.uchicago.edu/press/contact.html

For Maria Leonor, my fellow traveling companion

Contents

List of Figures and Tables * ix

INTRODUCTION
The Various Travels of Nicolaus Steno * 1
- The Worlds of Traveling Scholars
- The Intersection of Disciplines
- Mathematizing the Body
- Across the Roads of Europe

ONE
The Uses of Chaos; or, What Did Nicolaus Steno Know? * 19
- The Cathedral School, Mathematics, and Goldsmithing
- Travels, Note-Taking, and the Making of "Chaos"
- A Pious Search for Certainty and Focus
- Post-Harveian Anatomy at the Center
- Learning Interdisciplinary Methods

TWO
The Making of a Scholarly Anatomist * 63
- Authorship and the Blasius Controversy
- The *Historia* of the Glands
- "The Pen Following the Knife": The Making of a Scholar

THREE
Dissecting with Numbers, Machines, and Mixtures * 94
- The Mathematics and Mechanics of Glands and Muscles
- The (Non-Cartesian) Epistemology of Nicolaus Steno
- Observations in Practice: The Lymphatics Controversy
- Galen, God, and Gassendi
- The Mechanical and Chymical Worlds of the Netherlands

FOUR
In the Cradle of the Académie des Sciences * 134
- Monsieur Stenon, a *Sçavant Danois* in Paris
- The Académie de Physique de Caen
- Breathing Physico-Mathematics: The Friendship Between Steno and Swammerdam
- The Benefits of Being Anti-Cartesian

FIVE
Anatomy and Mathematics at the Medici Court * 166
- Travels Across the Networks of Florence
- The First Italian Month: Dissecting and Searching for Certainty
- Mathematics Mattered in Rome
- Intersecting Muscles with Mathematics in Florence
- Friendships, Patronage, and the Accademia del Cimento

SIX
Thinking the Earth with the Body * 207
- The Delayed Manuscript
- Comparisons in Anatomy
- Comparisons in Earth History
- The Search for Certainty in Anatomy
- Models in Anatomy
- The Search for Certainty in Earth History
- Models in Earth History

SEVEN
Anatomy of a Conversion * 241
- Religious Conversions in Seventeenth-Century Europe
- Friendships, from Collaboration to Conversion
- The Search for Certainty in Religion
- Steno and the Devout Women

EPILOGUE * 269

Acknowledgments * 283
Bibliography * 287
Index * 315

Figures and Tables

Figures

Figure I.1. Map of Nicolaus Steno's travels from 1659 to 1667 with arrival dates (except for the 1659 departure). Marks represent stops longer than a month and travels to Rome. Designed by Luís Ribeiro. 5

Figure I.2. Portrait of Nicolaus Steno, c. 1666–1675. Courtesy of the Gallerie degli Uffizi, Florence. 17

Figure 1.1. Steno's drawing of Ole Borch's microscope, in the Biblioteca Nazionale Centrale di Firenze (BNCF), Gal. 291, fol. 59v. Reproduced with the permission of the Ministero della Cultura/Biblioteca Nazionale Centrale di Firenze. 29

Figure 1.2. A personal note by Nicolaus Steno enclosed in parentheses "(N ... N)" and with a "D" marked in the margins, in BNCF, Gal. 291, fol. 67r. Reproduced with the permission of the Ministero della Cultura/Biblioteca Nazionale Centrale di Firenze. 36

Figure 1.3. Notes by Steno on Matthias Untzer, *Anatomia mercurii spagirica* (Halle, 1620), with no marginalia. BNCF, Gal. 291, fol. 49r. Reproduced with the permission of the Ministero della Cultura/Biblioteca Nazionale Centrale di Firenze. 52

Figure 1.4. Notes by Steno on Jean Pecquet, *Experimenta nova anatomica* (Paris, 1654), with headings and other notes in the margins. BNCF, Gal. 291, fol. 68r. Reproduced with the permission of the Ministero della Cultura/Biblioteca Nazionale Centrale di Firenze. 53

Figure 1.5. Roberval's bladder experiment. Pecquet, *Experimenta nova anatomica* (Paris, 1651), 51. Reproduced with the permission of Balliol College, University of Oxford (ref. 905 f 7 (5)). 59

Figure 2.1. Title page of Nicolaus Steno, *Disputatio physica de thermis* (Amsterdam, 1660). Reproduced with the permission of the Library Company of Philadelphia. 71

Figure 2.2. Blasius reading with students in his Amsterdam rooms; from Blasius, *Medicina generalis* (Amsterdam, 1661), frontispiece. The Bodleian Libraries, University of Oxford, 8° N 41 Med. 73

Figure 2.3. A probe inside the pulmonary artery of the heart in Cæcilius Folius, *Sanguinis a dextro in sinistrum cordis ventriculum ... reperta via* (Venice, 1639); shown here as reproduced in Thomas Bartholin, *Anatomia, ex Caspari Bartholini ... reformata* (The Hague, 1663). Courtesy of the Institute of the History of Medicine, Johns Hopkins University. 79

Figure 2.4. The palate viewed from below: (a) holes in the palate, (b) tonsils; in Steno, *Observationes anatomicæ* (Leiden, 1662). Reproduced with the permission of The President and Fellows of Magdalen College, Oxford. 80

Figure 2.5. Parotid glands in the head of a calf. The largest, on the right, is the conglomerate gland (a), with the salivary duct exiting downwards (c); the bean shape on the left is the conglobate gland (b). From Steno, *Observationes anatomicæ* (Leiden, 1662), 21. Reproduced with the permission of The President and Fellows of Magdalen College, Oxford. 82

Figure 2.6. The broadsheet *Apologiæ prodromus* (Leiden, 1663). From the British Library Archive, General Collection 548.e.32.(2.). 88

Figure 2.7. Paragraph with many quotations, where "D. B." refers to Hoboken's [*Novus*] *ductus blasianus* (1662) and "Med. G." to Blasius's *Medicina generalis* (1661), in Steno, *Apologiæ prodromus* (Leiden, 1663). From the British Library Archive, General Collection 548.e.32.(2.). 89

Figure 2.8. Table comparing Steno's and Blasius's claims in Steno, *Apologiæ prodromus* (Leiden, 1663). From the British Library Archive, General Collection 548.e.32.(2.). 90

Figure 3.1. A machine designed to pull a boat over a flat surface with the help of a third element (grease or rolling objects) laid underneath; from Simon Stevin, *Les oeuvres mathématiques* (Leiden, 1634), 481. Reproduced with the permission of Special Collections, The Sheridan Libraries, Johns Hopkins University. 99

Figure 3.2. Illustrations of the eye glands. I. Left eye of a calf with the superior eye gland (A); II. Interior surface of the palpebra with the superior eye gland (b) and ducts (c) ending in (d); III. Lacrimal gland (A) and corresponding cartilage (B). From Nicolaus Steno, *Observationes anatomicæ* (Leiden, 1680), 100. Reproduced with the permission of The President and Fellows of Magdalen College, Oxford. 101

Figure 3.3. Frontispiece of Steno, *De musculis et glandulis* (Copenhagen, 1664), with a geometrical representation of muscle fibers in the upper left corner. Reproduced with the permission of the Wellcome Collection. 104

Figure 3.4. Seventeenth-century pile drivers. Steno's analogy uses the machine on the left (4), operated by men pulling ropes. The pulley is not visible. See Cornelis Meijer, *Traité des moyens de rendre les rivières navigables* (Amsterdam, 1696), 18. Reproduced with the permission of The Warden and Fellows of All Souls College, Oxford. 107

Figure 4.1. Frontispiece of Jan Swammerdam, *Tractatus physico-anatomico-medicus de respiratione* (Leiden, 1667). Courtesy of Balliol College, University of Oxford (ref. 905 f 5 (2)). 149

Figure 4.2. Diagram of the human heart. While Harvey associated the motion of the lungs with the flow of blood into the pulmonary vein, Steno associated it with the superior and inferior vena cava. © Wikimedia Commons. 151

Figure 4.3. Gravity-dependent distribution of blood in a human and a dog. Designed by Luís Ribeiro. Based on Furst, *The Heart and Circulation*, 2nd ed. (Cham: Springer, 2020), 324. 153

Figure 5.1. Steno's lightly drawn sketch with two muscles inside a lobster claw in Florence, BNCF, Gal. 291, fol. 98r. Reproduced with the permission of the Ministero della Cultura/Biblioteca Nazionale Centrale di Firenze. 179

Figure 5.2. Lobster claw muscles, in Steno, *Elementorum myologiæ specimen* (Florence, 1667), tabula III. Reproduced with the permission of the Ministero della Cultura/ Biblioteca Nazionale Centrale di Firenze. 180

Figure 5.3. Geometrical representation of the lobster claw muscles in Steno, *Elementorum myologiæ specimen* (Florence, 1667), 41. Reproduced with the permission of the Ministero della Cultura/Biblioteca Nazionale Centrale di Firenze. 181

Figure 5.4. An ordo of muscle fibers in Graindorge's letter to Huet. Copenhagen, Royal Library, NKS 4660 kvart, 47. 191

Figure 5.5. A muscle as an oblique parallelepiped, in Steno, *Elementorum myologiæ specimen*, 3. Courtesy of the Institute of the History of Medicine, Johns Hopkins University. 192

Figure 5.6. "The delicate structure of the deltoid [shoulder] muscle in which twelve single muscles are counted," in Steno, *Elementorum myologiæ specimen*, tabula III, fig. I. Quote from ibid., 37. Courtesy of the Institute of the History of Medicine, Johns Hopkins University. 193

Figure 5.7. "Figure 3 shows the rectilinear, unequally equal motor fibers . . . , figure 4 shows the same motor fibers equally inflected," in Steno, *Elementorum myologiæ specimen*, tabula I, figs. III and IV. Quote from ibid., 35. Courtesy of the Institute of the History of Medicine, Johns Hopkins University. 194

Figure 5.8. Title page of the manuscript of Steno's *Elementorum myologiæ specimen* in Copenhagen, Royal Library of Denmark, NKS 4019 kvart, fol. 1r. 197

Figure 6.1. Drawing of a fish's ovaries and uterus. Steno, *Elementorum myologiæ specimen* (Florence, 1667), tabula VII. Courtesy of the Institute of the History of Medicine, Johns Hopkins University. 215

Figure 6.2. Steno, *Elementorum myologiæ specimen*, tabula IV. Courtesy of the Institute of the History of Medicine, Johns Hopkins University. 217

Figure 6.3. Steno, *Elementorum myologiæ specimen*, tabula VI. Courtesy of the Institute of the History of Medicine, Johns Hopkins University. 220

Figure 6.4. Cross-section of the brain, in Steno, *Discours sur l'anatomie du cerveau* (Paris, 1669). From the British Library Archive, General Collection 1477.bbb.32. 221

Figure 6.5. Relaxed (*left*) and contracted (*right*) muscles as parallelograms. Steno, *Elementorum myologiæ specimen*, 25. Courtesy of the Institute of the History of Medicine, Johns Hopkins University. 224

Figure 6.6. References to lemmas in the margins of the text on muscles (*left*) and to conjectures in the margins of the text on fossils (*right*). Steno, *Elementorum myologiæ specimen*, 20, 93. Courtesy of The Linda Hall Library of Science, Engineering & Technology. 235

Figure 6.7. Diagram of the formation of Tuscan mountains. To be read from 25 to 20, where 25 is the original Earth and 20 is the Earth as seen by Steno. Steno, *De solido* (Florence, 1669). Reproduced with the permission of the Ministero della Cultura/Biblioteca Nazionale Centrale di Firenze. 236

Figure 6.8. Descartes's diagram of the formation of the Earth. Descartes, *Principia philosophiæ* [*sic*]. Reproduced with the permission of the Wellcome Collection. 237

Figure 6.9. Cross-section by Nicolaus Steno of a cave in northern Italy. Florence, BNCF, Gal. 286, fol. 61r. Reproduced with the permission of the Ministero della Cultura/Biblioteca Nazionale Centrale di Firenze. 238

Figure 8.1. Drawing in sanguine or red chalk of a large-headed calf. Florence, ASF, MdP, f. 1020, fol. 665r. Courtesy of the Ministero della Cultura/Archivio di Stato di Firenze. 279

Table

Table 3.1. Proportion of the weights of maxillary and parotid glands as measured by Thomas Wharton and Nicolaus Steno 97

INTRODUCTION

The Various Travels of Nicolaus Steno

This is a book about a traveling anatomist who wandered as much with his feet as he did with his mind and heart. Owing to these varieties of travels, his anatomies opened more than just bodies. His intellectual scalpel also cut through the interior of the Earth, the boundaries between mathematics and nature, the rising field of mechanical philosophies, and the divisions of early modern Christianity. Today he is known as the founder of modern geology and as an exemplar of how to navigate the worlds of science and religion.[1] In his time, however, he was known as a very skillful anatomist with new ideas about science.[2]

It all started in 1638—an important year in the history of Europe. That year, Galileo Galilei (1564–1642) published his most significant book in the history of physics; King Louis XIV of France (1638–1715), the Sun King, was born; and France and Sweden reaffirmed a military alliance that led to the end of the devastating Thirty Years' War (1618–1648).[3] These events are important because they represent much of what Europe came

---

1. Alan Cutler, *The Seashell on the Mountaintop: A Story of Science, Sainthood, and the Humble Genius Who Discovered a New History of the Earth* (New York: Dutton, 2003), esp. 187; and Hans Kermit, *Niels Stensen: The Scientist Who Was Beatified* (Leominster: Gracewing, 2003).

2. By "science," I mean the study of nature in its various early modern forms, including such disciplines as natural philosophy, natural history, medicine, mathematics, and "chymistry." For a similar definition, see Katharine Park and Lorraine Daston, "The Age of the New," in Park and Daston, *Early Modern Science*, 1–17, esp. 2–6. On the use of "chymistry," see William Newman and Lawrence Principe, "Alchemy vs. Chymistry: The Etymological Origins of a Historiographic Mistake," *Early Science and Medicine* 3, no. 1 (1998): 32–65.

3. Treaty of Hamburg, March 1638. See Michael Roberts, "The Swedish Dilemma," in *The Thirty Years' War*, ed. Geoffrey Parker, 2nd ed. (New York: Routledge, 1997), 140–145, esp. 143.

to be in the second half of the seventeenth century. Galileo's last book, *Discorsi e dimostrazioni matematiche intorno a due nuove scienze* (*Discourses and Mathematical Demonstrations Concerning Two New Sciences*; Leiden, 1638) became one of the foundational texts of the so-called new sciences that proliferated in the hands of mathematicians and natural philosophers. The court of Louis XIV supported multiple artistic, cultural, and scientific projects for years to come, including the Académie Royale des Sciences, one of the first scientific academies of Europe. And the Peace of Westphalia, which ended the Thirty Years' War in 1648, emphasized Europe's political and religious divide for the following generations. Also in 1638, away from the agitation of central Europe, in the city of Copenhagen, Nicolaus Steno (1638–1686) was born. As he grew to maturity, he too played a key role in the transformations of early modern science, the culture of absolutist courts, and the ongoing political and religious dilemmas of Europe.

But who was Nicolaus Steno? The son of a well-positioned goldsmith in Denmark, Steno studied medicine for three years at the University of Copenhagen before embarking on a long journey across Europe.[4] After attending classes at the University of Leiden, from which he received his medical degree in 1664, he traveled to Paris and joined the intellectual circles that soon gave birth to France's Académie Royale des Sciences. Two years later, in Florence, Steno landed in the good graces of Grand Duke Ferdinand II de' Medici (1610–1670)—Galileo's last patron—in whose court he started working as a scientist. Also in Florence he abandoned his Lutheran faith in exchange for Catholicism. One might question his sincerity, had he not become a priest eight years later, in 1675, and then a bishop in 1677. As a bishop, Steno returned to northern Europe, where, among other things, he befriended Gottfried Wilhelm Leibniz (1646–1716) in Hannover. But to Leibniz's disappointment, Steno's focus was no longer the study of the natural world. Instead, Bishop Steno was busy tending to minority Catholic communities in Germany. After more travels, he died in 1686 in Schwerin, sixty miles east of Hamburg. His attachment to Florence, however, did not die. At the personal request of

---

4. I use the Latinized spelling "Nicolaus Steno" of the original name "Niels Ste[e]nsen," which is the most common form in English. The best biographies of Steno are still Roberto Angeli, *Niels Stensen: Il beato Niccolò Stenone, uno scienzato innamorato del vangelo e dell'Italia* (Milan: San Paolo, 1996), originally published as *Niels Stensen: Anatomico, fondatore della geologia, servo di Dio* (Florence: Libreria Editrice Fiorentina, 1968); and Gustav Scherz, *Niels Stensen: Eine Biographie*, 2 vols. (Leipzig: St. Benno-Verlag, 1987). The first volume of Scherz's biography was translated into English in Troels Kardel and Paul Maquet, eds., *Nicolaus Steno: Biography and Original Papers of a Seventeenth-Century Scientist*, ed., 2nd ed. (Berlin: Springer, 2018), 1–410.

Grand Duke Cosimo III de' Medici (1642–1723), Steno's body was shipped back to Florence, where it has remained ever since.[5] Today pilgrims still flock to pray before Steno's tomb in the Basilica di San Lorenzo thanks to the reputation he earned as a man of great devotion and virtue.

Steno's most famous contribution to the history of science is his theory concerning the origin of fossils. He argued that by looking at the Earth's strata it is possible to learn its history, and his ideas are still taught in geology courses as Steno's laws or principles of stratigraphy.[6] Moreover, owing to the subsequent sanctity of his life, he has also been described as a "great scientist who recognized God as his supreme Lord," in the words of Pope John Paul II at Steno's beatification.[7] Yet, despite Steno's excellence in these two areas, he is surprisingly absent from most histories of early modern Europe.[8] His multiple works in anatomy, geology, and theology have also favored fragmented accounts of his life because of modern disciplinary boundaries. However, I argue that Steno's different research projects should be understood in dialogue with one another. Only then can historians fully appreciate his foundational contributions to the history of science and early modern Europe.

This book introduces Nicolaus Steno on his own terms and on the terms of those who knew him. It reconstructs Steno's context by combining recent research on early modern science with a close reading of his publications, his private correspondence, and the writings of those around him, some of which are examined here for the first time. This study shows that, to early modern eyes, Steno was neither a geologist nor a model of how to integrate science and religion. Instead, he was a promising anatomist, remarkably good at dissections, who successfully

---

5. Cosimo III de' Medici to Theodor Kerckring, 7 January 1687, ASF, MdP, f. 4495, fols. 820r–821v, esp. 820v.

6. Keith Thomson, *Fossils: A Very Short Introduction* (Oxford and New York: Oxford University Press, 2005), 39.

7. John Paul II, "Homilia in Vaticana basilica habita ob decretos Dei Servo Nicolao Stensen," 23 October 1988, in *Acta Apostolicae Sedis* (Vatican: Libreria Editrice Vaticana, 1989), 81:290–296, esp. 292.

8. Exceptions are books on the history of geology and the interactions between science and religion, such as Cutler, *The Seashell on the Mountaintop*, and Stefano Miniati, *Nicholas Steno's Challenge for Truth: Reconciling Science and Faith* (Milan: FrancoAngeli, 2009). See also John Henry, "Nicolas Steno (1638–1686): A Polymath Reassessed," *Isis* 111, no. 2 (2020): 365–367. Recently Steno also attracted attention in bioethics; see Frank Sobiech, *Ethos, Bioethics, and Sexual Ethics in Work and Reception of the Anatomist Niels Stensen (1638–1686): Circulation of Love* (Dordrecht: Springer, 2016). For Steno in histories of anatomy, see references below.

applied knowledge from other disciplines to his anatomical studies. As an anatomist, Steno contributed to a new mechanical understanding of glands; argued against ancients and moderns that the heart was a muscle; renamed the so-called female testicles as ovaries; and developed a mathematical model for muscle motion. His name is rarely mentioned by scientists today, but most of these discoveries are still used in modern anatomy and physiology—two disciplines that developed from early modern anatomy.[9]

As I have already suggested, Nicolaus Steno's popularity was not due to his knowledge of anatomy alone. His ability to expound on geometry, mechanics, hydrostatics, chymistry, and epistemology, and to focus all this knowledge on specific scientific problems, makes his career truly remarkable.[10] My main argument in this book is that Steno's life reveals the intellectual and cultural importance of interdisciplinary research in early modern intellectual life and the history of science. As the work of Steno and his colleagues shows, the cross-fertilization of knowledge from various disciplines fostered creativity, contributed to a new understanding of the laws of nature, and provided impetus to the scholars' social integration in various contexts. I read Steno's late publications on the history of the Earth and theology as mere outcomes of a breadth of knowledge that he had cultivated since his early years in Copenhagen. At the same time, I also avoid reading Steno's career from its end. A young anatomist, even one who read broadly, would not necessarily write on such a variety of topics without specific reasons. In this book I explain not only why Steno became interested in mathematics, anatomy, the Earth, and other topics, but also how he was able to write about them at all.[11] In this way I also address unresolved problems about the history of science and early modern culture. For example, why was interdisciplinarity important in science? How did different areas of knowledge intersect with and relate to each other in history? What shared assumptions, if any, underlie the various interests of early modern scholars? And, perhaps more important, how well do modern views of polymathy and disciplinary boundaries apply to the early modern period? Historians have called the seventeenth century "a golden age of polymaths," but very few have tried

---

9. On physiology as part of early modern anatomy, see Vivian Nutton, "*Physiologia* from Galen to Jacob Bording," in Horstmanshoff, King, and Zittel, *Blood, Sweat and Tears*, 27–40.

10. On the word "chymistry," see n. 2.

11. There was no discipline of geology in the early modern period; see chap. 6 for more details.

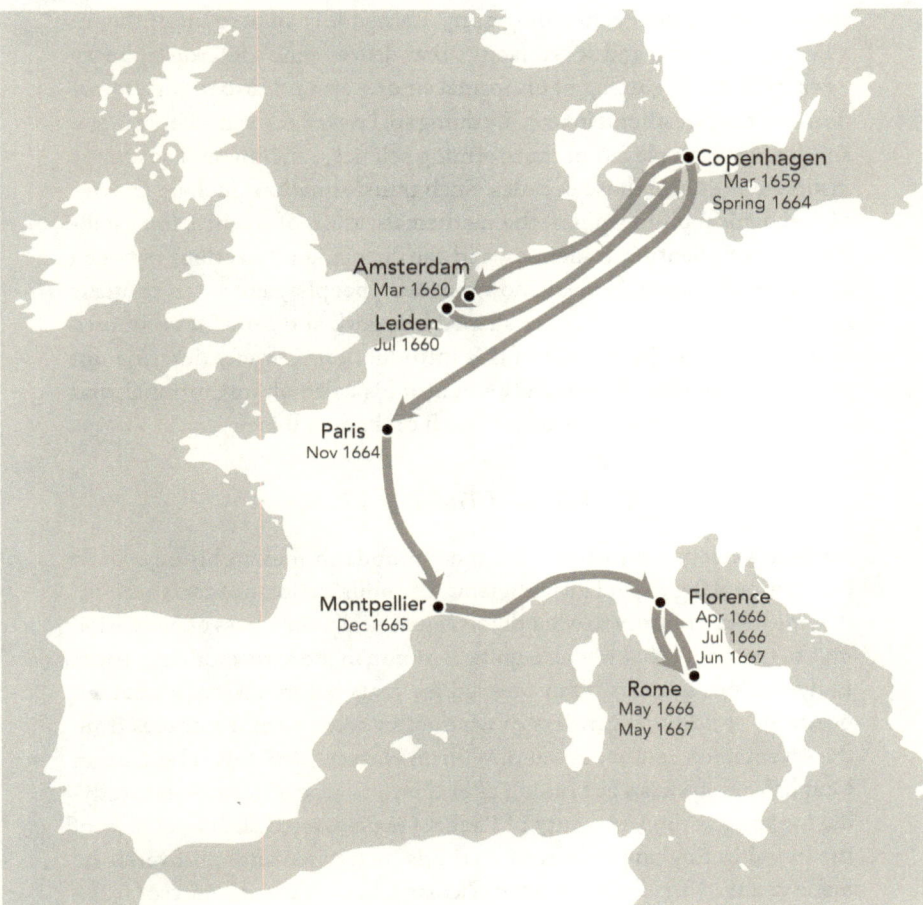

FIGURE I.1. Map of Nicolaus Steno's travels from 1659 to 1667 with arrival dates (except for the 1659 departure). Marks represent stops longer than a month and travels to Rome. Designed by Luís Ribeiro.

to problematize this category and analyze the intersection of disciplines in their historical context.[12]

In the following pages, I show that the intersection of disciplines in Nicolaus Steno's hands was especially successful due to the quantity and diversity of his travels (see fig. I.1). This book, however, is not a

---

12. Peter Burke, *The Polymath: A Cultural History from Leonardo da Vinci to Susan Sontag* (New Haven, CT: Yale University Press, 2020), esp. 4, 71, 255 for Steno. For polymathy in an academic context, see Neil Kenny, *The Uses of Curiosity in Early Modern France and Germany* (Oxford and New York: Oxford University Press, 2004), esp. 59–74.

comprehensive intellectual biography. Rather, it is an account of the intellectual, cultural, and social factors that drove Steno's interdisciplinary work from the beginning of his formal studies around 1650 to his ordination into the priesthood in 1675. By doing so, I unveil the central role Steno and his network played in transforming science, religion, and academic institutions during those decades. Such transformations include the rise of mechanical philosophies, the mathematization of science, the establishment of scientific academies, and the changing interactions between science and religion. I explain how the ideas, peoples, and social contexts that Steno met along the way shaped his work and how Steno in turn influenced them. In the rest of this introduction, I briefly describe not just Steno's geographical travels, but also his cultural, institutional, and intellectual itinerary. I thus turn to each of them in this order.

The Worlds of Traveling Scholars

Modern society seems to be the most mobile in human history. Daily international flights and instantaneous communication channels support that idea. Yet, contrary to what the technological uniqueness of our period may suggest, mobility was also quite common in the early modern period. Early modern scholars often traveled for years before taking permanent positions at home, regardless of where they were born. Examples from the seventeenth century abound. William Harvey (1578–1657) became an established physician in London after traveling across Europe and studying for years at the University of Padua. He discovered the circulation of the blood in England thanks to methods he learned and discoveries he made in Italy.[13] The French scholar Nicolas-Claude Fabri de Peiresc (1580–1637) traveled in Italy, Switzerland, England, and Flanders before settling down in his hometown of Aix-en-Provence. Peiresc possessed a widely known natural history collection and was at the center of a vast network of correspondents, both of which benefited from his travels.[14] Leibniz took a permanent position at the Brunswick court in Hannover only after visiting France, England, and the Netherlands for a few years. Meeting such scholars as Christiaan Huygens (1629–1695) and Henry Oldenburg (1619–1677) during this tour contributed to his ideas about mathematics

---

13. Roger French, *William Harvey's Natural Philosophy* (Cambridge: Cambridge University Press, 1994), 59–70.

14. Peter N. Miller, *Peiresc's Europe: Learning and Virtue in the Seventeenth Century* (New Haven, CT: Yale University Press, 2000), 7–9.

and philosophy.¹⁵ These academic travels were not unusual even among Steno's fellow Danes, who spoke of them as medical pilgrimages.¹⁶ A case in point is that of Ole Worm (1588–1654), owner of one of Europe's most famous cabinet of curiosities and whose early travels contributed to his collections and networks.¹⁷

Travel fundamentally shaped the lives and work of early modern scholars, perhaps even more than they do those of modern scientists. Therefore, it is not surprising that they also shaped Steno's career. But there is one striking difference between Steno and the travelers mentioned above. Most of the intellectual contributions of traveling scholars were made after their return home. Steno's scientific books, however, were written on the road, in between his stays in Copenhagen, Leiden, Paris, and Florence. Significantly, his scientific productivity reached its peak *before* he returned to Copenhagen as Royal Anatomist in 1672. Steno's ideas and publications drew from the acquaintances, locales, and social settings that he encountered on his travels. These encounters led him down unforeseen paths. His itineraries tied friends, patrons, and places together with the anxieties of early modern intellectual life and multiple fields of learning. The itinerant career described in this book thus displays with special clarity the connections between mobility and scientific interdisciplinarity in the seventeenth century.

The interactions between Steno and his world allow me to highlight important features of early modern culture and science that are not usually at the center of most historical accounts. One of them is the intrinsically collaborative nature of scientific enterprises.¹⁸ In the 1660s, when Steno was most active, collaboration was crucial to the success of the first

---

15. Maria Rosa Antognazza, *Leibniz: An Intellectual Biography* (Cambridge: Cambridge University Press, 2009), 141, 151.

16. Ole Peter Grell, "'Like the Bees, Who Neither Suck nor Generate Their Honey from One Flower': The Significance of the *peregrinatio academica* for Danish Medical Students in the Late Sixteenth and Early Seventeenth Centuries," in Grell, Cunningham, and Arrizabalaga, *Centers of Medical Excellence: Medical Travel and Education in Europe, 1500–1789*, 171–192. This was also the case in England; see Robert Iliffe, "Foreign Bodies: Travel, Empire and the Early Royal Society of London; Part 1, Englishmen on Tour," *Canadian Journal of History* 33, no. 3 (1998): 357–385.

17. Jole Shackelford, "Documenting the Factual and the Artifactual: Ole Worm and Public Knowledge," *Endeavour* 23, no. 2 (1999): 65–71.

18. Domenico Bertoloni Meli, "The Collaboration Between Anatomists and Mathematicians in the Mid-Seventeenth Century with a Study of Images as Experiments and Galileo's Role in Steno's *Myology*," *Early Science and Medicine* 13, no. 6 (2008): 665–709. See also Catherine Goldstein, "Routine Controversies: Mathematical Challenges in Mersenne's Correspondence," in "La guerre en lettres: La controverse

scientific academies and journals, whose stories will be mentioned in the following chapters.[19] A close look at collaborative work also reveals the complex nature of early modern authorship, an issue that was especially convoluted when professors disputed discoveries made by their students, as happened with Steno.[20] Moreover, my focus on scientific cooperation shows that friendships too were an essential part of early modern science. Steno made the distinction between friendship and collaboration come to light when conversations that he started about science led to more personal discussions about religion. Although the social functions of early modern friendships have been the subject of many studies, here I utilize such friendships to explore the personal depth of scholarly interactions.[21] Friendships that started because of scientific collaboration transformed Steno's medical pilgrimage into a religious one, contributing as they did to his subsequent conversion from Lutheranism to Catholicism. Other friendships that he made lasted for years despite the disparate opinions held by them on various topics. These complex relationships are especially fascinating because they portray science as a fully human enterprise.

This volume also elevates other forms of sociability as driving forces behind the scientific practices of Steno and his peers. In his travels, Steno befriended not only university scholars but also port masters, goldsmiths, surgeons, and independent scholars, some of whom were women. All these contacts reveal the various environments that nourished Steno's interests in mathematics, mechanics, and chymistry, as well as his late fascination with Catholicism. Such environments included artisanal workshops in Denmark and the Netherlands, the so-called radical circles of Baruch Spinoza (1632–1677) in Amsterdam, Jesuit colleges and libraries

---

scientifique dans les correspondences des Lumières," [special issue,] *Revue d'histoire des sciences* 66, no. 2 (2013): 249–273.

19. A comparative monograph about the rise of scientific institutions and journals in the seventeenth century is still lacking. On the Académie des sciences and the Accademia del Cimento, see chaps. 4 and 5.

20. Domenico Bertoloni Meli, "Authorship and Teamwork Around the Cimento Academy: Mathematics, Anatomy, Experimental Philosophy," *Early Science and Medicine* 6, no. 2 (2001): 65–95; and Adrian Johns, "The Ambivalence of Authorship in Early Modern Natural Philosophy," in *Scientific Authorship: Credit and Intellectual Property in Science*, ed. Mario Biagioli and Peter Galison (London: Routledge, 2003), 67–90.

21. See Paula Findlen, "Controlling the Experiment: Rhetoric, Court Patronage and the Experimental Method of Francesco Redi," *History of Science* 31, no. 1 (1993): 35–64; and Mario Biagioli, *Galileo, Courtier: The Practice of Science in the Culture of Absolutism* (Chicago: University of Chicago Press, 1993).

across Europe, the informal academies of Paris, and the princely and papal courts of Italy.

Steno's ability to navigate different social settings also speaks to larger matters of civility, which played an important role in the early modern Republic of Letters.[22] In particular, Steno cultivated the virtue of modesty in ways that may seem merely social, but that played an important intellectual role in his work. The same can be said about his reputation among various European patrons. As I show, rather than simply providing financial support, his patrons in Leiden, Paris, and Florence also allowed him to explore his broad interests, especially his fondness for mathematics. In a nutshell, collaborations, friendships, modesty, manners, and patronage enriched Steno's itinerant career in varying ways. But above all, they brought social dimensions to what I understand as his deepest intellectual ambition: the search for certainty.

The Intersection of Disciplines

Steno's breadth of knowledge—from anatomy and mathematics to the history of the Earth and theology—is not especially surprising for the early modern period. In past decades, historians have described the intellectual scope of early modern scholars as much wider than previously thought. Galileo studied astronomy and mechanics while actively discussing Dante's *Divine Comedy* and other literary works; Isaac Newton (1643–1727) was as interested in physics and mathematics as he was in alchemy and theology; and Leibniz tackled disciplines as different as history, mathematics, and law.[23] Polymathy is now an assumed of early modern intellectuals. However, rather than accepting that polymathy was widespread, this book complicates this claim through the evolving work of a seventeenth-century anatomist.

In the early modern period, polymaths, or polyhistors, were those who wanted to know all that there was to know, an attitude closely related to encyclopedism.[24] A classic example of a seventeenth-century polymath

---

22. Anne Goldgar, *Impolite Learning: Conduct and Community in the Republic of Letters, 1680–1750* (New Haven, CT: Yale University Press, 1995); and Miller, *Peiresc's Europe*, esp. 49–75.

23. John Heilbron, *Galileo* (Oxford and New York: Oxford University Press, 2010); Rob Iliffe, "Abstract Considerations: Disciplines and the Incoherence of Newton's Natural Philosophy," *Studies in History and Philosophy of Science* 35, no. 3 (2004): 427–454; and Antognazza, *Leibniz*.

24. Anthony Grafton, "The World of the Polyhistors: Humanism and Encyclopedism," in "The Culture of the Holy Roman Empire, 1540–1680," [special

is Athanasius Kircher, S.J. (1602–1680), whose pretensions to universal knowledge were widely known then.[25] Another famous polymath in his time was Leibniz, whose main goal in life was to create a "general science [*scientia generalis*]" of everything.[26] These cases resonate with the definition of polymathy provided by the German scholar Johann von Wower (c. 1574–1612) in his *De polymathia tractatio* (*Treatise on Polymathy*; Hamburg, 1603), the first book to have this word in its title. Polymathy was "the knowledge of various things ... spreading itself very widely gathered from all kinds of studies [*ex omni genere studiorum collectam, latissime sese effundentem*]."[27] Wower also remarked that the most important feature of the polymath was his mastery of grammar, one of the seven liberal arts most closely related to reading.[28] Since they had to read many books across a variety of fields, seventeenth-century polymaths used special note-taking techniques to capture and absorb most of the information that came their way.[29] Steno learned these reading skills, which had a significant impact on his scientific career. He also shared interests with Kircher and Leibniz, both of whom he met in his travels. But unlike them, Steno did not aspire to know everything.

In short, from an early modern perspective, Nicolaus Steno was no polymath. He read broadly like polymaths but applied his knowledge to focused intellectual problems instead. He even complained when new questions pulled him away from his projects. He feared that distractions would lead him to produce superficial scientific work—a criticism that some scholars applied to the writings of Kircher and other polymaths.[30] In this book, I interpret Steno's work method as *focused interdisciplinarity*.

---

issue,] *Central European History* 18, no. 1 (1985): 31–47, esp. 37; and Kathryn Murphy, "Robert Burton and the Problems of Polymathy," *Renaissance Studies* 28, no. 2 (2014): 279–297.

25. Paula Findlen, "Introduction: 'The Last Man Who Knew Everything ... or Did He? Athanasius Kircher, S.J. (1602–80) and His World,'" in *Athanasius Kircher: The Last Man Who Knew Everything*, ed. Paula Findlen (New York: Routledge, 2004), 1–48, esp. 25–26.

26. Antognazza, *Leibniz*, 6.

27. Johann von Wower, *De polymathia tractatio: Integri operis de studiis veterum* (Hamburg: Froben, 1603), 16. See also Ioannes Deitz, "Wower of Hamburg, Philologist and Polymath. A Preliminary Sketch of His Life and Works," *Journal of the Warburg and Courtauld Institutes* 58, no. 1 (1995): 132–151.

28. Deitz, "Wower of Hamburg, Philologist and Polymath," 144–148.

29. Ann Blair, *Too Much to Know: Managing Scholarly Information Before the Modern Age* (New Haven, CT: Yale University Press, 2010), 62–116.

30. Burke, *The Polymath*, 80–82. On Kircher, see Findlen, "Introduction: 'The Last Man Who Knew Everything ... or Did He?,'" 6–7, 26–29.

This term helps me to distinguish Steno from others who also worked in multiple disciplines at the time. Rob Iliffe, for instance, has argued that although Newton worked on alchemy, theology, and physics at about the same time, these interests arose independently of one another.[31] Unlike the work of Newton and the encyclopedists, Steno's cross-disciplinary knowledge flourished *within* his anatomical research. This does not necessarily mean that anatomy had priority over his other interests; Steno seemed to have an appreciation for mathematics from very early on in his life. But anatomy was conducive to the materialization of Steno's various interests.

Steno's focused interdisciplinarity was a response to specific intellectual problems of the early modern period. By the late 1650s several new systems had arisen that purported to explain natural phenomena, but that had certain core differences. For instance, the natural philosopher and Catholic priest Pierre Gassendi (1592–1655) believed in the existence of the vacuum, whereas René Descartes rejected it, and Jesuit mathematicians adopted a geo-heliocentric astronomy, in contrast to Galileo's heliocentrism.[32] These theories explained the available observations, but they were often opposite to one another and thus could not all be true. Steno faced similar disagreements in anatomy and the origin of fossils and became obsessed with solving them through what I call a search for certainty. Certainty was not necessarily the absolute certitude of geometry, but rather that of reliable knowledge which, because of its compelling and accurate descriptions, convinced others of its truth.[33] Steno thought this certainty could be achieved by using reason and observations not only from the discipline he was working on, but also from mathematics, mechanics, hydrostatics, and chymistry.

My treatment of Steno's focused interdisciplinary method benefits from recent studies of mechanisms in early modern science.[34] Broadly

---

31. Iliffe, "Abstract Considerations."

32. See Margaret Osler, *Divine Will and the Mechanical Philosophy: Gassendi and Descartes on Contingency and Necessity in the Created World* (Cambridge: Cambridge University Press, 1994), 182–188, 201–202, 208–210; and Christopher Graney, *Setting Aside All Authority: Giovanni Battista Riccioli and the Science Against Copernicus in the Age of Galileo* (Notre Dame, IN: University of Notre Dame Press, 2015).

33. Richard Serjeantson, "Proof and Persuasion," in Park and Daston, *Early Modern Science*, 132–178; Peter Dear, *The Intelligibility of Nature: How Science Makes Sense of the World* (Chicago: University of Chicago Press, 2006); and William Wallace, *Causality and Scientific Explanation*, 2 vols. (Ann Arbor: University of Michigan Press, 1972–1974).

34. See Domenico Bertoloni Meli, *Mechanism: A Visual, Lexical and Conceptual History* (Pittsburgh: University of Pittsburgh Press, 2019); Raphaële Andrault, *La raison des corps: Mécanisme et science médicale* (Paris: Vrin, 2016); and references therein.

conceived, mechanisms are descriptions of natural phenomena that rely not only on machines, but also on particles in motion, new concepts (e.g., the elasticity of the air), and chemical reactions.[35] I look at these mechanisms as historical instances of scientific interdisciplinarity that Steno's writings, reading practices, and collaborations make more visible. As I show in this book, Steno's interdisciplinary claims are an extension of his analogical reasoning in anatomy. They also resonate with what we today call scientific models.[36] Moreover, Steno's growing search for certainty in science also undergirds his rising religious anxieties. The fact that his conversion occurred at the peak of his scientific career was not mere coincidence. Rather, it speaks to a parallel between the early modern disjunctions within natural philosophy and Christianity.[37]

Mathematizing the Body

Nicolaus Steno's ability to apply his cross-disciplinary knowledge to studies of the body reveals the enormous flexibility of early modern anatomy. This disciplinary breadth may be hard to grasp today, when what we call "science" is made up of highly specialized and narrow disciplines. Modern scientists have grown so accustomed to this specialization that when different fields talk to each other, a new discipline is formed—for example, biophysics or computational neuroscience. But in Steno's time, things were not like this, and anatomy encompassed many things that fall outside its modern purview, such as humanist textual skills and mechanics. This book confirms that anatomy was the broad and compelling discipline that Katharine Park, Anita Guerrini, and other historians have written about.[38] But I add to their analyses not just anatomy's fascinating

---

35. Nuno Castel-Branco, "Physico-Mathematics and the Life Sciences: Experiencing the Mechanism of Venous Return, 1650s–1680s," *Annals of Science* 79, no. 4 (2022): 442–467, esp. 444–449.

36. William Wallace, *The Modeling of Nature: Philosophy of Science and Philosophy of Nature in Synthesis* (Washington, DC: Catholic University of America Press, 1996), 311–312.

37. Steno's late interest in theology, however, was more like Newton's, in the sense that it developed in parallel to his scientific work. See Rob Iliffe, *Priest of Nature: The Religious Worlds of Isaac Newton* (Oxford and New York: Oxford University Press, 2017), 14–15, 22–23.

38. Katharine Park, *Secrets of Women: Gender, Generation, and the Origins of Human Dissection* (New York: Zone Books, 2006); and Anita Guerrini, *The Courtiers' Anatomists: Animals and Humans in Louis XIV's Paris* (Chicago: University of Chicago Press, 2015). See also Cynthia Klestinec, *Theaters of Anatomy: Students, Teachers, and*

capacity to absorb major changes of the seventeenth-century new sciences, as Robert Frank and Domenico Bertoloni Meli have demonstrated, but also its power to change other disciplines, including what is today known as geology.[39] Indeed, Steno's consistent choice to publish his work in the eclectic genre of *observationes anatomicæ* (anatomical observations) intentionally linked the breadth of anatomy with his note-taking practices. This "epistemic genre," as Gianna Pomata called it, provided Steno with an opportunity to publish in the same book his empirical and textual notes alongside his interdisciplinary insights.[40] As such, this book is not only the first complete study of Steno's anatomical career, but also an account of its broader implications for the history of science.[41]

Thus, this book shows that, precisely because of anatomy, Steno played a fundamental role in the development of early modern science. His contributions were not just to geology, but also to the mathematization of life and the unification of all natural phenomena under a single set of laws. For example, science today relies so heavily on mathematics that it can be difficult to recognize how innovative Steno's blend of anatomy with mathematics was. Indeed, in the seventeenth century and shortly before, there was an ongoing battle over the uses of mathematics in science.[42] In the same decade in which Steno's research flourished, Isaac Barrow (1633–1677)—Newton's professor of mathematics—was claiming in

---

Traditions of Dissection in Renaissance Venice (Baltimore, MD: Johns Hopkins University Press, 2011).

39. Robert Frank, *Harvey and the Oxford Physiologists: A Study of Scientific Ideas* (Berkeley and Los Angeles: University of California Press, 1980); and Domenico Bertoloni Meli, *Mechanism, Experiment, Disease: Marcello Malpighi and Seventeenth-Century Anatomy* (Baltimore, MD: Johns Hopkins University Press, 2011).

40. Gianna Pomata, "Sharing Cases: The *Observationes* in Early Modern Medicine," *Early Science and Medicine* 15, no. 3 (2010): 193–236, esp. 203. For further references, see chap. 2.

41. Steno as anatomist is only a secondary character in most recent accounts. See Bertoloni Meli, *Mechanism, Experiment, Disease*; Guerrini, *The Courtiers' Anatomists*; Evan R. Ragland, "Mechanism, the Senses, and Reason: Franciscus Sylvius and Leiden Debates over Anatomical Knowledge After Harvey and Descartes," in Distelzweig, Goldberg, and Ragland, *Early Modern Medicine and Natural Philosophy*, 173–206; Antonio Clericuzio, "Meccanismo ed empirismo nell'opera di Stensen," in Vitoria and Gómez, *Scienza, filosofia e religione nell'opera di Niels Steensen*, 123–138; and Tim Huisman, *The Finger of God: Anatomical Practice in 17th-Century Leiden* (Leiden: Primavera Press, 2009).

42. For a recent summary, see Dmitri Levitin, *The Kingdom of Darkness: Bayle, Newton, and the Emancipation of the European Mind from Philosophy* (Cambridge: Cambridge University Press, 2022), 60–82.

Cambridge that "there is no branch of natural science that may not arrogate the Title [of the mathematical] to itself."[43] Barrow was building on the writings of Jesuit mathematicians about the boundaries between natural philosophy and mathematics, some of which Steno also read.[44] But despite the disciplinary breadth of Barrow's claim, most of his argument was about mathematics and natural philosophy. Steno, however, went far beyond natural philosophy, claiming that even the human body, through the study of muscles, had to become "part of mathematics."[45] Although Steno had been using mathematics in his work from the start of his career (as this book shows for the first time), this epistemic claim was written in Florence alongside the mathematician Vincenzo Viviani (1622–1703). Strikingly, Giovanni Alfonso Borelli (1608–1679), another Italian mathematician, was also working at the same time on his *De motu animalium* (Rome, 1680–1681), which is still the standard example of the early modern mathematization of life. But Borelli's book was not published until 1680, more than ten years after Steno wrote his mathematical myology (myology being the study of muscles).

Early modern actors thought of the mathematization of disciplines in terms of mixed mathematics or physico-mathematics. Physico-mathematics described natural phenomena using the pure mathematics (geometry and arithmetic) and experiments.[46] In the seventeenth century, these included topics as varied as astronomy, optics, mechanics, and hydrostatics. Seventeenth-century hydrostatics, for example, involved demonstrating the existence of air pressure and measuring it with newly invented instruments.[47] For this reason, Steno understood the mathematization of the body as being the same thing as its mechanization.[48] This disciplinary link between mechanics and mathematics indicates that Cartesian mechanical philosophy was not as influential on Steno's work as previously thought. On the contrary, my historical analysis of Steno's early writings and close circles reveals that Steno was never a committed Cartesian. This book thus contributes to a series of recent works that rightly avoids labeling every

---

43. Barrow, *Lectiones mathematicæ*, as quoted in Peter Dear, *Discipline and Experience: The Mathematical Way in the Scientific Revolution* (Chicago: University of Chicago Press, 1995), 223.

44. Dear, *Discipline and Experience*, 222–227.

45. Nicolaus Steno, *Elementorum myologiæ specimen* (Florence: Stella, 1667), preface.

46. Peter Dear, "Mixed Mathematics," in *Wrestling with Nature: From Omens to Science*, ed. Peter Harrison, Ronald L. Numbers, and Michael H. Shank (Chicago: University of Chicago Press, 2011), 149–172.

47. Dear, *Discipline and Experience*, 180–209.

48. Castel-Branco, "Physico-Mathematics and the Life Sciences," 465–467.

seventeenth-century mechanization of the body as Cartesian.[49] Steno's sources were much broader and more centered on the writings of Gassendi, making it hard to label him with a specific worldview.[50]

This focused interdisciplinary method was especially popular among early modern anatomists. As many historians of science have shown, anatomists had been using mechanical analogies since Galen.[51] Moreover, many of them, such as Harvey and Marcello Malpighi (1629–1694), were also scholars who were trained in a university humanist tradition and thus read broadly. Yet, owing to his unique career, Steno was more explicit about the epistemic value of his interdisciplinary endeavors than most of his fellows. Steno was different not only because of his various travels, but also because he evaded medical practice and teaching altogether. In Steno's day most physicians earned their income through university teaching and medical practice. Strikingly, Steno admitted in a 1670 letter to Malpighi that he "did not have a medical practice."[52] The reasons for this are twofold. First, Steno argued that medicine was in a poor state due to a lack of rigorous knowledge about the body.[53] Second, he was supported financially by different patrons throughout his life and did not need to worry about making ends meet by means of the lucrative practice of medicine. This exclusive dedication to research is what explains his incredible productivity—he wrote five books in seven years—his broad and successful interdisciplinary work, and his high reputation in the Republic of Letters.

Across the Roads of Europe

The structure of this book follows Steno's geographical itinerary as a framework, and mostly along chronological lines. Sometimes his sojourns in a European city were so rich that I address different aspects of his scientific

---

49. See Ragland, "Mechanism, the Senses, and Reason"; Bertoloni Meli, *Mechanism*, esp. 23–24; and Sophie Roux, "An Empire Divided: French Natural Philosophy (1670–1690)," in *The Mechanization of Natural Philosophy*, ed. Daniel Garber and Sophie Roux (Dordrecht: Springer, 2013), 55–98.

50. For similar claims, see Ann Blair, *The Theater of Nature: Jean Bodin and Renaissance Science* (Princeton, NJ: Princeton University Press, 1997), 107–109; and Bertoloni Meli, *Mechanism, Experiment, Disease*, 43–44.

51. See Sylvia Berryman, "Galen and the Mechanical Philosophy," *Apeiron* 35, no. 3 (2011): 235–253; and Bertoloni Meli, *Mechanism*, 11–16.

52. Nicolaus Steno to Marcello Malpighi, 30 September 1670, in *Epistolae*, 1:248–250, esp. 248.

53. Steno, *Elementorum myologiæ specimen*, preface.

production and social life in a specific city in more than one chapter. Chapter 1 recovers the broad interests of the twenty-one-year-old Steno in his final year at the University of Copenhagen through a notebook that he filled and carried with him throughout his life. It shows that, from the start, anatomy in Steno's hands was linked with other disciplines, owing especially to the influence of the focused interdisciplinary works of Pierre Borel (1620–1671) and Jean Pecquet (1622–1674). I also demonstrate that Steno's broad interests, and especially his interest in mathematics, came from his artisanal upbringing in goldsmith workshops and a solid pre-university training in the liberal arts.

Chapter 2 explores Steno's first years of research in the Netherlands through his famous discovery of the parotid salivary duct, a vessel that still bears his name.[54] During these years, Steno cultivated the image of a learned anatomist by turning a priority dispute into a display of anatomical knowledge, dissecting skills, and prestige among peer anatomists. I also show how Steno expanded the editorial practices of anatomists such as Pecquet and Gaspare Aselli (1581–1625) by writing *historiæ anatomicæ* (anatomical descriptions) of glands in his first book of *observationes anatomicæ*. In chapter 3 I summarize Steno's focused interdisciplinary research method by describing his increasing use of chymistry, mathematics, and mechanics in anatomical research. I argue that the people he met and the places he visited in the Low Countries strongly encouraged his eclectic work in anatomy.

In 1664, after a short return trip to Copenhagen, Steno traveled to Paris, where he lived for about a year. In chapter 4 I trace Steno's activities in Paris and argue that he was attractive to the intellectual elite there because of his anatomical skills and philosophical ideas. The chapter shows that Steno was more important to the foundation of the first royal scientific academies of France than has hitherto been noticed. It also describes the flourishing of Steno's friendship with Jan Swammerdam, which had a significant impact upon both their research projects and religious lives. Chapter 5 follows Steno to the Florentine Medici court, whose interests in mathematics strongly resonated with the methods Steno had developed up to that point. The friendships that he established in Florence, especially with Viviani, the city's leading mathematician, lie at the root of much of his work in Italy, including his geometrical theory of muscle motion. Strikingly, the Medici family liked Steno so much that they even

---

54. Henry Gray, *Anatomy of the Human Body*, 20th ed. (Philadelphia: Lea & Febiger, 1918), 1134.

FIGURE I.2. Portrait of Nicolaus Steno, c. 1666–1675. Courtesy of the Gallerie degli Uffizi, Florence.

had his portrait made. It now hangs in the Uffizi Galleries alongside portraits of other leading European scholars (see fig. I.2).

It was also in Florence that Steno developed his theory regarding the origin of fossils and the history of the Earth. In chapter 6 I demonstrate that it was Steno's search for certainty and his successful use of interdisciplinary scientific models that led him to begin his research on fossils. The chapter uses new manuscript evidence to show when and why Steno started his work on the history of the Earth after arriving in Florence.

After an intense year of research, in the fall of 1667 he also entered into full communion with the Catholic Church. Chapter 7 places Steno's religious conversion to Catholicism in its historical context. I argue that Steno turned his search for certainty from science to religion owing to his deep friendships with fellow scientists and, most important, with devout women throughout the intellectual circles of Europe.

Steno's scientific research did not stop after he left Italy in 1669, but he did not publish any other scientific book after *De solido* (Florence, 1669). In the epilogue I show that the book's main themes—his itinerant life, friendships, focused interdisciplinarity, and search for certainty—continued to be relevant throughout his life by describing Steno's career after his conversion. I also address the significance of Steno and his networks to the history of early modern Europe and science.

In a nutshell, this book shows that Steno's career was itinerant and interdisciplinary from the start. Through the various scholars, institutions, and cultures that he encountered, Steno's pilgrimage shaped his scientific and religious worldviews, giving birth to a sociable personality that connected well with others. At the same time, Steno's search for certainty helped him contain the complex and connected world that he saw in his travels within the research questions that most interested him. This consistent interdisciplinary focus, in turn, led him to successfully turn his work method from the body to the Earth, generating new techniques, insights, and theories about science. In the same way that Galileo applied geometry to the fall of projectiles and Newton studied the motion of planets as if they were falling earthly objects, so too did Steno describe muscle contraction with parallelograms and use anatomical, chemical, and mechanical principles to study the origin of fossils. This interdisciplinary research was not uncommon in the middle of the seventeenth century. In Steno's case, however, it was more intensive because his specific circumstances tied his research to travels, transnational scholarly networks, collaborations, and friendships. Nonetheless, despite their historical contingency, the themes that unite Steno's itinerary—collaboration, mobility, concern with certainty—remain of timeless importance for the history of science. His various travels were so significant that, even though Steno never left Europe, Francesco Redi (1626–1697) wrote of his "native curiosity that made him a Pilgrim of the World."[55]

---

55. Francesco Redi to Lavinia Arnolfni, 11 April 1667, in Florence, Biblioteca Marucelliana, Redi 8, fol. 119r: "quella nativa curiosità che lo ha fatto Pellegrino del Mondo."

CHAPTER ONE

The Uses of Chaos; or, What Did Nicolaus Steno Know?

At the National Library of Florence, near the autograph writings of Galileo, is a manuscript that has little to do with Galileo. It is a series of notes taken every day from 8 March to 3 July 1659 in Copenhagen by the young university student Nicolaus Steno.[1] These ninety-two quarto-sized pages include textual excerpts, experimental notes, thoughts about life, and personal prayers. Steno was only twenty-one years old when he took these notes. For this reason, these writings have been regarded as juvenilia by Steno scholars, who mention them only in the first pages of their extensive studies.[2] However, much more than mere student annotations, this manuscript had a lasting impact on Steno's life and career, even if only because he carried it with him across the roads of Europe.

Steno wrote the "Chaos" manuscript, as these notes came to be known, while Copenhagen was under military siege by Swedish troops.[3] Like a response to a pandemic, the city's energies were focused on holding off the threat at hand. Ordinary life was interrupted, including classes at the

---

1. Florence, BNCF, Gal. 291, fols. 28r–75v (hereafter Gal. 291). This manuscript is edited, translated, and thoroughly studied in Nicolaus Steno, *Chaos: Niels Stensen's Chaos-Manuscript, Copenhagen, 1659*, complete ed. with introduction, notes and commentary by August Ziggelaar, SJ, Acta Historica Scientiarum Naturalium et Medicinalium 44 (Copenhagen: Danish Library of Science and Medicine, 1997). See also Egill Snorrason, *Niels Stensen: En students notater fra 1659* (Copenhagen: Mölnlycke, 1966); and Nicolaus Steno, *Niels Stensen: A Danish Student in His Chaos-Manuscript 1659*, ed. H. D. Schepelern (Copenhagen: University Library, 1987).

2. See Scherz's biography in Kardel and Maquet, *Nicolaus Steno: Biography and Original Papers*, 49–64. See also Cutler, *The Seashell on the Mountaintop*, 17–32, and Miniati, *Nicholas Steno's Challenge for Truth*, 41–70.

3. The siege lasted from August 1658 to the fall of 1659; see Robert Frost, *The Northern Wars: War, State and Society in Northeastern Europe, 1558–1721* (New York: Pearson, 2000), 180–183.

University of Copenhagen, where Steno had enrolled three years before.[4] Despite this break in academic studies and the likely start of military responsibilities, Steno kept working hard on his intellectual life.[5] He took notes of his activities every day for four months, making this notebook a clear window into his daily life, social communities, and scientific practices in those days. They speak of the remarkable resources that were available in a besieged Scandinavian city in the middle of the seventeenth century, including books and new instruments. Over these months Steno also had inspiring conversations with friends and acquaintances and regular access to artisanal workshops. Therefore, "Chaos" is the result of a scholar deep at work in the library and the laboratory.

I begin this chapter by describing Steno's upbringing in a goldsmith workshop and his intellectual formation in a liberal arts school. These two environments shaped the roots of Steno's attachment to mathematics, which took pride of place among his interests. His early fascination with geometry and measurements is visible throughout the "Chaos" manuscript. I then explain how and why Steno wrote "Chaos," by contextualizing it, for the first time, with recent studies on early modern note-taking and reading practices.[6] Most of Steno's notes seem disorganized and hard to understand. He also wrote in Greek, French, and German, besides his usual Latin. Nonetheless, I argue that this apparent disorganization is due to how early modern authors produced books of commonplaces. By delving into how Steno studied and took notes, this chapter explains how the "Chaos" manuscript became useful for a student about to undertake a long journey through Europe.

The chaotic information in "Chaos" is also a symptom of the state of seventeenth-century science. In the months when Steno wrote this notebook, he encountered seventeenth-century authors on subjects as varied as chymistry, medicine, anatomy, and philosophy. Not unlike to other readers, these studies exhausted him—there was simply too much to be learned. However, rather than trying to master everything, as polymaths did, Steno understood the advantages of focusing on single

---

4. See Scherz's biography in Kardel and Maquet, *Nicolaus Steno: Biography and Original Papers*, 45–46.

5. On Steno's military responsibilities, see Holger Rørdam, *De danske og norske Studenters Deltagelse i Kjøbenhavns Forsvar mod Karl Gustav* (Copenhagen: C. G. Iversen, 1855), 203–214, which mentions Steno on "the participation of Danish and Norwegian Students in the defense of Copenhagen."

6. The literature on early modern note-taking is vast; see Blair, *Too Much to Know*, 62–116; and Richard Yeo, *Notebooks, English Virtuosi, and Early Modern Science* (Chicago: University of Chicago Press, 2014).

subjects. His internal but manifest fight against distractions created the intellectual muscles of his focused interdisciplinary method that would serve his career for years. Steno's attempts to focus also encompassed an intense search for certainty, which involved rejecting speculative claims and promoting knowledge grounded in observations. Strikingly, these efforts took on a personal dimension for Steno, who infused his devotional life with intellectual concerns. Steno's prayer life in Copenhagen is crucial to understanding why he would later fall under the influence of other devout people across Europe.

Finally, "Chaos" is the clear product of a university student of medicine. Thus, the chapter concludes with showing that medicine and anatomy were central to Steno's work by 1659. It also reveals that these disciplines were already in dialogue with other disciplines, especially chymistry, mathematics, and mechanics. I show that Steno's interdisciplinary method owed much to two previously unnoticed influences on his thought: the French physicians Pierre Borel and Jean Pecquet, who taught Steno how to incorporate ideas and concepts from other disciplines into his studies of the body. In short, the "Chaos" manuscript is a unique document that illuminates Steno's encounter with the ideas and practices of mid-seventeenth-century science. It provides a remarkable look into the rise of the new sciences in a corner of Europe far from Galileo's Italy in the 1650s.

The Cathedral School, Mathematics, and Goldsmithing

Born on New Year's Day of 1638 (11 January in the modern Gregorian calendar), Nicolaus Steno was raised in a Danish family who owned a goldsmith workshop in Copenhagen.[7] At around ten years of age, he entered the Copenhagen Cathedral School, and when he turned eighteen he started his medical studies at the University of Copenhagen.[8] It was before university, however, that Steno acquired his various intellectual interests, in part because of humanities classes that he took at the Cathedral School, classes that trained him to read broadly.

---

7. See Scherz's biography in Kardel and Maquet, *Nicolaus Steno: Biography and Original Papers*, 11–15.

8. See Scherz's biography in Kardel and Maquet, *Nicolaus Steno: Biography and Original Papers*, 25, esp. n. 82. The university's enrollment mentions that Steno had previously studied at the "Copenhagen School"; see Anon., *Kjøbenhavns Universitets Matrikel*, vol. 1, 1611–1667, ed. S. Birket Smith (Copenhagen: Gyldendalske Bøghandels Forlag, 1890), 265.

Like most seventeenth-century scholars, Steno likely understood his humanist formation as an instrument rather than a goal. In the words of Anthony Grafton, humanism "meant the cluster of disciplines that trained a scholar to interpret and produce literary texts in Latin."[9] In the Cathedral School Steno mastered Latin and Greek, vital elements for the work of early modern physicians, as stressed in a book about medical studies written by Caspar Bartholin the Elder (1585–1629), a physician and theologian famous in his native Denmark and the father of Steno's mentor, Thomas Bartholin (1616–1680).[10] Not surprisingly, one of Steno's first university exercises was in humanist rhetorical skills, as shown by a previously unknown thesis he defended on "methods related to the beginning of disputations," now held at the Royal Library of Denmark.[11] Interestingly, perhaps owing to its emphasis on a devotional life based on the Scriptures, the Copenhagen Cathedral School also taught its students to read Hebrew.[12] This knowledge of ancient languages, especially Hebrew, became an essential tool for Steno's later theological studies and religious conversion in Florence, as he would later recall.

Unlike the modern understanding of humanities, Steno's classical training also included significant studies in mathematics, one of the seven classical liberal arts. The subject became young Steno's preferred discipline. As he wrote years later, he had "dedicated many hours in the past" to the study of geometry, which he planned to treat "not as my primary, but as

---

9. Grafton, "The World of the Polyhistors," esp. 37.

10. Caspar Bartholin, *De studio medico* (Copenhagen: Georg Hantzsch, 1628), 3. See Jole Shackelford, "To Be or Not to Be a Paracelsian: Something Spagyric in the State of Denmark," in *Paracelsian Moments: Science, Medicine, and Astrology in Early Modern Europe*, ed. Gerhild Scholz Williams and Charles D. Gunnoe, Jr. (Kirksville, MO: Truman State University Press, 2002), 35–69, esp. 47. On Latin and medicine, see Vivian Nutton, "The Rise of Medical Humanism: Ferrara, 1464–1555," *Renaissance Studies* 11, no. 1 (1997): 2–19.

11. Nicolaus Steno, *Disputatio II de mediis nonnullis, genesin disputationis concernentibus . . . præside Christiano Schioldborg, defendet Nicolaus Stenonis* (Copenhagen: Petrus Morsingi, 1657), in Copenhagen, Royal Library of Denmark, 14,-143 4° 01108. This rhetorical disputation was Steno's first printed work alongside a short poem, published in Jacob Paulli, *Supremus honor funeri Reginæ Kallenbachiæ . . . exhibitus* (Copenhagen, 1657), trans. in *Nicolaus Steno and His Indice*, ed. Gustav Scherz (Copenhagen: Munksgaard, 1958), 290.

12. See Copenhagen, Danish National Archives, Metropolitanskolen (1646–1737), "Liber scholæ Hafniensis Rectore scholæ Georgio Hilario 1666," as referred to in Scherz's biography in Kardel and Maquet, *Nicolaus Steno: Biography and Original Papers*, 28–29.

my only work if straitened circumstances at home had not so much convinced as forced me to prefer the useful to the pleasant."[13] Because Steno went to university medical school and continued to work as an anatomist after that, his study of geometry must have happened at the Copenhagen Cathedral School, before his university years. Steno's comment also means that he saw his medical studies as less pleasant (but more useful) than mathematics.

What exactly happened that made Steno decide to study medicine at the university? I suggest that his ties to the Cathedral School's headmaster lie at the root of this decision. The headmaster was the geometer, astronomer, and theologian Georgius Hilarius (1616–1686).[14] Hilarius had a degree in theology from the University of Copenhagen, but Steno would remember his master, in the dedication of his first anatomical publication, as a "mathematician and a man of letters."[15] Hilarius was also appointed professor of mathematics at the University of Copenhagen in 1654, although he did not begin teaching there until 1672.[16] He was also writing books on mathematics by the time Steno was his student. Hilarius's textbook *Progymnasmatum mathematicorum enchiridion* (Copenhagen, 1656), published when Steno graduated from the Cathedral School, was probably where Steno learned Euclid's *Elements*. From it Steno would have learned basic definitions like that of a rhomboid, a figure whose opposite sides and angles are equal but is not a rectangle or a square.[17] Hardly did Steno know that he would use rhomboids to describe muscle fibers a decade later. Strikingly, when Hilarius was younger, he enrolled in medicine at the University of Copenhagen but changed his focus and obtained a theology degree instead. This headmaster may thus have recommended that Steno pursue a medical degree, thinking it would be a better option for Steno's broad interests, especially those in mathematics. Indeed, among all the useful early modern university degrees—theology, law, and medicine—only medicine was related to the natural sciences.[18]

---

13. Steno to Bartholin, 5 September 1662, in EMC IV, 103.

14. S. M. Gjellerup, "Eilersen, Jørgen," in *Dansk biografisk Lexikon, tillige omfattende Norge for tidsrummet 1537–1814*, ed. C. F. Bricka, 19 vols. (Copenhagen: Gyldendal, 1887–1905), 4:464–465.

15. Nicolaus Steno, *Observationes anatomicæ* (Leiden: Jacobus Chouët, 1662), 80.

16. Gjellerup, "Eilersen, Jørgen." See also Georgius Hilarius, *Progymnasmatum mathematicorum enchiridion* (Copenhagen: Petrus Morsingi, 1656).

17. Hilarius, *Progymnasmatum mathematicorum enchiridion*, 16. See Euclid's *Elements*, book 1, definition 33.

18. They were useful because all three degrees led to the job of clergyman, lawyer, or physician.

The relationship between mathematics and the natural sciences in the early modern period was complex. Everyone knew there were no disciplines as certain as geometry and arithmetic, the pure mathematics.[19] But geometry was an abstract subject and not originally intended to be applied to the natural world. In the generation before Steno, authors such as Galileo Galilei, René Descartes, and various Jesuit mathematicians explored ways to use mathematics in their descriptions of natural phenomena, often with great accuracy.[20] One of the leading mathematicians at the University of Copenhagen when Steno studied there was Erasmus Bartholin (1625–1698), the son of Caspar Bartholin the Elder and an admirer of Cartesian thought. During his European travels, in Leiden in the 1650s, Erasmus edited two books on Cartesian mathematics.[21] On his return to Copenhagen, he continued to correspond with European mathematicians such as Christiaan Huygens in the Netherlands and Vincenzo Viviani in Florence.[22] Some of his letters to Viviani, preserved in the same library collection that holds the "Chaos" manuscript, date from the years Steno studied at the University of Copenhagen.[23] In fact, Steno's interests in mathematics might have benefited from the lectures of Erasmus Bartholin, who, in a speech to university students in 1657, claimed that "the mathematical disciplines contain the fundaments of all arts and sciences."[24] Nevertheless, Steno never spoke about Erasmus Bartholin in his writings, making it hard to determine how much of an influence, if any, he had on Steno.

On the other hand, the "Chaos" manuscript reveals various authors on mathematics that Steno mastered and who therefore influenced him. Steno excerpted a section from Athanasius Kircher's *Magnes sive de arte magnetica* (Cologne, 1643) where Kircher argued that "mathematics has

---

19. Kirsti Andersen and Henk Bos, "Pure Mathematics," in Park and Daston, *Early Modern Science*, 696–723.

20. On the debates on mathematics, see Dear, *Discipline and Experience*.

21. Frans van Schooten, *Principia matheseos universalis seu introductio ad geometriæ methodum Renati des Cartes edita ab Er. Bartholino, Casp. Fil.* (Leiden: Elsevier, 1651); and Florimond de Beaune, *De æquationum natura, constitutione et limitibus opuscula duo . . . edita ab Erasmio Bartholino, Medicinæ et Mathematum in Regia Academia Hafniensi Professore publico* (Amsterdam: Ludovicus et Danielis Elsevier, 1659).

22. Kirsti Andersen, "An Impression of Mathematics in Denmark in the Period 1600–1800," *Centaurus* 24, no. 1 (1980): 316–334.

23. Florence, BNCF, Gal. 254, fols. 137r–138v.

24. Erasmus Bartholin, *De naturæ mirabilibus quæstiones academicæ* (Copenhagen: Georg Gödian, 1674), 146–170, esp. 157. The date of the speech is in the table of contents of "De studio lingua danicæ."

many ways in which it concludes true and absolutely certain things [*vera et certissima concludit*] from falsities and impossibilities."²⁵ Kircher explained that mathematics was so powerful that it accurately described astronomical phenomena with models built on wrong assumptions. He gave the example of astronomic models that had the Earth dislocated from the center of the universe (at a point called eccentric) and planets moving on orbits upon orbits (known as epicycles). After Steno excerpted this page from Kircher, he commented in the margin of his notebook that "only mathematics deduces truth from false principles."²⁶ Another mathematical book Steno read was *Deliciae physico-mathematicae* (Nuremberg, 1636), by Daniel Schwenter (1585–1636), on the construction of small mechanical devices. This German book was part of the textual tradition of the French *Récréations mathématiques* (Pont-à-Mousson, 1624).²⁷ Schwenter's book was later expanded by the German mathematician Georg Harsdörffer (1607–1658) into two volumes with newer mathematical devices described in books by such scholars as Kircher, Pecquet, and Santorio Santorio (1561–1636), all of whose works Steno knew well.²⁸ Steno transcribed only a short note from the third volume and none from the second, but from the first he took many notes, having read it in its entirety.²⁹ Although the authorship of the original French text has been questioned, it was thought at the time to be the work of the Jesuit mathematician Jean Leurechon (c. 1591–1670). Gaspar Schott (1608–1666) wrote in a printed volume that Schwenter translated Leurechon into German.³⁰ Schott was a famous

---

25. Athanasius Kircher, *Magnes sive de arte magnetica* (Cologne: Iodocum Kalkoven, 1643), 500. Steno read the 1643 edition, as can be inferred from the page numbers he copied.

26. Gal. 291, fol. 38v (22 March), in Steno, *Chaos*, 119. This is also printed in the margin of Kircher, *Magnes sive de arte magnetica*, 500.

27. Gal. 291, fols. 54rv (16 April), in Steno, *Chaos*, 251–253. Steno copied the preface from the 1636 edition. See also Albrecht Heeffer, "*Récréations mathématiques* (1624): A Study on Its Authorship, Sources and Influence," *Gibecière* 1 (Summer 2006), 1:77–167, esp. 86.

28. Georg Harsdörffer, *Delitiæ mathematicæ et physicæ*, 2 vols. (Nuremberg: Jeremia Dümler, 1651–1653). For the list of authors, see the cover of vol. 2. For Roberval's experiment as presented in Pecquet's *Experimenta nova anatomica*, see Harsdörffer, *Delitiae*, 2:658–659.

29. Gal. 291, fol. 54v (16 April), in Steno, *Chaos*, 253. Steno copied only nineteen of more than a hundred propositions from vol. 1, on arithmetical games, mechanics, music, optics, and astronomy.

30. Gaspar Schott, *Mechanica hydraulico-pneumatica* (Frankfurt: Henricus Pigrin, 1657), 222. On the questionable authorship of Leurechon, see Heeffer, "*Récréations mathématiques*," 86–95.

disciple of Kircher, and Steno briefly mentioned his books in "Chaos."[31] Steno also mentioned authors that he still wanted to read, such as the English mathematician Henry Briggs (1561–1630) and his German colleague Wilhelm Avianus (d. 1636), and took notes from Bernhard Varen's *Geographia generalis* (Amsterdam, 1650).[32] Varen opened his book saying that geography is a mixed mathematical science and expounding on what the significance of that status was, but Steno did not take notes from there.[33]

In the end, two of the mathematical treatises in "Chaos" on which Steno took the most notes from—Kircher's and Schwenter's books—came from the Jesuit tradition of mixed mathematics.[34] Even though the Jesuits had been explicitly forbidden to teach in Denmark since 1604, their intellectual influence reached the country's young students through their books on mathematics.[35] They would continue to be present in Steno's intellectual life as he traveled south in Europe.

Steno learned the importance of quantitative reasoning and rigorous measurements not only from books, but also from real-life problems encountered in his home's goldsmithing workshop. Steno's father and two stepfathers supported the family with their artisanal skills by making objects such as chalices and candleholders, polishing metals, and other similar activities.[36] Steno mentioned this workshop in the "Chaos" manuscript, noting that "for grinding lenses, one can use what is available in our workshop [*in officina nostra*]."[37] He was also familiar with other workshops around the city, such as that of a distiller named Niels, whose "laboratory [*laboratorium*], where there are often so many flies, is filled with vapors, essences, etc."[38] Not surprisingly, this artisanal environment be-

---

31. Steno was aware of Schott's hydraulics, but does not excerpt from it; see Gal. 291, fol. 54v (18 April), in Steno, *Chaos*, 253.

32. Correspondingly, Gal. 291, fols. 55v (24 April), 57v (13 May), in Steno, *Chaos*, 261, 277.

33. Varen, *Geographia generalis* (Amsterdam, 1650), "Epistola dedicatoria."

34. Besides Kircher, Schott, and Leurechon, Steno also mentioned the Jesuit mathematicians Niccolò Cabeo (1586–1650), Christoph Scheiner (c. 1575–1650), and Christoph Clavius (1538–1612), see, respectively, Gal. 291, fols. 38r (21 March) and 55v (24 April), in Steno, *Chaos*, 113, 261–262.

35. Henry Nielsen, Helge Kragh, Kristian Hvidfeldt Nielsen, and Peter Kjaergaard, eds., *Science in Denmark: A Thousand-Year History* (Aarhus: Aarhus University Press, 2008), 51–52. On the influence of Jesuit mathematics elsewhere in Europe, see Dear, *Discipline and Experience*, 32–42.

36. For an introduction to Steno's family workshop, see Scherz's biography in Kardel and Maquet, *Nicolaus Steno: Biography and Original Papers*, 35–38.

37. Gal. 291, fol. 60r (25 May), in Steno, *Chaos*, 290.

38. Gal. 291, fol. 60v (3 June), in Steno, *Chaos*, 296.

came part of his frame of reference in "Chaos." When rereading his notes from Matthias Untzer's (1581–1624) *Anatomia mercurii* (Halle, 1620), for instance, he underlined that for goldsmiths "life is short due to the oppression of the[ir] hot nature by the coldness of mercury," perhaps thinking of the early deaths of his father and stepfather.[39] More interestingly, Steno used his artisanal skills and knowledge from workshop acquaintances in his intellectual studies. One day, he said, "the examination of gold, pure or mixed with silver, might very well be practiced in our flask [*in phiala nostra*]."[40] On another day, he turned to goldsmiths to solve his inquiries about the epistemic role of colors, noting that "one should inquire from goldsmiths about the nature of colors."[41] The answer, which arrived one month later, was that "goldsmiths make a red color either from burnt vitriol or thin iron sheets," meaning that one color could have various origins, a conclusion that Steno often used in his late scientific research.[42] In short, Steno placed his artisanal background at the service of the natural sciences, which he explicitly acknowledged when he wrote that his chemical experiments "have not so much to do with craftsmanship as with the natural sciences [*non tam Mechanicam quam physicam spectantium*]."[43] These inquiries also reveal that Steno circulated widely in the social world of Danish goldsmiths, including during the military siege of Copenhagen.

In the days in which he wrote "Chaos," the overlap between Steno's intellectual interests and artisanal practices extended beyond chymistry to mechanics, glassmaking, and other crafts. With his forthcoming travels in mind, Steno dedicated several days to building an "instrument for scaling maps" in different proportions and accurate ratios.[44] His interest in grinding lenses was also associated with new seventeenth-century instruments such as telescopes and microscopes.[45] Steno considered

---

39. Gal. 291, fol. 48v (9 April), in Steno, *Chaos*, 200–201. Ziggelaar mentions in the notes that Steno's father and first stepfather, both goldsmiths, died very early in the son's life. See also Scherz's biography in Kardel and Maquet, *Nicolaus Steno: Biography and Original Papers*, 20–21. The sources do not reveal how either man died.

40. For this and another example of goldsmithing skills to investigate philosophical questions, see Gal. 291, fols. 60r (25 May), 36r (15 March), in Steno, *Chaos*, 290, 86.

41. Gal. 291, fol. 56v (2 May), in Steno, *Chaos*, 268.

42. Gal. 291, fol. 61r (5 June), in Steno, *Chaos*, 298.

43. Gal. 291, fol. 33v (14 March), in Steno, *Chaos*, 80.

44. Gal. 291, fol. 44v, (1 April), in Steno, *Chaos*, 162. Steno also mentions this instrument in Gal. 291, fols. 58r (13 May), 60r (28 May), 61r (7 June), in Steno, *Chaos*, 279–281, 292, 299.

45. On the telescope, see Gal. 291, fol. 59r (21 May), 60v (3 June), 75v (3 July), in Steno, *Chaos*, 286, 297, 448. On the microscope, see Gal. 291, fols. 59v–60v (24 May,

building "microscopes with a shape other than round [*explorandum anne microscopiis alia dari possit figura quam rotunda*]." However, he concluded that it "might be in vain because the glass [lens] should be adapted not to the shape of the object but to that of the eye." On that same day, while "inspecting the glasses of a telescope [*inspiciendo vitra . . . telescopii*]," he remarked that "the quality of the grinding [of the lenses] could be examined."[46] Once he even drew the microscope of his friend Ole Borch (1626–1690) in the middle of his notes (see fig. 1.1). Microscopes had not yet become the popular tool that they would be a decade later, in the hands of Malpighi, Jan Swammerdam (1637–1680), and Anton van Leeuwenhoek (1632–1723).[47] Thus, this early interest in microscopes reveals how up-to-date scientific instruments were in Copenhagen in 1659.

Besides learning chymical practices and maneuvering instruments, Steno also learned in his artisanal environment the need for rigor and accuracy, which involved mathematics. While writing "Chaos," Steno measured and compared the weights and volumes of many substances in different conditions. He compared the weight "of something which smells" with its weight once the smell disappeared; the weight of melted and burned wax; the "weight ratio of gold or any other thing to water"; the density of distilled and non-distilled rainwater; and so on.[48] Faithful to his goldsmithing training, Steno knew that accurate quantities mattered if one were to achieve good results, so he searched for ways to find better values for his measurements. He suggested measuring different spices, liquors, and salts by dissolving them in distilled water or wine and measuring the weight of the water or wine before and after making the solution. He hoped that "perhaps in this way we may achieve a more accurate knowledge [*accuratiorem cognitionem*]."[49] Steno also developed instructions on "how glass can be graduated accurately [*ut vitrum accurate dividatur*]" by attaching "paper strips to both sides" of the vessel, which could be

---

28 May, 3 June), in Steno, *Chaos*, 289–290, 292, 296. On the early history of these two instruments, see Christoph Lüthy, "Atomism, Lynceus, and the Fate of Seventeenth-Century Microscopy," *Early Science and Medicine* 1, no. 1 (1996): 1–27.

46. Gal. 291, fol. 61r (3 June), in Steno, *Chaos*, 296–297.

47. Edward Ruestow, *The Microscope in the Dutch Republic: The Shaping of Discovery* (Cambridge: Cambridge University Press, 1996); on Malpighi and the microscope, see Bertoloni Meli, *Mechanism, Experiment, and Disease*, 40–45.

48. Gal. 291, fols. 41v (28 March), 56r (29–30 April), in Steno, *Chaos*, 143, 266–267. See also Gal. 291, fol. 58v–59r (15 May, 18 May), 60r (25 May, 27 May), 61r (7 June), in Steno, *Chaos*, 281, 285, 290, 291, 299.

49. Gal. 291, fol. 58v (18 May), in Steno, *Chaos*, 284.

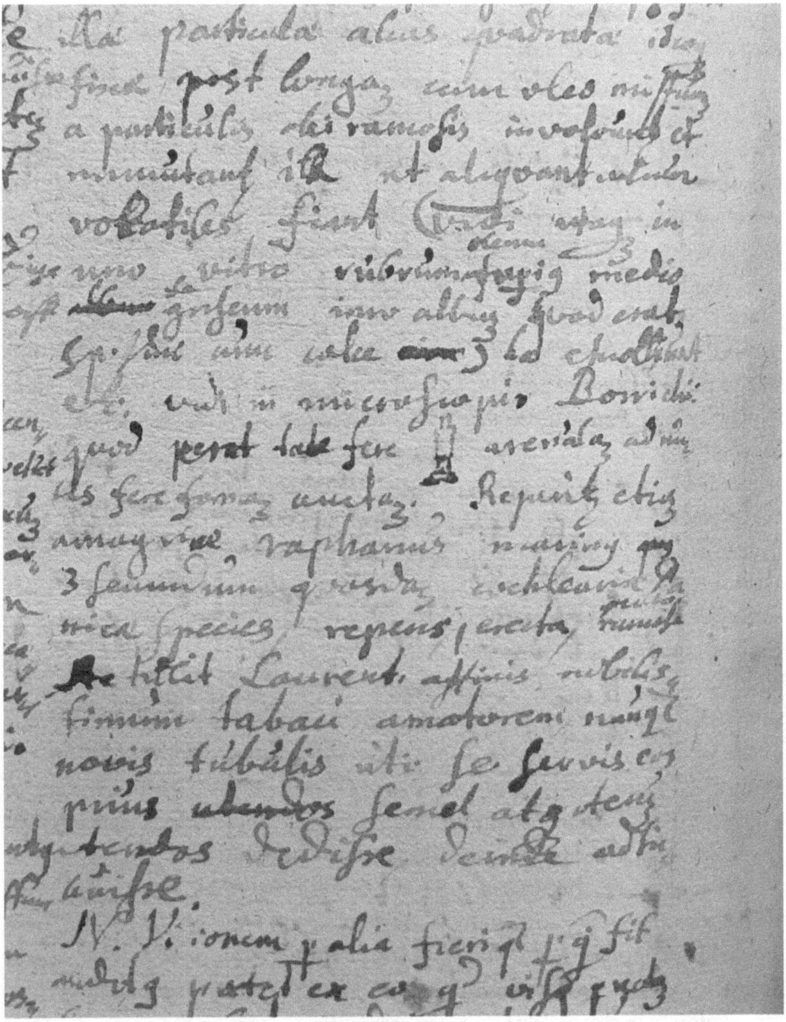

FIGURE 1.1. Steno's drawing of Ole Borch's microscope, in the Biblioteca Nazionale Centrale di Firenze (BNCF), Gal. 291, fol. 59v. Reproduced with the permission of the Ministero della Cultura/Biblioteca Nazionale Centrale di Firenze.

improved "if more strips were attached."[50] Steno's obsession with rigorous weighing and measurements speaks to the epistemic role he attributed to quantities. As he admitted, the differences between simple substances

---

50. Gal. 291, fol. 60r, (31 May), in Steno, *Chaos*, 293.

might be better known with the "accurate knowledge of arithmetic [*accurata Arithmetices cognitione*]."[51]

Measuring everything also became central to Steno's studies of the body. It led him to interact with the most famous book of quantitative medicine at the time: Santorio's *De statica medicina* (Venice, 1615). Santorio, a physician from the prestigious University of Padua, introduced in the same book a chairlike scale to weigh persons before and after they ate in order to determine "how much healthy food is convenient for each person [to take]" and "how much insensible perspiration [*perspiratio insensibilis*] there is in each body."[52] Insensible perspiration was a concept first mentioned by Hippocrates and Galen and was related to hidden evacuations of the body through the skin, and which Santorio wanted to weigh quantitatively.[53] Steno did not excerpt from Santorio's book in his notes, but he was fully aware of it since he wrote extensive comments on Santorio's work. He followed Santorio's quantitative approach closely and wrote, "In dietetics, the body should be weighed when you wish to urinate and after that." The goal of this measurement was to know "how much the body at that time gives off in exhalation." Santorio wrote his ideas in separate sentences, as aphorisms. Steno, on the other hand, suggested arranging the "observations in an ordered way [*rite observationes institutas*]" and laid out a program to "weigh the eaten food with exactitude [*cibum quem sumis exacte ponderabis*]" to compare its value with the weight difference of the body before and after eating. One of the reasons for comparing the weight difference in the body before and after eating with the weight of the food was to identify whether there was any weight loss unrelated to the food and linked to perspiration, and how much it was. That is, "if you observe here that the weight of your body increased exactly by the same amount as the weight of what you have taken in, you will judge that there has been no perspiration [*transpiratio*]." Steno even tried to go beyond Santorio, who "seems to be mistaken," he wrote, for not considering external effects in the production of perspiration, such "as the warmth of the

---

51. Gal. 291, fol. 60r, (30 May), in Steno, *Chaos*, 292. For Steno's instructions on how to build "a geometrical instrument" that transformed "an arithmetical operation" into "simple counting," see Gal. 291, fol. 41v (25 March), in Steno, *Chaos*, 141.

52. Santorio Santorio, *De statica medicina* (Leiden: David Lopes de Haro, 1642), "Aphorismi qui continentur." This excerpt was first published in Santorio, *Commentaria in primam fen primi libri Canonis Avicennae* (Venice: Jacob Sarcina, 1625), 557–558.

53. Teresa Hollerbach, *Sanctorius Sanctorius and the Origins of Health Measurement* (Cham: Springer, 2023), 49–62. See also Michael Stolberg, "Sweat: Learned Concepts and Popular Perceptions, 1500–1800," in Horstmanshoff, King, and Zittel, *Blood, Sweat and Tears*, 503–522.

air." Therefore, Steno suggested "always have a thermoscope in your office [*habeas in museo tuo semper thermoscopium*]" to control the temperature.[54] It is not clear whether Steno went so far as to try these Santorio-like measurements by himself. Most likely he did not try them, because at this time, Steno had no medical practice of his own. Nevertheless, he would return to perspiration when working on the anatomy of skin glands. In short, this engagement with Santorio's work illustrates how easily Steno adapted his interest in accurate measurements and quantitative reasoning to medicine and the life sciences. Mathematics and measurements were not the central theme of Steno's notebook. However, they testify to the long-term presence in "Chaos" of interests from his times at home, in the goldsmithing workshop, and at the Cathedral School.

Travels, Note-Taking, and the Making of "Chaos"

Early modern Denmark was an exciting place for the study of nature. The most famous Dane of the time was Tycho Brahe (1546–1601), the astronomer who performed the most accurate celestial observations without telescopes.[55] Yet, the generations after him were the ones who contributed to what is known as the "golden age for science in Denmark":[56] such scholars as Ole Worm, the Bartholins, and Ole Rømer (1644–1710), the mathematician whose observations of Jupiter's moons supported the first measurement of the speed of light.[57] One of these leading intellectuals was Thomas Bartholin, a professor of medicine at the University of Copenhagen and, for a period, the university's rector.[58] Bartholin became Steno's mentor when the latter matriculated in the fall of 1656, an event that was so significant that Steno compared being "adopted as a disciple" by Bartholin "to a [new] birthday."[59]

---

54. For this account and a similar criticism on disregarding the effects of perspiration on clothes, see correspondingly Gal. 291, fols. 57r (11 May), 56v (3 May), in Steno, *Chaos*, 274–275, 270.

55. Victor Thoren, *The Lord of Uraniborg: A Biography of Tycho Brahe* (Cambridge: Cambridge University Press, 1990).

56. Nielsen, Kragh, Nielsen, and Kjaergaard, *Science in Denmark*, 20.

57. Bernard Cohen, "Roemer and the First Determination of the Velocity of Light (1676)," *Isis* 31, no. 2 (1940): 327–379.

58. John H. Skavlem, "The Scientific Life of Thomas Bartholin," *Annals of Medical History* 3, no. 1 (1921): 67–81; and Jesper Andersen, *Thomas Bartholin: Laegen & anatomen; Fran enjhørninger til lymfekar* (Copenhagen: FADL's Forlag, 2017).

59. Nicolaus Steno to Thomas Bartholin, 22 April 1661, in EMC III.

Thomas Bartholin, just like his father, Caspar Bartholin the Elder, and his uncle Ole Worm, took part in the well-established Danish tradition of medical pilgrimages.[60] These pilgrimages were the main reason Danish science was on par with scientific developments elsewhere in Europe. They also explain how Danes became well-known in the Republic of Letters. According to Bartholin, who wrote an entire book called *De peregrinatione medica* (Copenhagen, 1676), such pilgrimages were essentially academic, since they were meant "to achieve the final step of academic studies [*ad studiorum metam assequendam*]."[61] In a nutshell, the pilgrimage was a journey a young man undertook in order to enable him to study and learn from scholars in different institutions across Europe and to obtain a medical degree from another university. After years abroad, Danish scholars returned home with knowledge of the newest ideas and practices developed across Europe and a vast network of correspondents.[62] Worm's museum of natural curiosities in Copenhagen became known across Europe precisely because of Worm's vast correspondence network.[63] All the Bartholin family members, including Thomas and Worm, took professorships at the university when they returned home, and many became university rectors for a period. The Bartholin family shaped the intellectual culture at the University of Copenhagen in the seventeenth century. Therefore, it is not surprising that Steno rejoiced in studying under Thomas.[64]

---

60. Andrew Cunningham, "The Bartholins, the Platters and Laurentius Gryllus: The *peregrinatio medica* in the Sixteenth and Seventeenth Centuries," in Grell, Cunningham, and Arrizabalaga, *Centers of Medical Excellence?*, 3–16; and Grell, "'Like the Bees, Who Neither Suck nor Generate Their Honey from One Flower,'" in Grell, Cunningham, and Arrizabalaga, *Centers of Medical Excellence?*, 171–192. See also Nielsen, Kragh, Nielsen, and Kjaergaard, *Science in Denmark*, 50–52. For Tycho Brahe undertaking similar travels, see Thoren, *The Lord of Uraniborg*, 13–14.

61. Thomas Bartholin, *De peregrinatione medica* (Copenhagen: Daniel Paull, 1674), 3. Bartholin further added that the final purpose of those studies was to serve God and the nation.

62. Ole Peter Grell, "In Search of True Knowledge: Ole Worm (1588–1654) and the New Philosophy," in Smith and Schmidt, *Making Knowledge in Early Modern Europe*, 214–232; see also Ole Peter Grell, "Caspar Bartholin and the Education of the Pious Physician," in Grell and Cunningham, *Medicine and the Reformation*, 78–100.

63. Ole Worm, *Museum wormianum, seu historia rerum rariorum* (Leiden: Johann Elsevier, 1655), 67–68. See also Ella Hoch, "Diagnosing Fossilization in the Nordic Renaissance: An Investigation into the Correspondence of Ole Worm (1588–1654)," in Duffin, Moody, and Gardner-Thorpe, *A History of Geology and Medicine*, 307–327.

64. J. F. C. Danneskiold-Samsøe, *Muses and Patrons: Cutures of Natural Philosophy in Seventeenth-Century Scandinavia*, Ugglan, Minervaserien 10 (Lund University, 2004), 207–210, at 257. See also Grell, "Caspar Bartholin and the Education of the Pious Physician," 91.

As a typical medical Danish student, Nicolaus Steno, too, started an extended European trip four years after entering the University of Copenhagen. The "Chaos" manuscript was written a few months before his departure and witnesses the preparations for such a trip. Steno recorded in his notebook having a conversation with a friend "about traveling [*de peregrinando*]," and the next day he wrote instructions on how "to scale maps."[65] In those days he also excerpted from travel literature such as the *Itinerarium frisio-hollandicum* (Leiden, 1630) and Vincent Le Blanc's *Les voyages* (Paris, 1649).[66] Notes from books unrelated to travels also included information relevant to his pilgrimage. When reading a section on types of soils from Varen's *Geographia generalis*, he noted that "the houses in Amsterdam are usually built on piles."[67] Not surprisingly, Amsterdam was the first place he would settle after leaving Copenhagen. When he started writing "Chaos," Steno also recorded that someone from a known noble family in Denmark had visited a glassmaking workshop in Amsterdam.[68] Steno started this record with a "N[ota] B[ene]," showing that he had personal interests in visiting glassmaking workshops abroad, which he did. Finally, when reading the final pages of Pecquet's *Experimenta nova anatomica* (Paris, 1654), Steno copied only the names of the scholars mentioned there, such as "Jacques Mentel," professor of medicine at the University of Paris, and "Pierre de Mercenne," another physician from Paris.[69] Paris would be Steno's first long stop after leaving the Netherlands.

There were, however, more profound reasons for Steno to take notes. As Ann Blair and others have shown, early modern note-taking emerged from the tradition of commonplace books.[70] Scholars copied excerpts

---

65. Gal. 291, fol. 44r (31 March–1 April), in Steno, *Chaos*, 161–162.

66. See correspondingly, Gal. 291, fols. 58r (13 May), 36r–37v (17–18 March), in Steno, *Chaos*, 278, 90–93. Steno read the 1630 edition of *Ititnerarium* as per the page numbers he copied.

67. Gal. 291, fol. 55v (24 April), Steno, *Chaos*, 263; excerpted from Varen, *Geographia generalis*, 69.

68. Gal. 291, fol. 28r (9 March), Steno, *Chaos*, 26.

69. Gal. 291, fols. 64v (19 June), 67v (23 June), in Steno, *Chaos*, 329, 360; and Jean Pecquet, *Experimenta nova anatomica . . . [et] dissertatio anatomica de circulatione sanguinis* (Paris: Sebastian & Gabriel Cramoisy, 1654), 19, 152.

70. See Anthony Grafton and Joanna Weinberg, "Johann Buxtorf Makes a Notebook," in *Canonical Texts and Scholarly Practices: A Global Comparative Approach*, ed. Anthony Grafton and Glenn W. Most (Cambridge: Cambridge University Press, 2016), 275–298; and Michael Stolberg, "Empiricism in Sixteenth-Century Medical Practice: The Notebooks of Georg Handsch," *Early Science and Medicine* 18, no. 6 (2013): 487–516.

from multiple sources for their own study in these books.[71] They also used notebooks to assist them in close reading. The Jesuit scholar Jeremias Drexel (1581–1638), author of one of the earliest and most popular books on note-taking, famously wrote that "it is no waste of time to take notes, but rather to read without taking notes."[72] Thomas Bartholin himself wrote a book "on books that should be read," *De libris legendis* (Copenhagen, 1676). He opened by stating that "the perfect summit of the natural and medical sciences" cannot be reached "without reading books" and devoted an entire chapter to reading and taking notes.[73]

Steno was familiar with various reading and note-taking methods because of his cultural environment. Learning how to manage books was part of his training in the humanities at the Cathedral School. Indeed, references to other notebooks that he used show that he knew a number of techniques of reading and note-taking. Besides "Chaos," Steno had a notebook of observations in which he recorded observational knowledge learned from other books. After copying excerpts from a book on the supernatural causes of diseases, Steno asked himself whether "[certain] accounts could be written in the book of observations [*liber historiarum*] without mentioning the authors."[74] At another time, after reading "[Daniel] Sennert's tables," Steno wanted to add "an ampler explanation or description in 'Chaos' or the tables if space permits."[75] These tables were likely another group of papers in which Steno organized information related to medical practice, such as lists of medical symptoms and *materia medica*.[76] Indeed, Thomas Bartholin recommended the "frequent use of tables in support of memory," citing as a good example Christian Wincklemann's *Institutiones medicinæ Danielis Sennerti in tabulas redactae*

---

71. See Blair, *Too Much to Know*, 63–64, 69–73, 84–85; and Richard Yeo, "Thinking with Excerpts: John Locke (1632–1704) and His Notebook," *Berichte zur Wissenschafts-Geschichte* 43, no. 2 (2020): 180–202.

72. As translated in Blair, *Too Much to Know*, 78.

73. Thomas Bartholin, *De libris legendis* (Copenhagen: Daniel Paull, 1676), preface and chap. 6.

74. For the quote and another reference to the *liber historiarum*, see correspondingly Gal. 291, fols. 37v (21 March), 58r (14 May), in Steno, *Chaos*, 100, 281. The book Steno read is Hieronymus Jordan, *De eo quod divinum aut supernaturale est in morbis humani corporis* (Frankfurt: Johann Gottfried Schönwetter, 1651).

75. Gal. 291, fol. 37r (21 March), in Steno, *Chaos*, 96–97.

76. Steno also mentioned these tables when quoting from Schylander's *Practica chyrurgiæ* in Gal. 291, fols. 28r (8 March), 36v (16 March), in Steno, *Chaos*, 22, 89. Steno wanted to include Henricus Regius's medicine (perhaps *Fundamenta medica* [Utrecht, 1647]) in the same tables; see Gal. 291, fol. 37r (19 March), Steno, *Chaos*, 95.

(Paris, 1637)—the book that Steno was working on.[77] Daniel Sennert (1572–1632) was a German physician whose works, especially the *Institutiones medicinæ* (Wittenberg, 1611), had "a distinctly corpuscularian emphasis," in the words of William Newman.[78] Thus, one of Steno's expositions to atomism was likely through chymical texts such as Sennert's.

Finally, Steno referred to a "book of commonplaces [*loci communes*]" and some "loose sheets [*folia libera*]" on which he was writing notes.[79] Steno wrote this comment as a personal note in "Chaos," which he distinguished from excerpts of his readings by bracketing them between two Ns "(N. . . . N.)," either meaning "Nota" or himself as "Nicolaus" (see fig. 1.2). In short, Steno seemed to have begun writing "Chaos" as a draft for writings that would later be organized in other notebooks, as was common practice at the time. Scholars with multiple notes first wrote a notebook with unedited information that would later be compiled in another document organized by headings, just like a commonplace book.[80] Steno continued to write notes in multiple places until he died, as his confusing papers held in Florence show.[81]

The "Chaos" manuscript takes pride of place among Steno's papers because it is the only one from his early years in Copenhagen to have survived. Steno probably learned the advantages of keeping all information in one document. On the surface, this surviving notebook looks like a chaos of information—hence the name he gave it.[82] But Steno brought order to the chaos through various textual techniques and marginalia. Note-taking authors like Drexel suggested writing passages in a single notebook only once and in the order in which they were read, adding headings to the margins later.[83] The final product of the "Chaos" manuscript employs the techniques advocated by Drexel. Most of the manuscript has a second layer of notes in Steno's hand, such as cross-referenced notes, numbered propositions, headings, underlining, and specific marking notes. One of these marking notes was the letter "D," which Steno used to signal "both reasons and things occurring

---

77. Bartholin, *De libris legendis*, 188–189.

78. William Newman, *Atoms and Alchemy: Chymistry and the Experimental Origins of the Scientific Revolution* (Chicago: University of Chicago Press, 2006), 90, 157–189.

79. For the quote and more on commonplace books, see correspondingly Gal. 291, fols. 58r (14 May), 37v (21 March), 56r (28 April), in Steno, *Chaos*, 281, 98, 266.

80. Blair, *Too Much to Know*, 73, 88–89.

81. Gal. 291, fols. 76r–245v.

82. On Steno's naming of the manuscript, see Ziggelaar's introduction in Steno, *Chaos*, 15.

83. Blair, *Too Much to Know*, 77, 89; for a similar case, see Grafton and Weinberg, "Johann Buxtorf Makes a Notebook," 278–279. Bartholin advocated a similar practice for contents that did not fit under a heading in Bartholin, *De libris legendis*, 191.

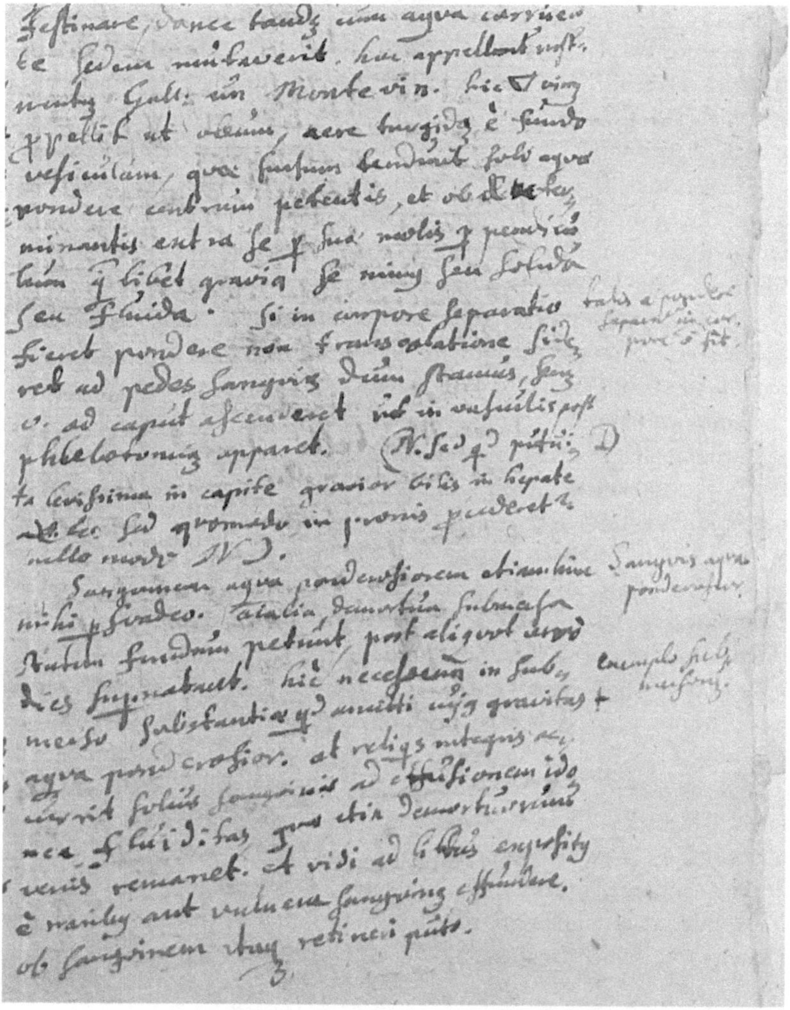

FIGURE 1.2. A personal note by Nicolaus Steno enclosed in parentheses "(N . . . N)" and with a "D" marked in the margins, in BNCF, Gal. 291, fol. 67r. Reproduced with the permission of the Ministero della Cultura/Biblioteca Nazionale Centrale di Firenze.

spontaneously [*cum rationibus tum extempore occurrentibus*]" (see again fig 1.1).[84] These marginalia show Steno actively rereading what he had written, confirming that "Chaos" was the axis of his intellectual life in those months.

---

84. Steno explains using the letter "D" as marginalia in Gal. 291, fol. 42r (28 March), in Steno, *Chaos*, 146.

Steno, however, could only write this notebook because of his friendship with Ole Borch. Borch is known today as the author of one of the first histories of chymistry, but during his lifetime he was known as "a very good philosopher as well as an excellent humanist," according to an English diplomat.[85] The University of Copenhagen hired him as a professor of philology in 1660.[86] Borch took his first job after finishing his studies at the university in 1650 as a teacher at the Cathedral School, where he met the young Steno for the first time. Five years later he became the tutor to two sons of Joachim Gersdorff (1611–1660), steward of the Royal Court and a learned Danish nobleman.[87] The tutoring continued even after Borch started his European travels in 1660 because the Gersdorff brothers accompanied him.[88] The income from this private teaching likely helped him cover the cost of his journeys. He once wrote that he "enjoyed a courtly life" at the Gersdorffs' in Copenhagen, where he had access to an extensive library and a chymical laboratory that was also useful to Steno.[89]

Almost all authors named in the "Chaos" manuscript are mentioned in the three-volume catalog of the Gersdorff library, an indication that Borch was one of the primary sources for Steno's books.[90] There were good libraries in Copenhagen in 1659, including the university library, which received about 2800 volumes of medical literature that year alone.[91] However, because of the Swedish military siege of the city, it was

---

85. As translated in Nielse, Kragh, Nielsen, and Kjaergaard *Science in Denmark*, 53.
86. On Borch, see OBI, 1:xv–xxi.
87. Joachim Gersdorff was the "magister aulae" at this time; see C. G. Hoffmann, *Scriptores rerum lusaticarum antiqui et recentiores* (Leipzig: David Richter, 1719), Pars altera, 158.
88. Steno mentions Borch saying that "the Court Steward has seen the experiment himself"; see Gal. 291, fol. 37v (21 March), in Steno, *Chaos*, 99. See also OBI, 1:xxxi.
89. OBI, 1:xvi.
90. See Copenhagen, Royal Library of Denmark, KBs arkiv (indtil 1943) E 2: Catalogi Bibliothecae Gerstorffianae. Strangely, Egil Snorrason and Gustav Scherz did not consider the Gersdorff library, which was only mentioned later by Schepelern and Ziggelaar. Some books read by Steno were neither at the University Library nor at Thomas Bang's library but were in Gersdorff's, including Drexel's *Joseph egyptiæ prorex* (Antwerp: Cornelius Leysser, 1641), Christian Nold's *Leges distinguendi* (Frankfurt: Johann Wilhelm Ammonius, 1657), and Francis Bacon's *De dignitate et augmentis scientiarum* (Leiden: Franciscus Moyardus & Adrianus Wijngaerde, 1645).
91. These volumes came from the collection of Henrik Fuiren (1614–1659), a cousin of Thomas Bartholin who died on 8 January 1659. See Steno, *Niels Stensen*, ed. Scheperlen, xii; and Egill Snorrason, "The Studies of Nicolaus Steno 1659 in Copenhagen Libraries," in Scherz, *Steno and Brain Research in the Seventeenth Century*, 69–93, esp. 75–76.

hard, if not impossible, to access them.⁹² Steno's notes also speak of advice that Borch gave Steno on what and how to read, further confirming Borch's influence on Steno's reading program. Borch once recommended "the doctrine of purges, . . . [which] should be read in Simon Paulli's *Quadripartitum* [*de simplicium medicamentorum facultatibus* (Rostock, 1640)] and that all dosages in it should be noted down from it."⁹³ Borch spoke with Steno either when they went "for a walk outside" or while they were at the laboratory making observations with "Borch's microscope [*microscopio*]" (see again fig. 1.1).⁹⁴ Borch even shared his reading notes with Steno, including notes on Gassendi's philosophy. Although Steno studied Gassendi from Borch's notes, he concluded his notes as if they were excerpts from another printed book, with the original book's title underlined: "Ex Epicuri Philosophia Gassendi. O[laus] B[orrichius]."⁹⁵ Borch was one of many friends with whom Steno shared his scientific research, to great intellectual and personal profit. More important, through Borch's friendship Steno first experienced the advantages of researching under courtly patronage, something he would seek, with success, during his European tour.

A Pious Search for Certainty and Focus

Steno took his notes in small handwriting and a two-column format, numbering all 184 columns. Most of his notes are excerpts of book sections; he rarely copied entire books. When Steno read Kircher's *Magnes sive de arte magnetica*, he took notes from the third and last part, which was quite long in itself.⁹⁶ On the contrary, from Galileo's *Sidereus nuncius* (Venice, 1610), he took a few notes only on the dedicatory preface and on the magnification power of telescopes.⁹⁷ Overall, Steno carefully read and excerpted more than thirty scientific books, "first to understand and

---

92. The university library was in the Trinity Church building; the building housed the Round Tower, which had an astronomical observatory and was most likely used for military purposes during the siege. See Snorrason, "The Studies of Nicolaus Steno."

93. Gal. 291, fol. 61v (12 June), in Steno, *Chaos*, 313.

94. Gal. 291, fol. 28v (9 March), in Steno, *Chaos*, 32. Borch's name is explicitly mentioned at least nineteen times throughout the *Chaos* manuscript.

95. Gal. 291, fol. 75v (28 June), in Steno, *Chaos*, 447.

96. Kircher, *Magnes sive de arte magnetica*, 463. Steno wrote several pages with excerpts from Kircher's book 3, but he dedicated only one page to both books 1 and 2; see gal. 291, fol. 55r (20–22 April), in Steno, *Chaos*, 257, 259, 260.

97. Gal. 291, fol. 61v (11 June), in Steno, *Chaos*, 301–302.

then to become more familiarized," as he wrote.[98] Most books were about chymistry, medicine, natural philosophy, mechanics, or mathematics—all topics that are part of modern science today—but he also read travel literature, moral theology, and even poetry.[99] Interestingly, there is no mention of traditional textbooks, such as books on Aristotle's *Physics* or Euclid's *Elements*, which means he was reading books outside the university curriculum. In addition, most of the books had been published very recently. Borel's *Historiarum et observationum medicophysicarum centuriae IV* (Paris, 1656) was published only three years before Steno read it. Moreover, the oldest books he read were less than a century old: Cornelius Schylander's *Practica chirurgiae* (Antwerp, 1577) and Tycho Brahe's *Epistolarum astronomicarum liber* (Uraniborg, 1596), which he read for medical information.[100]

Steno's choice of recent books speaks to the status of science in the 1650s. As new theories about the natural world emerged across Europe, there were many textbooks to follow and little agreement between them.[101] Thus, "Chaos" became something like Steno's own textbook, where he quoted from authors as varied as Gassendi on natural philosophy, Untzer on chymical theories of matter, Kircher on physico-mathematics, Borel on medical cases, and Pecquet on anatomy. In true humanist fashion, the "Chaos" manuscript became a commonplace book of the new sciences.

Unsurprisingly, all this information eventually overwhelmed Steno. Soon he was to write that a "harmful hastening should be avoided [*noxia volitatio vitanda*] in conversations, especially with Borch." The solution should be to "stick to one topic [*una materia persistendum est*]" instead.[102] At another time he stopped reading a book because "these things

---

98. Gal. 291, fol. 37r (20 March), in Steno, *Chaos*, 96–97. I consider books Steno read in passing as those on which he wrote fewer than five lines of notes.

99. On moral theology, he read Drexel's *Joseph ægypti prorex*; see, e.g., Gal. 291, fol. 28r (8 March), in Steno, *Chaos*, 22–23; for poetry, Steno said that the verses of Jacques Favereau, a Frenchman mentioned in Borel's *Historiarum*, were communicated to him by Borch; see *Chaos*, fol. 29r (10 March), in Steno, *Chaos*, 35.

100. See correspondingly, Gal. 219, fols. 28r (8 March), 36r (15 March), in Steno, *Chaos*, 21, 86. Steno also copied from a 1641 edition of Marsilio Ficino's *Opera omnia*, in Gal. 291, fol. 42r (28 March); Steno, *Chaos*, 145. Tycho's book is a collection of astronomical letters, but Steno copied twelve lines on failed healing attempts.

101. See Ann Blair, "Natural Philosophy," in Park and Daston, *Early Modern Science*, 365–406, esp. 393–403; Mark A. Waddell, *Jesuit Science and the End of Nature's Secrets* (Burlington, VT: Ashgate, 2015), 17–28; and Robert Pasnau, *After Certainty: A History of Our Epistemic Ideals and Illusions* (Oxford and New York: Oxford University Press, 2017).

102. Gal. 291, fol. 33v (12 March), in Steno, *Chaos*, 77.

considered in such a superficial way [and] under such limits of time seem useless [*haec superficialiter ita cogente angustia percurrere inutile videtur*]."[103] When studying the tables of Sennert's medicine, he wrote that his mind became "distracted with various thoughts for nearly the entire day."[104] This temptation of distractions and superficiality affected him so much that this struggle acquired an interior dimension, leading him to write down personal resolutions to overcome distractions at work. Steno made prayerful resolutions to "take away from me, O God, such plague and grant that I may free the soul from all distractions, work on one thing alone [*unum agere*], and make familiar to me those tables of medicine."[105] In practice, he blocked out his mornings to focus on medicine, writing that "before noon nothing must be done except medical things [*præter medica nihil agendum*]."[106] But a few days later Steno was distracted again and found himself copying several chapters from Kircher's book on magnetism, ending the day with the realization that he had "hardly done anything else today [*parum etiam qua cætera actum est hodie*]." He then asked God to guide him and "grant that I may work in good order on something sound and persistently."[107] This combination of work habits with his interior life shows that Steno worked hard to focus, an ability that would become central to his interdisciplinary method.

In the world of ideas, Steno was also frustrated with the disagreement between seventeenth-century authors on questions such as the origin of colors and the existence of the vacuum. After reading Herman Conring's *De hermetica medicina* (Helmstedt, 1648), Steno wrote that "neither Conring, nor Sennert, nor [Johann] Sperling is reliable" on the theory that placed sulfur as the origin of all color.[108] A few months later he copied from Gassendi's natural philosophy that "color is not truly in bodies" but is a result of different combinations of particles.[109] This disparity contributed to Steno's skepticism, mentioned earlier, about the epistemic value of colors, which became crucial for his work on the body and fossils. He also copied ideas from Gassendi and Pecquet regarding the vacuum, whose

---

103. Gal. 291, fol. 37v, (21 March), in Steno, *Chaos*, 100. He was reading Jordan, *De eo quod divinum aut supernaturale est*.
104. On the quote and Sennert's tables, see correspondingly Gal. 291, fol. 37r (19 and 21 March), in Steno, *Chaos*, 95–97.
105. Gal. 291, fol. 37r (19 March), in Steno, *Chaos*, 95.
106. Gal. 291, fol. 36v (16 March), in Steno, *Chaos*, 89.
107. Gal. 291, fol. 39r (22 March), in Steno, *Chaos*, 123.
108. Gal. 291, fol. 42r, (28 March), in Steno, *Chaos*, 145.
109. Gal. 291, fol. 71r, (27 June), in Steno, *Chaos*, 401.

existence both Aristotle and Descartes denied.[110] In a series of personal notes unrelated to a book but written after he read Pecquet, Steno wrote that "all observations and objections against the vacuum" prove that a vacuum does not exist in ample space. However, they say nothing against "small empty spaces [*vacua spaciola*]."[111] Moreover, he also noticed that certain disagreements originated in poor argumentation. When reading Varen's *Geographia*, one of the books that contributed to the spread of the Copernican model after the 1650s, Steno wrote down the inconsistencies of the geocentric model.[112] According to Steno's notes (and Varen's book), Aristotle assumed that the Earth was at the center of the universe to prove its sphericity, "not without a logical fallacy [*non sine paralogismo*]." Christoph Clavius (1538–1612) and Willebrord Snellius (1580–1626), on the other hand, used Archimedes' argument on the sphericity of water to prove that the Earth was at the center of the universe. However, according to Steno's excerpt, Archimedes also assumed that the Earth had a center to which all bodies tend; thus, Clavius's and Snellius's argument seemed circular.[113] On the other hand, the young Steno had likely learned that the Earth was at the center of the universe, according to the textbook of his Cathedral School mentor Hilarius on the "principles of spherical doctrine."[114] Contradictory ideas advanced by reputable authors with good and not-so-good arguments abounded in the seventeenth century.

Despite these intellectual discrepancies, seventeenth-century scholars were convinced that a proper explanation of natural phenomena could be found. One of them was Gassendi,[115] who opted for theories that matched observations with significant probability but could not be proved with certainty.[116] Gassendi admitted that he wanted to find a "middle way between skeptics . . . and the dogmatics." Skeptics thought reality was "so completely unknown that no criteria can be found for determining it." In contrast, dogmatics were those who "do not really know everything they believe they know."[117] Steno would use a strikingly similar definition

110. For a summary of ideas on the vacuum in "Chaos," see Steno, *Chaos*, 473–474.
111. Gal. 291, fol. 68v, (23 June), in Steno, *Chaos*, 369.
112. William Warntz, "Newton, the Newtonians, and the *Geographia Generalis Varenii*," *Annals of the Association of American Geographers* 79, no. 2 (1989): 165–191.
113. Gal. 291, fol. 55v (24 April), in Steno, *Chaos*, 261. See also Varen, *Geographia generalis*, 20–23.
114. Hilarius, *Progymnasmatum mathematicorum enchiridion*, 92.
115. Pasnau, *After Certainty*, 186.
116. Levitin, *The Kingdom of Darkness*, 98–101.
117. Gassendi, *Syntagma philosophicum*, as quoted in Osler, *Divine Will and the Mechanical Philosophy*, 106.

of this middle way in his last scientific publication about the Earth. Historians have also called Gassendi's epistemology a science or knowledge of appearances because it was built on analogies between imagined mechanisms and observed reality.[118] To draw an example from Steno's notes, Gassendi described light rays as "shapes of corpuscles continuously succeeding each other with unutterable agility [*per radios lucis intelligo . . . species corpusculorum sibi continenter pernicitate ineffabili succedentium*]." He then used this corpuscular description of light to explain its "reflection out of hindering solid bodies, and refraction from those [bodies] that allow for some, but not entirely free, traffic [*ex iis qvae transitum qvidem sed non omnino liberum praebent*]."[119]

Steno also encountered another probabilistic approach to the study of nature in those days. Recent studies on Kircher showed that the spectacular devices in his Roman museum were intended to display probable descriptions of nature rather than to promote certain knowledge of causes.[120] In his work on magnetism, Kircher explicitly commented that, as he contemplated "the theater of the world covered in its infinite variety of things," he found no effect so marvelous or a force so hidden to which "at least a probable or verisimilar cause could be assigned to, if not a certain or evident one [*causa probabilis saltem, aut verisimilis, si non certa aut evidens, assignari possit*]."[121] Steno did not copy this specific quote. However, he definitely read it, because he took and underlined notes from this section of Kircher's book, including that many "assert that nothing can be known, but the author inveighs against these extensively."[122] Steno learned and found interest in Gassendi's and Kircher's probabilistic epistemologies, but it is hard to know what he thought about them. Nonetheless, he grew closer to epistemic probabilism as he relied increasingly on interdisciplinary analogies, especially in his publications about the Earth.

Steno's personal view in "Chaos" was that there was a need for accurate knowledge in the natural sciences. For him, useful knowledge either improved epistemic certainty or helped human life: "In all physical problems, as well as in mathematical, it must be seen whether something useful drawn from it may either extend life or [extend] an accurate knowledge

---

118. Osler, *Divine Will and the Mechanical Philosophy*, 102–118.
119. Gal. 291, fol. 71r (27 June), in Steno, *Chaos*, 402–403.
120. Waddell, *Jesuit Science and the End of Nature's Secrets*, 91.
121. Kircher, *Magnes sive de arte magnetica*, 467–468. See also Waddell, *Jesuit Science and the End of Nature's Secrets*, 119–159.
122. Gal. 291, fol. 38r (22 March), in Steno, *Chaos*, 114. See Kircher, *Magnes sive de arte magnetica*, 463–469, esp. 467. Underlining in the original.

[*cognitio accurata*] of nature."[123] Steno perceived that early modern medicine lacked depth and that worried him significantly. In his notebook he wrote that "in medicine, one can make progress by looking not so much at different humors and [their] causes alone, but at the different qualities [of a disease] and their causes." He gave the example of a "burning and hard tumor [*tumor ... fervidus durus*]" that should be treated by looking at its qualities. Even though not everyone agrees "on the [general] natures of humors [*naturæ humorum*]," they should at least "agree on the cure."[124] He concluded that "the doctrine of faculties and dosages of medicines seems far from perfect [*videtur doctrina ... parum perfecta*]."[125] As a testament to his unity of life—the coherence between the worlds of work and prayer—, Steno united his searches for reliable knowledge and focus. One Sunday Steno asked God "that an abstinence from ... a hasty and ill-considered judgment or opinion about something ... may rule over me, unless known with great exactitude [*nisi exactissime noto*]."[126]

The "Chaos" manuscript also shows that the young Steno placed great weight on scientific observations, in contrast to overly bookish and speculative practices. In yet another religious reference, he wrote that those "who are not willing to inspect the very works of nature [*qui non ipsa naturæ opera inspicere volunt*]" but instead prefer "reading others ... sin against the majesty of God." "In this way," he continued, "they not only do not enjoy by inspection the beauty of God's wonders, but they also lose time which is to be given to necessities and the good of our neighbor [*necessariis et proximo commodo dandum*]." This charitable principle was likely related to the improvement of medical knowledge. Therefore, Steno made yet another resolution to spend his time "not on meditations, but only on the search, Experience, and register of natural things and similar reports observed by the Ancients, as well as in investigating them if possible."[127] Perhaps he had Descartes's *Meditationes de prima philosophia* (Paris, 1641) in mind when rejecting meditations. Steno's intellectual quest for natural observations also included travel books, which should be read so that "all experiments concerning science etc. should be considered."[128]

---

123. Gal. 291, fol. 54v (17 April), in Steno, *Chaos*, 253.
124. Gal. 291, fol. 56v (5 May), in Steno, *Chaos*, 271.
125. Gal. 291, fol. 58v (17 May), in Steno, *Chaos* 283.
126. Gal. 291, fol. 41v (27 March), in Steno, *Chaos*, 142.
127. Gal. 291, fol. 44r (30 March), in Steno, *Chaos*, 160.
128. Gal. 291, fol. 37r, (20 March), in Steno, *Chaos*, 95. Steno wrote this after reading Vincent Le Blanc, *Les Voyages* (Paris: Gervais Clousier, 1649).

Finally, these religious comments on his work resolutions speak to Steno's personal piety, manifested in "Chaos" in more explicit ways. He often wrote of his efforts to keep "the cross of Christ always before the eyes" and to "expect death at any moment."[129] This religious sensibility resonates with the pious environment of Lutheran Denmark, which affected academic studies, especially medical studies, in that country. Caspar Bartholin the Elder began his book *De studio medico*, on "advice about the beginning, continuation, and conclusion of medical studies," by recommending medical students to foster "piety, an unblemished life, daily prayers, love of the Word of God, and heed for it."[130] This integration of Steno's religious devotion with his scientific work is specific to that moment of his life. But it was not the last time it would happen, and it may explain why his religious conversion happened amid an intensive work moment in Florence.

Post-Harveian Anatomy at the Center

By 1659 Nicolaus Steno was well informed of the scientific developments of his time in natural philosophy, medicine, and chymistry. However, the discipline that mattered the most for Steno then was anatomy, probably owing to almost three years of anatomical studies under Thomas Bartholin at the University of Copenhagen. Early modern anatomy, perhaps better than any other discipline, showed Steno that he could obtain precise knowledge about the body and rely on ideas and methods from other disciplines. Nevertheless, to fully understand the significance of Steno's early anatomical studies and practices, it is essential to know what anatomy was like in the 1650s.

In the early modern period, the knowledge of the human body was based on the works of Galen of Pergamon (129–216 AD), which all early modern physicians had to master, including Steno. Although Vesalius and other sixteenth-century physicians highlighted a few errors in Galen's human anatomy, they continued to follow its general outlines.[131] This attachment to Galen was not because of some conservative attitude but

---

129. Gal. 291, fol. 37r (18 March), in Steno, *Chaos*, 94.
130. Caspar Bartholin, *Opuscula quatuor singularia* (Copenhagen: Georg Hantzsch, 1628), as quoted in Grell, "Caspar Bartholin and the Education of the Pious Physician," 79.
131. Andrew Cunningham, *The Anatomical Renaissance: The Resurrection of Anatomical Projects of the Ancients*, 2nd ed. (London: Routledge, 2016), esp. 116–121. On Galenic physiology, see Rudolph E. Siegel, *Galen's System of Physiology and Medicine: An Analysis of His Doctrines and Observations on Blood Flow, Respirations, Humors and Internal Diseases* (New York: S. Karger, 1968).

because Galenic medicine and anatomy proved to be remarkably adaptable to new ideas and practices.[132] However, when Steno started his university studies in 1656, anatomy was undergoing dramatic changes away from Galen's understanding of the body. Two new vital discoveries were at the forefront of these changes: the circulation of the blood, published in 1628 by William Harvey, and the observations of the lacteals or milky vessels, as described by Gaspare Aselli (c. 1581–1625) in a book published in 1627.[133] Their books were often bound together in the early modern period as a testament to their joint significance, as Domenico Bertoloni Meli has pointed out.[134]

Among other things, Harvey challenged the basic understanding of Galenic anatomy by uniting veins and arteries into one circulatory system.[135] He discovered that blood circulated in the body by observing the blood flow in the arteries of living animals, often with ligatures, and associating it with the heart's systole (contraction) rather than its diastole (distension). By measuring the timings of the pulse, Harvey concluded that so much blood could only flow so rapidly in a circular motion.[136] The lacteals, in turn, were observed by Aselli when vivisecting animals shortly after they ate. Aselli saw a white fluid in the vessels of the mesentery, a region inside the abdomen. However, Aselli interpreted his observations within a Galenic framework, concluding that chyle (the fluid produced in the intestines during digestion) moved from the intestines to the liver through the newly discovered lacteals. According to Galen, the liver used the chyle to make blood, which then moved from the liver to the heart. Therefore, Aselli's and Harvey's discoveries did not affect this Galenic understanding of the liver as the producer of blood in the body.[137] They did, however, reinforce the role of vivisection as an essential form of observation in

---

132. Evan R. Ragland, *Making Physicians: Tradition, Teaching, and Trials at Leiden University, 1575–1639*, Clio Medica 106 (Leiden: Brill, 2022), 7–18, 113–129, 298–304.

133. Bertoloni Meli, *Mechanism, Experiment, Disease*, 1–3, 34–40. Gabriele Falloppio (c. 1522–1562) also observed the lacteals in the 1550s; see Michael Stolberg, *Gabrielle Falloppia, 1522/23–1562: The Life and Work of a Renaissance Anatomist* (Abingdon: Routledge, 2023), 50–51.

134. Bertoloni Meli, *Mechanism, Experiment, Disease*, 2.

135. Frank, *Harvey and the Oxford Physiologists*, 1–20.

136. Measurement was one of the arguments he provided. See Andrew Cunningham, *"I Follow Aristotle": How William Harvey Discovered the Circulation of the Blood* (London: Routledge, 2022), 102–115, esp. 110–113.

137. Thomas S. N. Chen and Peter S. Y. Chen, "William Harvey as Hepatologist," *The American Journal of Gastroenterology* 83, no. 11 (1988): 1274–1277; and Guerrini, *The Courtiers' Anatomists*, 67.

anatomy.[138] Harvey explicitly wrote that he discovered "the motion and use of the heart [*motum et usum cordis*]" by "collecting many observations by means of frequently examining many and various living animals [*multa frequenter et varia animalia viva introspiciendo multis observationibus collatis*]."[139] The anatomists who built on their legacy throughout the seventeenth century thus also vivisected many animals. For example, Pecquet, the French anatomist whose work was foundational for Steno, vivisected more than a hundred animals in just three years.[140]

Thomas Bartholin learned about these discoveries and vivisection practices during his medical pilgrimage in the 1640s. In Leiden he befriended the anatomists Franciscus Sylvius (1614–1672) and Johannes Walaeus (1604–1649), who were making new observations that further confirmed that blood circulated within the body.[141] Bartholin, too, made his contribution by producing a new edition of his father's famous anatomy textbook—the *Anatomicæ institutiones* (Wittenberg, 1611). It became the first anatomy manual that mentioned the circulation of the blood.[142] Back home in Copenhagen, Bartholin promoted regular dissections as a standard practice at the new university anatomical theater. One of his students, the German Michael Lyser (1626–1660), published a dissection manual while studying under Bartholin.[143] *Culter anatomicus* (*The Anatomical Knife*; Copenhagen, 1653) was republished through the eighteenth century and translated into English and German—a testament to the influence of Danish anatomy in the second half of the seventeenth century.[144]

---

138. Anita Guerrini, "Experiments, Causation, and the Uses of Vivisection in the First Half of the Seventeenth Century," *Journal of the History of Biology* 46, no. 2 (2013): 227–254; and Domenico Bertoloni Meli, "Early Modern Experimentation on Live Animals," *Journal of the History of Biology* 46, no. 2 (2013): 199–226.

139. William Harvey, *Exercitatio anatomica de motu cordis et sanguinis in animalibus* (Frankfurt: Gulielmus Fitzeri, 1628), 20–21 (hereafter cited as *De motu cordis*). See also Andrew Wear, "William Harvey and the 'Way of the Anatomists,'" *History of Science* 21, no. 3 (1983): 223–249, esp. 239.

140. Bertoloni Meli, "Early Modern Experimentation in Live Animals," 209–210.

141. For a recent account, see Andersen, *Thomas Bartholin: Laegen & anatomen*, 30–142.

142. Thomas Bartholin, *Anatomia parentis Caspari Bartholini* (Leiden: Franciscus Hack, 1641). The circulation was in the two attached epistles by Johannes Walaeus; see Bertoloni Meli, *Mechanism, Experiment, Disease*, 35.

143. Skavlem, "The Scientific Life of Thomas Bartholin," 69–70.

144. Michael Lyser, *Culter anatomicus* (Copenhagen: Georg Lamprecht, 1653). See also Axel Garboe, "Michael Lyser, a 17th Century Anatomist," *Acta Medica Scandinavica* 142, no. S266 (1952): 63–73. On the impact of Lyser's manual, see Guerrini, *The Courtiers' Anatomists*, 38–39.

The main anatomical breakthrough after Harvey and Aselli happened when Pecquet discovered a new vessel called the thoracic duct and published an account of it in *Experimenta nova anatomica* (Paris, 1651).[145] Unlike Aselli, who claimed that the chyle in the lacteals moved to the liver, Pecquet argued that the chyle sidestepped the liver and moved toward the thoracic duct and from there to the heart. This was a dramatic claim since, in Galenic anatomy, the liver produced blood from ingested food. If the chyle, the product of digestion, did not flow to the liver, then the liver would remain without its material to make blood, thus losing its central function. Thomas Bartholin confirmed Pecquet's discovery by finding the same thoracic duct in animal and human bodies but initially disagreed with the claim that the chyle did not move to the liver.[146] However, after a few more observations, he sided with Pecquet that the chyle sidestepped the liver altogether.[147] From then on, Bartholin became one of Pecquet's most preeminent supporters and wrote a book about the dethroning of "the poor liver from the job of sanguification [*sanguificatio*]."[148] Part of this support was directed at himself, because Bartholin also discovered a new type of vessel, like the lacteals, linked to the thoracic duct throughout the body. He called these vessels the lymphatics and became an apologist for this discovery in various controversies.[149] Modern anatomy still describes this group of vessels as the lymphatic system. Bartholin published it first in a book named the *Vasa lymphatica . . . et hepatis exequiæ* (*Lymphatic Vessels . . . and the Exequies of the Liver*; Copenhagen, 1653), where he wrote an obituary poem for the liver, a sign of the significance of his anatomical discovery.

By the time Steno entered the University of Copenhagen, Bartholin had already performed his last public dissection.[150] However, he contin-

---

145. Bartolomeo Eustachi (1520–1574) also observed this in 1656; see Charles Ambrose, "Immunology's First Priority Dispute: An Account of the 17th-Century Rudbeck-Bartholin Feud," *Cellular Immunology* 242, no. 1 (2006): 1–8, esp. 3.

146. Thomas Bartholin, *De lacteis thoracicis . . . historia anatomica* (Copenhagen: Melchior Martzan, 1652), 60–62.

147. Raphael Suy, Sarah Thomis, and Inge Fourneau, "The Discovery of the Lymphatics in the Seventeenth Century. Part III: The Dethroning of the Liver," *Acta Chirurgica Belgica* 116, no. 6 (2016): 360–397, esp. 395.

148. Thomas Bartholin, *Vasa lymphatica . . . et hepatis exequiae* (Copenhagen: Petrus Hakius, 1653), esp. 54.

149. Raphael Suy, Sarah Thomis, and Inge Fourneau, "The Discovery of the Lymphatic System in the Seventeenth Century. Part IV: The Controversy," *Acta Chirurgica Belgica* 117, no. 4 (2017): 270–278.

150. Skavlem, "The Scientific Life of Thomas Bartholin," 80.

ued to perform them privately at various homes and the university's anatomical theater.[151] As his student, Steno most likely witnessed most if not all of them. Because Bartholin kept writing about the lymphatics amidst ensuing controversies, Steno probably learned about these new vessels directly from Bartholin's dissections. Steno's days with Bartholin ended with the 1659 siege of Copenhagen, which put a stop to all academic activities. Bartholin escaped the city and went to his home in the countryside before it began.[152] Perhaps because he was alone, Steno began stretching the boundaries of what kinds of dissections and vivisections he could do when writing "Chaos." The craftsmanship skills he probably acquired in his family's goldsmith workshop undoubtedly contributed to his dexterity with the anatomical knife. There are various references in "Chaos" to dissections that Steno may have done during the siege of Copenhagen. The longest one is a guideline for the vivisection of a dog. It is worth quoting in full because it speaks to Steno's familiarity with new techniques for the injection of fluid into the blood vessels.[153]

> Take a dog. It should be placed vertically if you like. Open two different veins. Out of one of them, let the blood flow continuously; through the other [vein], infuse just as much hot water (which can be easily done through a little pipe with a bladder attached, such that it may be introduced through incision in the vein and closed with tightenings [*adstringentes*] all around). Let this continue until pure water begins to flow through the other vein. Let it continue for as long as you want until you observe either that the dog died or fell into lipothymia [i.e., fainted] or something else. Then, take another dog, and once tied to the table, open a vein and then through a suitable channel, having placed in between very hot water or fire, let the blood flow into the dead dog. The water is extracted through the other channel by the same force. Here, what follows must be noted. If you observe that the previously dead

---

151. Bartholin dissected an African lion at the anatomical theatre on 19 December 1656; see Thomas Bartholin, *Historiarum anatomicarum rariorum centuria III & IV* (The Hague: Adriaan Vlacq, 1657), 276–280. He also dissected a muscular man with eleven ribs on one side and, at Simon Paulli's house, an owl frozen to death, both in 1658; see Thomas Bartholin, *Historiarum anatomicarum rariorum centuria V et VI* (Copenhagen: Henric Gödian, 1661), 1–3, 105–107. I am thankful to Jesper Andersen for pointing these references to me.

152. See Scherz's biography in Kardel and Maquet, *Nicolaus Steno: Biography and Original Papers*, 47.

153. Bertoloni Meli, *Mechanism, Experiment, Disease*, 36–37, 97. See also Henricus a Moinichen to Thomas Bartholin, 22 April 1655, in EMC I, 600–601.

[dog] comes back to life, tighten and close the bandage, and it will live as long as it can without it.[154]

Anatomists used injection techniques in the 1650s to confirm Harvey's theory of blood circulation. The English anatomist Francis Glisson (1597–1677) also mentioned, in 1654, that he injected hot water into blood vessels using a bladder similar to that employed by Steno.[155] Yet, Steno did not mention Glisson in the "Chaos" manuscript and may have been unaware of his work.[156] Moreover, unlike Glisson, Steno was interested in seeing the external effects of injecting other fluids into the blood flow rather than studying the vessels themselves. Therefore, Steno included the extra step of transfusing blood from a living dog to a dead one to see whether it came back to life.[157] Blood transfusions were rarely performed in the 1650s and were not published about until 1666, in the English journal *Philosophical Transactions*.[158] When the Académie des Sciences in Paris began a blood transfusion project that same year, there were so many difficulties that it was soon abandoned.[159] Therefore, given the speculative style of Steno's guidelines, it is unlikely that Steno conducted this specific one in Copenhagen. But he did perform dissections during the siege of Copenhagen, such as when he dissected another dog that had drowned.[160] Regardless, it is remarkable that he was so familiar with blood injections that he wrote about the possibility of transfusions in so much detail.

The German physician Andreas Libavius (1555–1616) had already written about blood transfusions, suggesting that transfusion could be an excellent method by which to heal sick persons by injecting them with the blood of a healthy one.[161] Steno may have read Libavius's speculations on

---

154. Gal. 291, fol. 56v (3 May), in Steno, *Chaos*, 269–270.
155. Francis Glisson, *Anatomia hepatis* (London: Du-Gard for Octavian Pulleyn, 1654), 314.
156. The 1655, 1660, and 1663 editions of Thomas Bartholin's *Anatomia reformata* do not yet include Glisson's treatise on the liver.
157. Gal. 291, fol. 56v (3 May), in Steno, *Chaos*, 270. Steno probably did this experiment to test the ancients' and Harvey's idea that life is contained in the blood; French, *William Harvey's Natural Philosophy*, 298.
158. N. S. R. Maluf, "History of Blood Transfusion," *Journal of the History of Medicine and Allied Sciences* 9, no. 1 (1954): 59–107.
159. Guerrini, *The Courtiers' Anatomists*, 99–105.
160. Gal. 291, fol. 68v (23 June), in Steno, *Chaos*, 369.
161. Maluf, "History of Blood Transfusion," 59–60.

this topic, because he mentions Libavius's works in "Chaos."[162] But Steno seemed more interested in using blood transfusions to study the physiological process of digestion than for the healing purposes mentioned by Libavius. After the guidelines on injecting fluid into a living dog, Steno suggested opening "the [dog's] abdomen, observ[ing] the stomach" and "introduc[ing] well-chewed food" in it. He wanted to see "whether new blood or excrements could still be generated" while warm water circulated in the animal's veins.[163] Steno's medical interests seemed to lie in studying the causes of things, even if they were not immediately applicable to the world. A week after writing on transfusion, Steno suggested raising a dog, "accustom[ing] it from birth to human food and see[ing] if its excrements are like those of humans."[164] It was also after writing these notes that Steno turned to the studies of Santorio and his measurements of discharged fluids, as mentioned earlier. Steno wanted to understand the invisible digestion processes in greater detail.

During the months Steno was compiling his notebook, he was also interested in the physiology of respiration and the heart's motion, topics that would occupy him in the future.[165] How he wrote about them in "Chaos" reveals that his anatomical interests at the time were already in dialogue with knowledge from other disciplines. In one of his personal notes, he wrote that "in living beings, air enters the lungs in such a way that it goes out simultaneously with soot [*fuligo*]." Then, "through the holes left by the soot, subtle matter [*materia subtilis*] penetrates the [blood] vessels" and perhaps reaches the heart. Steno was interested in the processes that transformed venous blood, often described as sooty, into arterial blood, thought to contain vital spirits.[166] These ideas, however, "should be inquired [into] more accurately and in the order according to the method of Descartes [*accuratius et ordine ad methodum Cartesii*], or simply considering what enters the pores of blood, what its particles [*particulae*] are, how they move," and so on.[167] Steno wrote

---

162. Gal. 291, fol. 35r (undated), 48r (8 April), in Steno, *Chaos*, 82, 199. Steno mentions Libavius, *Singularum partes quatuor* (Frankfurt, 1601), and an unknown Libavius book called *Apocal[ypsi] Hermet[ica]*.

163. Gal. 291, fol. 56v (3 May), in Steno, *Chaos*, 270.

164. For this and another quantitative experiment on digestion, see Gal. 291, fols. 58v (16 May), 61v (9 June), in Steno, *Chaos*, 282, 300–301.

165. On respiration after William Harvey, see Frank, *Harvey and the Oxford Physiologists*.

166. Referring to this problem, Harvey wrote of the venous blood's soot [*fuligines*] and the arterial blood's spirit [*spiritus*]; see Harvey, *De motu cordis*, 16. See also Frank, *Harvey and the Oxford Physiologists*, 14–16.

167. On respiration, see *Chaos*, fol. 39r (22 March), 58v (17 May), in Steno, *Chaos*, 123, 284.

THE USES OF CHAOS; OR, WHAT DID NICOLAUS STENO KNOW? 51

these notes a few days after reading Henricus Regius (1598–1679), a Dutch physician and renowned disciple of Descartes.[168] A few days later he also copied into "Chaos" a few propositions from Descartes's *Principia philosophiae* (Amsterdam, 1644) about the constitution of sulfur, bitumen, and clay, perhaps because of his chymical interests. Interestingly, in Descartes's book, these propositions come after those on the origins of mountains, which Steno must have read too.[169] He would indeed write about the history of the Earth and quote these propositions in his later writings. However, my point here is that Steno's consideration of Cartesian philosophy when thinking about respiration shows that he incorporated corpuscular natural philosophy into his anatomical thinking.

Natural philosophy was not the only interest that Steno placed in dialogue with his anatomical studies. He also did so with his artisanal skills. Upon reading Kircher's lengthy explanation of a hydraulic device, Steno noted only that this device also "showed the systole and diastole of the heart," as mentioned by Kircher.[170] Then, about a month later, Steno proposed "to imitate the motion of the heart" by building himself a device that would do just that. He suggested taking "a little bladder, [which,] using some bottlenecks and bandages, can be shaped into two ventricles." Then he mentioned adding "the arteries or nerves [and] if a certain instrument is made like the clock's pendulum, the bladder [can be] squeezed regularly after a defined interval of time, then in this way, the inserted blood is expelled into the arteries."[171] Anatomy was thus a fertile field for engagement with other disciplines and mechanical gadgets. Not surprisingly, the most-studied books in "Chaos" were templates of the intersection of other disciplines with medicine and anatomy.

Learning Interdisciplinary Methods

There are three sections of "Chaos" that Steno annotated most thoroughly. The first is at the beginning of the notebook, where he excerpted from Borel's *Historiarum et observationum medicophysicarum centuriæ iv*

---

168. Gal. 291, fol. 36r (15 March), Steno, *Chaos*, 87, 89.
169. Gal. 291, fol. 45v (4 April), in Steno, *Chaos*, 177–178. Steno took brief notes from part 4 (On the Earth), propositions 62, 70, 76, 77, and 81. He read the 1644 edition as per the page numbers he copied.
170. Gal. 291, fol. 39r, 22 March, in Steno, *Chaos*, 122; and Kircher, *Magnes sive de arte magnetica*, 527–529.
171. Gal. 291, fol. 59r, 18 May, in Steno, *Chaos*, 285. On the motion of the heart, see also *Chaos*, fol. 39r (22 March), 62v (17 June), in Steno, *Chaos*, 122–123, 322. It is not clear whether Steno built this device.

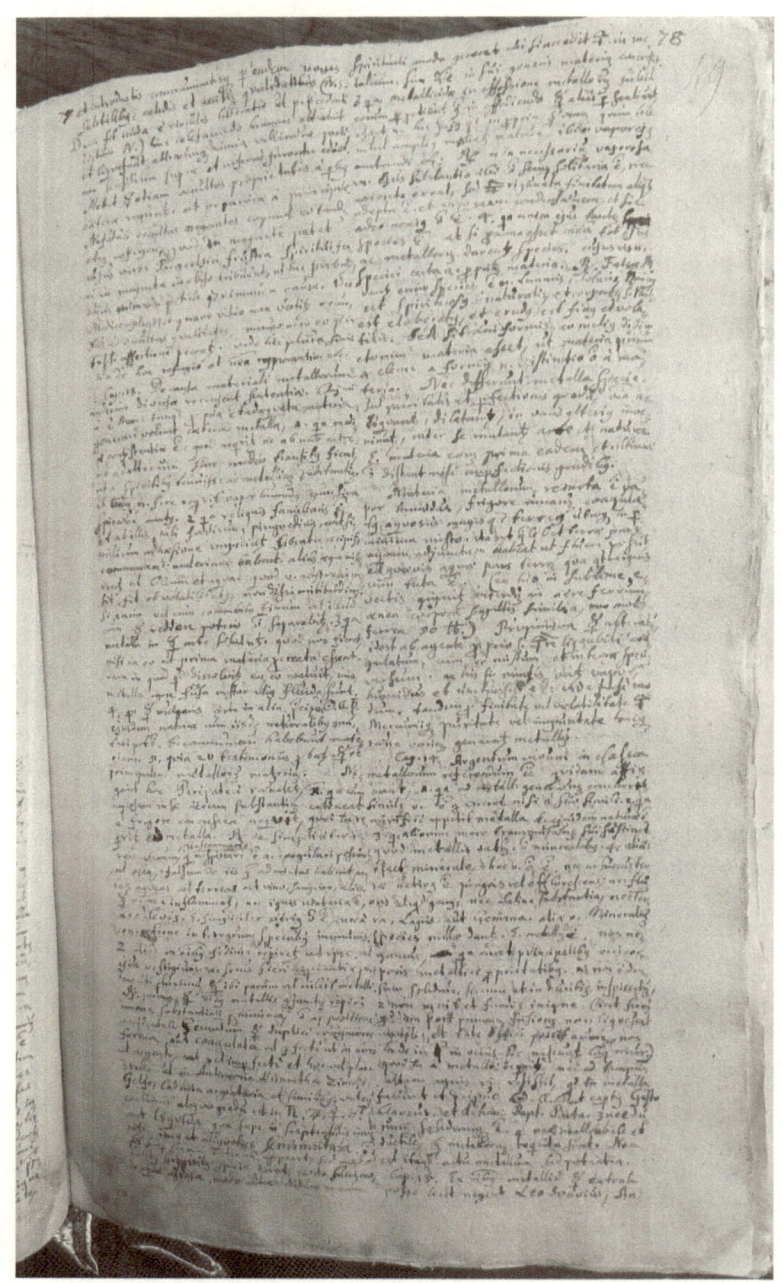

FIGURE 1.3. Notes by Steno on Matthias Untzer, *Anatomia mercurii spagirica* (Halle, 1620), with no marginalia. BNCF, Gal. 291, fol. 49r. Reproduced with the permission of the Ministero della Cultura/Biblioteca Nazionale Centrale di Firenze.

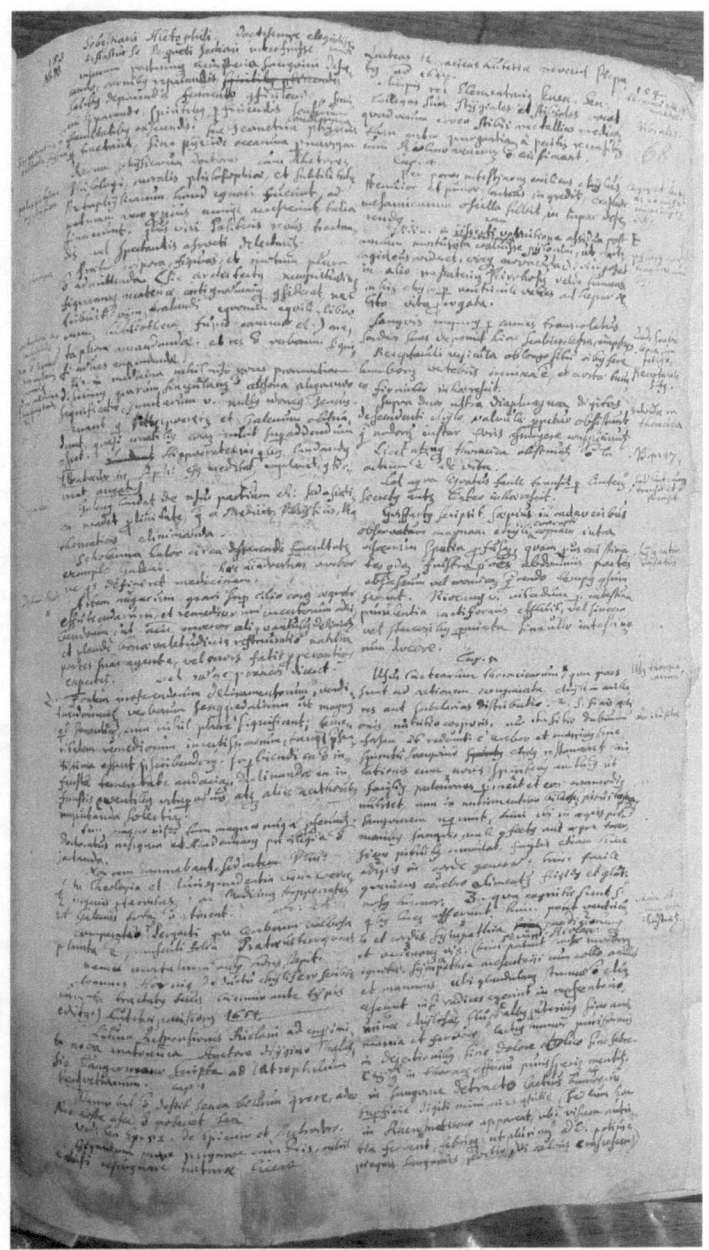

FIGURE 1.4. Notes by Steno on Jean Pecquet, *Experimenta nova anatomica* (Paris, 1654), with headings and other notes in the margins. BNCF, Gal. 291, fol. 68r. Reproduced with the permission of the Ministero della Cultura/Biblioteca Nazionale Centrale di Firenze.

(Paris, 1656). The other two are at the end, when Steno excerpted from Pecquet's *Experimenta nova anatomica* (Paris, 1654) and Gassendi's *Animadversiones* (Lyon, 1649) (see figs. 1.3 and 1.4).[172] These marginalia speak to which parts of the notebook were most helpful to Steno in his subsequent research. They also happen to match the books that mattered the most to Steno at the time of writing, because Borel's and Pecquet's works were the only two books that Steno copied in full into the "Chaos" manuscript.

Borel's *Centuriæ iv* was part of the epistemic genre of *observationes medicæ*, a genre that was growing in popularity in the seventeenth century.[173] As Gianna Pomata has argued, epistemic genres were textual tools for communicating and creating new knowledge. In particular, the *observationes medicæ* was a genre that physicians used to describe and organize empirical data. Rather than focusing on general aspects of diagnosis and therapeutics, these books were lists of individual observations of symptoms and rare phenomena, often with no direct relation to one another.[174] Steno mastered this epistemic genre precisely because, in his first days of writing "Chaos," he copied the complete list of Borel's observations.[175] Borel's *Historiarum et observationum medicophysicarum centuriae IV (Four Volumes of a Hundred Medico-Physical Observations and Descriptions Each)* contained descriptions of diseases, rare anatomical cases, and other strange things such as "meat that glows at night," with no apparent connection.[176] Indeed, it was typical of books of *observationes* that the order in which they were presented was the order in which the data were gathered. Borel's first two *centuriæ* were published as soon as they were ready, three years before his four-volume book came out.[177] Since books of

---

172. Steno also wrote marginalia to his notes on Drexel's *Joseph ægypti prorex*; see Gal. 291, fols. 28r (8 March), 29v (10 March), 33v (14 March), and 36r–36v (16 March).

173. See Pomata, "Sharing Cases"; and Gianna Pomata, "Observation Rising: Birth of an Epistemic Genre, ca. 1500–1650," in Daston and Lunbeck, *Histories of Scientific Observation*, 45–80.

174. Pomata, "Sharing Cases," 206, 219. On the definition of epistemic genre, see ibid., 195, 198–199.

175. Gal. 291, fols. 28v–33r (9 March–12 March), in Steno, *Chaos*, 26–74. These notes are interspersed with notes on other books he was also reading. Steno copied at least the titles of each of Borel's four hundred cases, if not more.

176. Pierre Borel, *Historiarum et observationum medico-physicarum centuriae IV* (Paris: Jean Billaine & Veuve Mathurin Dupuis, 1656), 5. For Steno's notes on this work, see Steno, *Chaos*, 27.

177. Pierre Borel, *Historiarum, et observationum medico-physicarum centuria prima et secunda* (Toulouse: Arnald Colomeri, 1653). Steno read the complete 1656 edition, as attested in Gal. 291, fol. 28v (9 March), in Steno, *Chaos*, 32.

observations were often published as works-in-progress, there is a strong link between the genre of *observationes* and note-taking practices.

This bond between books of *observationes* and note-taking lends greater significance to the "Chaos" manuscript and the role that writing played in Steno's long-term career.[178] Indeed, "Chaos" is just a collection of things that Steno read in books (primarily observations), ideas that he had, and experiments he performed. Observational accounts appear intercalated with reading excerpts in the manuscript in the order in which they were seen or read, revealing a profound union between empirical and textual approaches in Steno's work methods. This union was not uncommon in the practices of early modern scholars, who, in the words of Anthony Grafton and Joanna Weinberg, "transformed the rusty old tools of erudition into the shiny new ones of empirical science, founded on observation."[179] Steno himself spoke of observations in the same terms in which he spoke of textual notes. He wrote that "in the study of nature, it is good to be bound by no discipline at all [*in physicis nulli scientiæ alligari præstat*] but rather to assign every single thing that can be observed under specific headings." This statement confirms not only that Steno liked to read broadly but also that he was selectively searching for empirical data that would then be organized under different headings. This was a direct inheritance from commonplace books that allowed Steno to "elicit [from this data], if nothing else, exact knowledge of something [*elicere, si nihil aliud, aliqualis certa cognitio*]."[180] Thus, the "Chaos" manuscript was the middle step between observing natural reality and writing a book of *observationes*. In this light, the act of writing "Chaos" was more than just gathering information. It was scientific practice in itself.[181]

The genre of *observationes medicæ* also allowed greater inclusion of other topics. Borel alludes to this diversity of observations in his book's preface, where he assumes that his topic was "accounts [not just] of animals, minerals, and plants" but of "almost every art and curious science, especially the hermetic and Galenic disciplines."[182] As such, an apparent list of observations allowed Borel to write about his medical (Galenic)

---

178. On *observationes* and note-taking, see Pomata, "Sharing Cases," 198, 214–215, 221.

179. Grafton and Weinberg, "Johann Buxtorf Makes a Notebook," 277. See also Ann Blair, "An Early Modernist's Perspective," *Isis* 95, no. 3 (2004): 420–430.

180. For this and previous quotes, see Gal. 291, fol. 37v (21 March), in Steno, *Chaos*, 100.

181. See also Pomata, "Sharing Cases," 230–232.

182. Borel, *Historiarum et observationum medico-physicarum centuriæ IV*, iii. Steno did not take notes from the preface, but there is no reason to think he did not read it.

and alchemical (hermetic) interests.[183] This combination of medicine and chymistry might indeed have made Borel attractive to Steno, whose interest in chymistry will become increasingly evident in the following chapters. Borel also published the first biography of Descartes at the end of his book, which Steno read, and a book of observations made with telescopes and microscopes.[184] Interestingly, all these themes converged in the eclectic genre of *observationes*. Therefore, it is not surprising that when Steno started publishing his anatomical research, he also used the genre of *observationes*.

If Borel's book speaks to Steno's interest in scientific observations, Pecquet's *Experimenta nova anatomica* presented attractive, detailed, first-person narratives of anatomical dissections. Moreover, given Bartholin's role in promoting Pecquet's discovery of the thoracic duct and its anatomy, Steno probably considered Pecquet one of the leading anatomists of his time. Nevertheless, observations and medicine were not the only reasons Pecquet's book mattered to Steno; the book also spoke directly to Steno's cross-disciplinary interests in mathematics, since Pecquet was one of the first authors to publish experiments on the vacuum carried out by mathematicians interested in mechanics and hydrostatics.[185] Steno read all five parts of the second edition of Pecquet's *Experimenta nova anatomica* (Paris, 1654). Reading Pecquet was also Steno's most focused time while writing "Chaos." He read it in four consecutive days with almost no interruptions, unlike his approach to other books on which he took notes, including those by Borel and Kircher. The 1654 edition begins with Pecquet's discovery of the thoracic duct, followed by a long second part with mechanical explanations of blood circulation and the motion of chyle.[186] There is then a third part, not included in the original 1651 edition, which he called a "new dissertation on the thoracic lacteals," wherein Pecquet responded to objections he received from Jean Riolan (1580–1657), a leading physician from the University of Paris and prominent critic of Harvey's theory of the circulation of the blood.[187] The fourth part contains letters

183. On the relationship between hermeticism and alchemy, see Lawrence Principe, *The Secrets of Alchemy* (Chicago: University of Chicago Press, 2013), 31, 218 n. 10.

184. Pierre Borel, *Vitæ Renati Cartessi summi philosophi compendium* (Paris, 1656). Steno excerpted it in Gal. 291, fol. 33r (12 March); see in Steno, *Chaos*, 75–76. See also Borel, *De vero telescopii inventore, cum brevi omnium conspiciliorum historia . . . accessit etiam centuria observationum microcospicarum* [sic] (The Hague: Adriaan Vlacq, 1655).

185. Dear, *Discipline and Experience*, 32–62, 151–179.

186. Pecquet, *Experimenta nova anatomica*, 25–50.

187. Pecquet, *Experimenta nova anatomica*, 93–138. See also French, *William Harvey's Natural Philosophy*, 265–279.

from physicians and mathematicians addressed to Pecquet and supporting his work. A new letter by Samuel Sorbière (1615–1670), on the importance of geometry in anatomy, was also included in the 1654 edition.[188] The book concludes with an essay by the physician Pierre de Mercenne (fl. 1650) refuting Riolan's objections.[189]

Steno took only short notes from the first part, probably because he already knew about the discovery of the thoracic duct from his studies with Bartholin. In these few notes he wrote that a "plucked-out heart" continued to beat with "about ninety systoles with the same force," which was not central to Pecquet's argument but spoke to Steno's interests in measurements and the motion of the heart. Moreover, when Pecquet claimed that the lacteals do not merge in the pancreas or the liver, Steno wrote only a note on "valves of the lacteal veins."[190] In that part, Pecquet described a valve in the thoracic duct that stopped the chyle from moving downward to the liver.[191] As to whether there were valves in other lymphatics, this was still an open question, to which Steno would have much to add after arriving in Leiden.[192]

Steno was also interested in Pecquet's statement that "the principle of blood motion is not its weight, . . . neither systole alone . . . nor the diastole's attraction."[193] Although Harvey explained that the heart's systole pumped the blood through the arteries, it was unclear just how it returned to the heart through the veins. This question was especially important in the case of the human body, where most of the blood and lymphatic flow moves against the natural downward tendency of fluids or, in modern terms, against the force of gravity.[194] In the second part of his book Pecquet proposed a mechanical solution to this problem by integrating anatomy with physico-mathematics. His arguments relied on analogies between

---

188. Pecquet, *Experimenta nova anatomica*, 139–180.

189. Pecquet, *Experimenta nova anatomica*, 181–252. This essay was followed by a short letter to Pecquet from Matheus Chatelain, a physician from Montpellier, and an anagram on Riolan, which Steno copied. See *Chaos*, fol. 68v (23 June), in Steno, *Chaos*, 368.

190. Gal. 291, fol. 63v, 19 June, in Steno, *Chaos*, 329. On the beating heart, see Pecquet, *Experimenta nova anatomica*, 5. Interest in valves of the vein had also been critical for Harvey's discovery of blood circulation.

191. Pecquet, *Experimenta nova anatomica*, 12–21.

192. For now, see Suy, Thomis, and Fourneau, "The Discovery of the Lymphatic System in the Seventeenth Century: Part IV."

193. Gal. 291, fol. 64v (20 June), in Steno, *Chaos*, 334–335.

194. Pecquet, *Experimenta nova anatomica*, 43–46. See also, Castel-Branco, "Physico-Mathematics and the Life Sciences," esp. 449–452.

bodily vessels (the veins and thoracic duct) and siphons and bladders. Pecquet rejected the opinion that chyle and blood moved due to an invisible attraction. He had in mind the Aristotelian explanation that water moved upward in siphons because of "fear of the vacuum [*metus vacui*]."[195] On the contrary, he argued, chyle moved upward due to the expansion of air, a property that Pecquet called the "elasticity of the air [*elater aëri*]."[196] Pecquet demonstrated the weight and elasticity of the air with a section entitled "Physico-Mathematical Experiments on the Vacuum," where he mentioned recent experiments performed by mathematicians in France, including the famous Puy-de-Dôme experiment, with which Blaise Pascal (1623–1662) demonstrated that atmospheric air had weight.[197] Pecquet even reported, for the first time in print, new experiments developed by his mathematician friends Gilles Personne de Roberval (1602–1675) and Adrien Auzout (1622–1691). One of these experiments was a bladder placed within Torricellian tubes that was seen to expand owing to the vacuum that formed at the top of the tube (see fig. 1.5).[198] Pecquet's anatomy book thus became a leading vehicle through which to divulge new physico-mathematical experiments on the vacuum and to show that air has an elastic nature.[199]

Steno read the physico-mathematical part of Pecquet's book very carefully. He copied all the relevant details into the "Chaos" manuscript, including the names of the mathematicians who performed the vacuum experiments, the relevant locations of where the experiments happened, and the exact measurements indicated.[200] These notes were also thoroughly annotated. He added many headings in the margins; numbered the text in order to organize it better, almost as a list of observations; underlined certain parts on the new properties of the air, such as that "the *air has weight even in its own sphere*," or that elasticity is like a "sponge or rather wool in a stack." He even added personal notes on other things to be investigated with Torricellian tubes, such as to what height mercury would rise to if the Torricellian tube were made of silver instead of glass, in light of mercury's

---

195. Pecquet, *Experimenta nova anatomica*, 62–73, esp. 62.

196. Pecquet, *Experimenta nova anatomica*, 49–50; see also Bertoloni Meli, "The Collaboration Between Anatomists and Mathematicians," esp. 672.

197. Gal. 291, fol. 65r (20 June), in Steno, *Chaos*, 338; and Pecquet, *Experimenta nova anatomica*, 55–56.

198. For a more detailed explanation see Bertoloni Meli, "The Collaboration Between Anatomists and Mathematicians," 670–677.

199. Dear, *Discipline and Experience*, 202–203.

200. Gal. 291, fol. 64v–66r (20–21 June), in Steno, *Chaos*, 336–349.

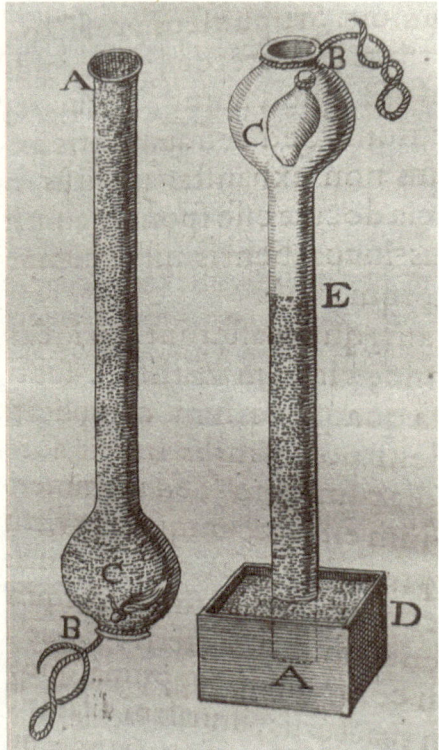

FIGURE 1.5. Roberval's bladder experiment. Pecquet, *Experimenta nova anatomica* (Paris, 1651), 51. Reproduced with the permission of Balliol College, University of Oxford (ref. 905 f 7 (5)).

reaction with that element.[201] Steno also copied and underlined many anatomical details that appeared toward the end of this physico-mathematical section and in the book's third part on thoracic lacteals.[202]

More significant, two letters addressed to Pecquet and published at the end of his book speak about the advantages of using methods of physico-mathematics in anatomy. When he excerpted from the letter by Sebastian Alethophilus, a pseudonym of Samuel Sorbière, Steno wrote that "he who deals with the physics of medicine without Geometry [*sine*

---

201. For the two emphasized sentences and the silver Torricellian tube, see correspondingly Gal. 291, fols. 64v (20 June), 65r (21 June), and 64v (20 June), in Steno, *Chaos*, 336, 339, 337.

202. Gal. 291, fols. 66r–67v (21–22 June), in Steno, *Chaos*, 349–359.

Geometria medicinæ physicæ qui tractarit], sails through the ocean without a compass." Sorbière was one of the leading disciples of Gassendi, whose mechanical philosophy was built on the principles that "besides bodies, shapes, and motions nothing else should be admitted," as Steno copied in "Chaos" from Sorbière's epistle.[203] Pecquet himself was a member of the circles frequented by Gassendi in Paris when writing *Experimenta nova anatomica*. Interestingly, Steno's excerpts do not necessarily say what he thought about these letters, but his marginalia do. When Steno reread his notes, he added to the start of Sorbière's epistle one of the few double "Nota Bene [*NB.NB.*]" marks in the "Chaos" manuscript, followed by the heading that "without geometry, natural studies should not be dealt with [*sine geometria non tractanda physica*]."[204] In another epistle to Pecquet, Auzout repeated similar ideas about mechanical analogies, such as when claiming that "the circulation should be studied mechanically [*inquirendum in circulationem mecha[nice]]*."[205] Auzout also mistrusted a few medical practices (he spoke of "the harm of bleeding," i.e., he criticized bloodletting) and spoke of the solid epistemic value of "experiments and observations and that which is deduced from them together with the principles of metaphysics and mechanics."[206]

In several ways, Pecquet's *Experimenta nova anatomica* contained much of what was important about seventeenth-century anatomy. Pecquet spoke of the epistemic value of repeated dissections and vivisections, the discovery of new body parts and processes, and the usefulness of physico-mathematics. It also illustrated to Steno the successful interdisciplinary outcome of interactions between anatomists and mathematicians, such as between Pecquet, Roberval, and Auzout.

❋

Nicolaus Steno finished his notebook with excerpts from Gassendi's mechanical philosophy, the other of the three most heavily annotated parts of "Chaos." Steno excerpted from the natural philosophy section

---

203. Gal. 291, fol. 68r (23 June), in Steno, *Chaos*, 362–363. On Sorbière and Gassendi, see Guerrini, "Experiments, Causation, and the Uses of Vivisection," 239–240. On Alethophilus as the pseudonym of Sorbière, see Vincent Placcius, *De scriptis et scriptoribus anonymis atque pseudonymis* (Hamburg: Christianus Guthius, 1674), 141.

204. Gal. 291, fol. 68r (23 June), in Steno, *Chaos*, 362.

205. Gal. 291, fol. 67v (23 June), in Steno, *Chaos*, 361.

206. For the quotes, see correspondingly Gal. 291, fol. 68r (23 June), 67v (23 June), in Steno, *Chaos*, 362, 361, 360, 361.

of Gassendi's book on Epicureanism, one of the most important works of early modern atomism.²⁰⁷ As always, these notes by themselves do not say much about Steno's thoughts, but his textual interaction with them in marginalia shows that they were relevant to him. An early part of the "Chaos" manuscript also indicates that Steno viewed natural phenomena, especially hidden attractions, in terms of particles in motion, as did Descartes and Gassendi. His written decision to study an "electrical attraction" and the "particles [*particulæ*]" involved in it speaks to this view.²⁰⁸ His note that Kircher, an eclectic Aristotelian, did not "totally deny Epicurean Atoms [*non totaliter negamus Atomos Epicureos*]," also confirms this opinion.²⁰⁹ By the time he finished "Chaos," Steno also seemed to think the vacuum did exist. As he wrote in a personal note, "The vacuum exists because there is no motion possible unless in a void."²¹⁰ Belief in the vacuum meant that Steno's mechanical philosophy of particles in motion was closer to the natural philosophy of Gassendi than to that of Descartes, who rejected the notion of vacuum.²¹¹

Steno's own views of Cartesian ideas also seemed to change over the course of the months in which he wrote "Chaos."²¹² At the beginning of the notebook, written in early March 1659, Steno mentions Borel's biography of Descartes and several books by the Cartesian Henricus Regius.²¹³ Shortly after that, and since he had also read in those days Francis Bacon's *De augmentis scientiarum* (Leiden, 1645), Steno proposed to do research "according to Verulam [i.e., Bacon] or Descartes."²¹⁴ However, later in May, while working with microscope lenses in his workshop, he started noticing Descartes's mistakes on various fronts. Steno noted that "less work is required for grinding lenses than Descartes thought."²¹⁵ In early June, when

207. Gal. 291, fols. 70r–75v (26–28 June), in Steno, *Chaos*, 393–447. Steno jumped from the astronomy section directly to the meteor parts, in Steno, *Chaos*, 438.

208. Gal. 291, fol. 40r (23 March), in Steno, *Chaos*, 132. Kircher described a series of attractions as electrical due to the example of amber (*electrum* in Greek), a material famous for attracting metals after being rubbed; this effect is known today as static electricity. Kircher, *Magnes sive de arte magnetica*, 563–571.

209. Gal. 291, fol. 38r (22 March), in Steno, *Chaos*, 116. See Kircher, *Magnes sive de arte magnetica*, 470.

210. Gal. 291, fol. 68v (23 June), in Steno, *Chaos*, 368.

211. See Osler, *Divine Will and the Mechanical Philosophy*, 176–188.

212. For the opposite argument, see Miniati, *Nicholas Steno's Challenge to Truth*, 41–46.

213. Gal. 291, fol. 33r (12 March), 36r–36v (15–16 March), in Steno, *Chaos*, 76, 87, 89.

214. Gal. 291, fol. 39r (23 March), in Steno, *Chaos*, 124; see also fol. 39r (22 March), in Steno, *Chaos*, 123.

215. Gal. 291, fol. 60r (30 May), in Steno, *Chaos*, 293.

he read Descartes's *Meteora* (Amsterdam, 1644), he "doubted [*dubito*]" the Cartesian explanation of what happens to air when heated.[216] By the end of June 1659, as he was finishing the "Chaos" manuscript, Steno was explicitly critical of Cartesian philosophy and Descartes's failures in medicine. He remarked that some people, "after an accurate examination of Cartesian philosophy, noted some things in which it cannot be accepted." Moreover, when Descartes had a fever in Sweden, "he wanted to cure himself according to the norm of his own philosophy, and thus killed himself by continually drinking water."[217] As Steno started his travels, his antagonism toward the writings of Descartes and his followers developed in this same direction.

When Steno left Copenhagen to start his medical pilgrimage, he showed all the signs of the scholar he would become. He was a learned and well-read young student interested in applying interdisciplinary knowledge in order to find certainty in his anatomical studies. The handicraft skills that he developed at home served him in performing delicate dissections of animal bodies and in making precise measurements in workshops. His friendship with Borch, who joined Steno for part of his travels, would also bear fruit as they met other scholars together. Thankfully, Borch noted everything in his journal. Steno's note-taking techniques would also form the basis of almost all his scientific publications. In short, Nicolaus Steno kept applying his humanist training, high sociability, and artisanal hands to the problems he encountered, especially problems about the body. To put it another way, the intellectual, artisanal, and social roots that made Steno a traveling anatomist were established by 1659. The time had come for the fertile environment of the Netherlands to make them grow.

---

216. Gal. 291, fol. 60v (1 June), in Steno, *Chaos*, 294.
217. Gal. 291, fol. 75v (29 June), in Steno, *Chaos*, 447, for both quotes.

CHAPTER TWO

The Making of a Scholarly Anatomist

It was Christmas 1660, and twenty-two-year-old Nicolaus Steno had been living in Leiden since the summer. Now he returned to Amsterdam to meet his friend Ole Borch, whom he had not seen for almost a year.[1] Borch had left Copenhagen in October and arrived in the Netherlands to join Steno. His goal was to pursue medical studies at the University of Leiden, which Steno had already started.[2] After the Christmas holidays, Steno received Borch in Leiden, introduced him to the city's academic environment, and immediately showed him his recent anatomical discovery—a new salivary duct in humans and animals. At this time, Steno did not predict the controversy that this discovery would trigger. A few months later Gerard Blasius (1625–1682), an Amsterdam anatomist and a prolific editor of medical books in the Netherlands, tried to claim the discovery for himself.[3] Blasius had taught Steno for a few months in Amsterdam in the spring of 1660 before Steno settled down in Leiden. After an intense dispute in the Republic of Letters, most scholars recognized Steno as the true discoverer of the parotid salivary duct, still called the *ductus stenonianus*.[4]

Most historians who have written about Steno's Dutch sojourn invariably mention this dispute, often describing it as unfair to the young Steno. However, many things in these accounts do not make sense. For instance, why did Blasius, a respected scholar in the Netherlands and abroad, decide

---

1. OBI, 1:7, 25 December 1660.
2. Steno enrolled on 27 July 1660 and Borch on 21 February 1661; see Leiden University Library, ASF 10, fols. 585, 601, as quoted in *Album studiosorum Academiæ Ludguno Batavæ* (The Hague: Martin Nijhoff, 1875), 482, 486.
3. Harald Moe, "When Steno Brought New Esteem to Glands," in *Nicolaus Steno 1638–1686: A Re-consideration by Danish Scientists*, ed. Jacob Poulsen and Egill Snorrason (Gentofte, Denmark: Nordisk Insulinlaboratorium, 1986), 51–96.
4. Gray, *Anatomy of the Human Body*, 1134.

to claim credit for himself if Steno was the one who really discovered it? And why did Thomas Bartholin, Steno's former mentor from Copenhagen, never take Steno's side publicly? These questions have not yet been answered because previous studies have focused mostly on Steno's side of the controversy and not on what early modern scholars wrote about it. Thomas Bartholin, for example, wrote more than ten letters on the subject to several people; so did Borch and other physicians. Blasius and his friends also wrote many pages against Steno. This chapter uses this correspondence to examine the wider context behind this controversy and understand its full significance to Steno's career. I argue that the Blasius controversy played a crucial role in inserting Steno in the medical Republic of Letters and in shaping him into an anatomist and scholar.[5] This was only possible, however, because of the humanist education that Steno received at the Copenhagen Cathedral School, and the way it formed his writing.

This chapter starts by examining Steno's discovery of the salivary duct in the spring of 1660 and the possible reasons why Blasius claimed priority over it. I show that at stake were practices of authorship that differed between institutions as well as different methods of doing anatomical research. Thomas Bartholin, a good friend of both Steno and Blasius, followed the dispute carefully through his epistolary network and tried to moderate its divisiveness. Although he did not take sides publicly, behind the scenes he acknowledged Steno's remarkable anatomical skills. Bartholin was happy with his former student because, as the controversy ensued, Steno's position evolved from questioning *who* discovered the salivary duct to *how* it was discovered, as well as *where* it was situated. As I explain in the second part of the chapter, this controversy reinforced Steno's search for certainty in anatomy. He employed several techniques to convince others of the reliability of his observations, such as writing vivid narratives of his multiple dissections. To leave a mark in the medical Republic of Letters, he also expanded his work from a study of the salivary duct into a full anatomical description of the glands of the mouth. By all accounts, he wrote what early modern anatomists called a *historia* of the glands.[6] Finally, the chapter turns to Steno's editorial efforts with respect to his own texts and his choice of the eclectic genre of *observationes anatomicæ*. As Steno made use of rhetorical tools and

---

5. Ian Maclean, "The Medical Republic of Letters Before the Thirty Years War," *Intellectual History Review* 18, no. 1 (2008): 15–30.

6. See Gianna Pomata, "*Praxis Historialis*: The Uses of *Historia* in Early Modern Medicine," in Pomata and Siraisi, *Historia: Empiricism and Erudition in Early Modern Europe*, 105–146, esp. 118–121.

textual techniques to better articulate his points against Blasius, he was developing his own style not just as an anatomist, but as a scholar. Unlike what we moderns like to think, Steno's works show that books and texts were deeply intertwined with the early modern culture of empiricism.[7] Ultimately, the role of this controversy in establishing Steno's scientific persona was much deeper than making him the acknowledged winner of the dispute.[8]

Authorship and the Blasius Controversy

Steno arrived in Amsterdam from Copenhagen in March 1660 and "around the feast of Easter" was received in Blasius's lodgings.[9] Blasius was a new professor in the Amsterdam Athenæum, where he received a full position only later in September.[10] Thomas Bartholin, who probably knew Blasius from the latter's family connections in Denmark, wrote Steno a letter of recommendation to present to Blasius.[11] Steno would later thank Bartholin as his *mecænas*, indicating that Bartholin helped make Steno's travels financially viable.[12] As soon as he arrived in Amsterdam, Steno began classes with Blasius, and when the Easter holidays started he decided to perform dissections by himself, since he had already done some in Copenhagen. Thus, on 7 April, as Steno later wrote, "when I was dissecting alone in the small cabinet, I found in a sheep's head a duct which, as far as I know, nobody had described [before].... Amazed by the novelty of the thing, I called my host to hear his opinion." But his host was skeptical. The most recent book on the glands by Thomas

---

7. Anthony Grafton, *Defenders of the Text: The Traditions of Scholarship in an Age of Science, 1450–1800* (Cambridge, MA: Harvard University Press, 1991), esp. 178–203.

8. Lorraine Daston and H. Otto Sibum, "Introduction: Scientific Personæ and Their Histories," *Science in Context* 16, nos. 1–2 (2003): 1–8.

9. Steno, *Observationes anatomicæ*, 11–12. On the year, see Steno's dissertation defense in the Amsterdam Athenæum on 8 July 1660 in Nicolaus Steno, *Disputatio physica de thermis* (Amsterdam, 1660), frontispiece.

10. See Dirk van Miert, *Humanism in an Age of Science: The Amsterdam Athenæum in the Golden Age, 1632–1704*, trans. Michiel Welema, with Anthony Ossa-Richardson, Brill's Studies in Intellectual History 179 (Leiden and Boston: Brill, 2009), 92–93, 319–324.

11. Nicolaus Steno, *Apologiæ prodromus* (Leiden, 1663), in OPH, 1:147. This letter is now lost.

12. Nicolaus Steno to Thomas Bartholin, 22 April 1661, in EMC III, 88. In a dedication, Steno also acknowledged the patronage of Hans Svane (1606–1668), a Danish archbishop, and Otto Krag, a Danish ambassador in the Netherlands.

Wharton (1614–1673) had made no mention of this duct.[13] Blasius could also not see it himself, owing to "rather carelessly treated vessels [*tractata negligentius vasa*]," as Steno admitted.[14] Steno found the duct by inserting a probe inside an opening in the parotid gland, which he then pushed until it was "no longer retained inside the straits of the tunics [i.e., the vessel's walls] but roamed in a large cavity, and soon I heard the protruding iron make the teeth themselves resound."[15] However, Blasius, aware of the risk of ripping the duct, thought that Steno had forced the probe through the flesh into the mouth.[16] Undeterred, Steno continued searching, and a few days later he saw the same duct in a dog, although "it was more obscure [*obscurius*]."[17] Regardless, Steno became convinced of the existence of this new duct, whose function was to bring salivary fluid into the mouth, as he wrote to a Danish friend.[18]

According to this first account, this duct clearly seemed difficult to observe. At first Steno refrained from making his discovery public, probably remembering Gassendi's warning against "wanting to acquire fame by telling something new," which he had copied and underlined in the "Chaos" manuscript.[19] Ascertaining the novelty of something was a main preoccupation of seventeenth-century anatomists. It required not only repeated observations in dissection but also careful scrutiny of the existing anatomical literature since antiquity.[20] Moreover, another salivary duct had been recently discovered—the *ductus whartonianus*—connecting the maxillary gland to the mouth. There was no reason another salivary duct might be needed. Thus, Steno decided to "remain silent" until he consulted the "famous Sylvius on the matter."[21] Steno moved to Leiden in July 1660 to pursue medical studies at the university under the renowned professors Franciscus Sylvius and Johannes van Horne (1621–1670), both

---

13. Steno to Bartholin, 22 April 1661, in EMC III, 88, 89. See also Thomas Wharton, *Adenographia, sive glandularum totius corporis descriptio* (London: J. G., 1656), 124–127.

14. Steno to Bartholin, 22 April 1661, in EMC III, 89.

15. Steno to Bartholin, 22 April 1661, in EMC III, 88–89. "Tunics" were the membranes or walls of vessels inside the body.

16. Steno to Bartholin, 22 April 1661, in EMC III, 89.

17. Steno to Bartholin, 22 April 1661, in EMC III, 89.

18. Steno to Bartholin, 22 April 1661 in EMC III, 89. In this letter Steno mentioned writing about this discovery to his friend Jakob Henrik Paulli in April of 1660.

19. Gal. 291, fol. 74v, 28 June, in Steno, *Chaos*, 436.

20. Later, when studying the heart, Steno asked Bartholin whether his discovery "was noticed by others," considering Bartholin's "abundant reading." See Steno to Bartholin, 30 April 1663, in EMC IV, 414.

21. Steno to Bartholin, 22 April 1661, in EMC III, 89.

of whom were old friends of Thomas Bartholin.[22] After Steno mentioned the newly discovered duct to his professors, Sylvius searched for it himself and, after finding it in a human cadaver, "demonstrated [it] to many spectators."[23] Van Horne was also impressed with Steno's discovery: within a few months he was showing it to students and calling it the *ductus stenonianus* in honor of Steno, the name it bears to this day.[24] The observation and confirmation of the duct's existence by his Leiden professors allowed the young Dane to be certain of the duct's existence.[25]

Steno kept on studying the new salivary duct and its place within the head. Since dissections were rarely carried out in isolation, Steno showed his progress to those around him in Leiden. These small audiences also served as witnesses who helped to confirm the reliability of Steno's observation. Robert Sibbald (1641–1722), a young Scottish medical student who was in Leiden from March 1660 to September 1661, wrote in his autobiography that Steno "dissected in my chambers sometymes, and showed me there, the *ductus salivalis superior*, he had discovered."[26] We can also imagine the excitement with which Steno showed these recent discoveries to Borch, in their first days together in Leiden. Early in January 1661 Borch described in his diary the vivisection of a dog in which he saw lymphatic vessels connected to the thoracic duct, the motion of the heart in the living dog, and the production of pancreatic juice. Borch did not mention Steno by name, but it was most likely Steno showing him all of this, because he mentioned some medical opinions of Sylvius who had been teaching Steno. But the strongest evidence that Steno was the dissector is Borch's reference to seeing the "salivary ducts, including the superior ducts departing quickly from the jaws and the cheek, and the ducts spread out in the mouth itself, close to the [lingual] papillae."[27] In fact, by then, Steno

---

22. Bartholin participated in Sylvius's dissections around the year 1640; see Johannes Walaeus to Thomas Bartholin, 10 October 1640, in Thomas Bartholin, *Institutiones anatomicæ*, ed. Caspar Bartholin (Leiden: Franciscus Hack, 1641), 408. Johannes van Horne published "De aneurysmate epistola" in Thomas Bartholin, *Anatomica aneurysmatis dissecti historia* (Palermo: Alfonso dell'Isola, 1644).

23. Steno to Bartholin, 22 April 1661, in EMC III, 89.

24. Steno to Bartholin, 22 April 1661, and Ole Borch to Bartholin, 3 March 1661, in EMC III, 89, 362. It was printed for the first time in Johannes van Horne, *Mikrokosmos seu brevis manuductio ad historiam corporis humani* (Leiden: Jacob Chouët, 1662), 23.

25. Steno to Bartholin, 22 April 1661, in EMC III, 89.

26. Robert Sibbald, *The Autobiography of Robert Sibbald, Knt., M.D., to Which Is Appended Some Account of His MSS*, ed. James Maidment (Edinburgh: Thomas Stevenson and John Wilson, 1833), 15–16.

27. OBI, 1:18, 5–7 January 1661.

was already pursuing a full study of all the glands in the mouth which formed the basis of his first anatomical book, published a few months later.

On 3 March 1661 the news of Steno's discovery of the salivary duct entered the Republic of Letters, when Ole Borch broke it to Thomas Bartholin. Bartholin was an important node of the medical Republic of Letters in northern Europe. Just as Oldenburg and Marin Mersenne (1588–1648) corresponded with natural philosophers out of Paris and London, Bartholin received news from the medical world and shared it with the scholarly community.[28] Borch said that the new salivary duct was "shown publicly twice" and that its name was *ductus stenonianus*, as "the most illustrious van Horne is accustomed to call them in honor of the discoverer."[29] Within a week Bartholin responded that "the new salivary ducts observed by yourself and Steno are worthy of all praise."[30] Until then, Bartholin had corresponded only with Borch to learn about the situation in the Netherlands, not with Steno.[31] But around this time Bartholin also wrote to Steno a letter encouraging him to "produce an image of the external salivary duct [*ductus salivalis exterioris iconem edam*]," as Steno mentioned in his reply.[32] Surprisingly, the original letter from Bartholin to Steno is missing from all four volumes of Bartholin's *Epistolarum medicinalium centuria* (Copenhagen, 1663–1667), the source for most of the letters mentioned. Its absence is most likely intentional and related to the ensuing controversy.

To everyone's surprise, at the end of March 1661, Gerard Blasius published the book *Medicina generalis*, where he claimed to have discovered the upper salivary duct, "which a year ago showed itself discovered by me, when I was occupied with private anatomical exercises . . . in a calf's head."[33] Although Blasius's book was a general introduction to medicine, not

---

28. Justin Grosslight, "Small Skills, Big Networks: Marin Mersenne as Mathematical Intelligencer," *History of Science* 51, no. 3 (2003): 337–374; and Marie Boas Hall, *Henry Oldenburg: Shaping the Royal Society* (Oxford and New York: Oxford University Press, 2002), 125–156.

29. Borch to Bartholin, 3 March 1661, in EMC III, 362. Borch uses the plural, possibly thinking of the tiny salivary ducts of the cheek.

30. Bartholin to Borch, 9 March 1661, in EMC III, 370.

31. Before Steno responded to Bartholin on 22 April 1661, Borch had already sent Bartholin at least four letters that year, see EMC III, 360–369, 374–377, 382–387, 390–395. In an earlier letter on Steno's discovery, Bartholin repeats Borch's account; thus, it does not seem that the account came from Steno himself. See Bartholin to Georg Seger, 9 March 1661, in EMC III, 48–52.

32. Steno to Bartholin, 22 April 1661, in EMC III, 87.

33. Gerard Blasius, *Medicina generalis, nova accurataque methodo fundamenta exhibens* (Amsterdam: Pieter van den Berge, 1661), vi.

anatomy, Blasius mentioned the newly discovered duct in the preface, illustrating the importance he attributed to it. When Steno learned about this, he was shocked and felt as one who "attained the hateful mark of ignominy [*turpis ignominiæ nota*]," accusing Blasius of committing the "envious crime of plagiary."[34] Considering that Steno alone discovered the salivary duct, which he tried to demonstrate with letters announcing it to friends before Blasius's publication, these strong words are reasonable.[35] What is harder to understand is the reaction of other anatomists, especially of Thomas Bartholin. Borch immediately reported to Bartholin that "Mr. Blasius attributes to himself the discovery of the upper salivary duct" and that Steno was preparing a response.[36] Bartholin, however, did not even address this matter in his response to Borch.[37] Hence, when Borch replied back, he said that he "did not want to touch [*non attingo*]" it either.[38] Overall, Bartholin never took sides in his letters to Steno. At most, he tried to keep the dispute at arm's length. A few months later, in September 1661, Bartholin asked the Amsterdam surgeon and medical author Paul Barbette (1620–1666) for help in solving this dispute.[39] Barbette acknowledged Steno's great anatomical skills, but he also knew Blasius was an important and respected physician.[40] Historians have explained Bartholin's reaction as part of the academic diplomacy and politeness required of the Republic of Letters.[41]

However, there was more to Bartholin's reaction than just politeness. The truth is, Bartholin and Blasius were friends. Indeed, Blasius's *Medicina generalis* begins with a poem by Bartholin to the "most excellent man Gerard Blasius, friend, and honored as family," a fact that has rarely been mentioned in accounts of this controversy.[42] Blasius was also deeply involved in producing the Amsterdam edition of Bartholin's books on the lymphatics, in the midst of a larger controversy on the topic between Bartholin and the anatomist Lodewijk de Bils (1624–1671).[43] Johannes Blasius (1639–1672), brother

34. Steno to Bartholin, 22 April 1661, in EMC III, 87.
35. Steno to Paulli, May 1660, and Johannes Blasius to Eyssonius, March 1661, both mentioned in EMC III, 89–90; Borch to Bartholin, 3 March 1661, in ibid., 362.
36. Borch to Bartholin, 20 March 1661, in EMC III, 376.
37. Borch to Bartholin, 20 March 1661 and 29 March 1661, in EMC III, 376, 378–381.
38. Borch to Bartholin, 21 April 1661, in EMC III, 394.
39. Bartholin to Paul Barbette, 28 August 1661, in EMC III, 200.
40. Barbette to Bartholin, 30 July 1661, in EMC III, 196–197.
41. See Scherz's biography in Kardel and Maquet, *Nicolaus Steno: Biography and Original Papers*, 88; and Moe, "When Steno Brought New Esteem to Glands," 60.
42. Blasius, *Medicina generalis*, vii.
43. Bartholin to Georg Seger, 9 March 1661, in EMC III, 48–49. See also Thomas Bartholin, *Spicilegia bina ex vasis lymphaticis* (Amsterdam: Pieter van den Berge, 1661).

of Gerard and a friend of Steno, also wrote a poem to Gerard that was published in the same book.[44] In the early modern period it was customary for brothers and friends to praise each other in poems published before the prefaces in medical books. The problem is that both Bartholin and Johannes Blasius were aware of Steno's discovery of the duct *before* Blasius's book was printed. Bartholin had been informed of the discovery of the salivary duct by Borch, and Steno said that "in a letter to [the Groningen physician] Eyssonius [1620–1690] his own brother [Blasius] attributed the discovery of this duct to me some days before these reproaches" appeared.[45] Therefore, they must have been surprised to find that Gerard claimed Steno's discovery for himself. They were probably unaware of Gerard Blasius's text when they wrote their poems and were taken by surprise when it came out.

Considering that Gerard Blasius, a respected physician, was close to both Bartholin and Steno, it is hard to believe that he proceeded out of malice or plagiarism, Steno's accusation notwithstanding. The main reason for Blasius's claim for credit lies, instead, in the complex practices of authorship in the early modern period.[46] In the seventeenth century the propositions of most academic disputations were written by professors, whereas students' contributions appeared only in the final corollaries.[47] This means that Steno wrote only the two final corollaries of what has erroneously been considered his first original work, the *Disputatio physica de thermis* (*Physical Disputation on Hot Springs*; Amsterdam, 1660) (see fig. 2.1).[48] The theme of hot springs (*de thermis*) and its relation to geology has tempted historians to conclude that Steno worked in geological topics from an early age. However, this disputation's corollaries, the ones that Steno really authored, were not about hot springs but rather about the constitution of gold and iron.[49] These chymical themes thus speak to his

---

44. Steno was close to the Blasius brothers; see Eric Jorink, "*Modus politicus vivendi*: Nicolaus Steno and the Dutch (Swammerdam, Spinoza and Other Friends), 1660–1664," in Andrault and Lærke, *Steno and the Philosophers* (Leiden: Brill, 2018), 13–44. Blasius's poem was written on 6 March 1661, while Steno was living in his house in Leiden.

45. Steno to Bartholin, 22 April 1661, in EMC III, 90.

46. Bertoloni Meli, "Authorship and Teamwork Around the Cimento Academy"; and Biagioli and Galison, *Scientific Authorship*.

47. On early modern dissertations, see Kevin Chang, "From Oral Disputation to Written Text: The Transformation of the Dissertation in Early Modern Europe," *History of Universities* 19, no. 2 (2004): 129–187, esp. 150–152.

48. Gustav Scherz, "Niels Stensen's First Dissertation," *Journal of the History of Medicine and Allied Sciences* 15, no. 3 (1960): 247–264.

49. Steno, *Disputatio physica de thermis*, 7.

FIGURE 2.1. Title page of Nicolaus Steno, *Disputatio physica de thermis* (Amsterdam, 1660). Reproduced with the permission of the Library Company of Philadelphia.

long-held interest in goldsmithing and show how he brought his artisanal upbringing to his academic training and interests.

In the Amsterdam Athenæum it was even standard practice for professors to publish their students' scholarly disputations under their own names, especially if they had a private group of students, known as the *collegium*, which Blasius had.[50] Both Gerard Blasius and Arnold Senguerdus (1610–1667), the professors who presided over Steno's disputation on hot springs on 8 July 1660 in Amsterdam, published their students' disputations under their own names.[51] Senguerdus's book on bones, the *Osteologia corporis humani* (Amsterdam, 1662), for example, was a compilation of twenty-two disputations by his students. As the historian Dirk van Miert explained, "Blasius followed the same method for his *Compendium* [Amsterdam, 1667], based on the text of twelve [student] disputations."[52]

This inversion of authorship, as it seems to a modern eye, was not uncommon elsewhere in Europe. Thomas Bartholin, whose discovery of the lymphatic vessels was made in close collaboration with his student Michael Lyser, was himself accused of plagiarism.[53] However, unlike Steno, Lyser rejected any claim to honors, saying that he did not want to deprive Bartholin of "the immortal honor of the discovery which rightfully belongs to thee, or in any way try to diminish it."[54] As further confirmation of Blasius's attitude, the book where he claimed credit for the duct was an introduction to medical students, to the point that the frontispiece showed Blasius reading with students in his private rooms—perhaps one of them Steno (see fig. 2.2). In this book, Blasius even mentioned the "bright and young Nicolaus Steno," who had also seen the duct in "a human's head a few months before, exhibiting it to spectators."[55] That is, Blasius wrote as if he had made the discovery of the salivary duct and of Steno as his "diligent disciple" who was spreading its glory.[56] It is thus possible that Bartholin was taken by surprise not by Blasius's claim in his book, but by Steno's strong response. Steno and Blasius did not share Bartholin and

50. See Collegium privatum Amstelodamense, *Observationes anatomicæ selectiores* (Nieuwkoop, 1667, 1673). On professors writing students' disputations, see van Miert, *Humanism in an Age of Science*, 156. Senguerdus was the main author of Steno's disputation, as further confirmed by disputations defended by other students at about the same time, almost all of them on similar topics; see Scherz, "Niels Stensen's First Dissertation," 253.

51. For the presiders, see title page of Steno, *Disputatio physica de thermis*.

52. van Miert, *Humanism in an Age of Science*, 153–154.

53. Garboe, "Michael Lyser, a 17th Century Anatomist," esp. 65.

54. Michael Lyser to Thomas Bartholin, 10 April 1656, in EMC I, 638; translation from Garboe, "Michael Lyser, a 17th Century Anatomist," 65. The Swedish physician Olaf Rudbeck (1630–1702), who also claimed priority for the lymphatics, was the one accusing Bartholin.

55. Blasius, *Medicina generalis*, vi. Blasius likely meant before the publication of his book.

56. Blasius, *Medicina generalis*, vi.

FIGURE 2.2. Blasius reading with students in his Amsterdam rooms; from Blasius, *Medicina generalis* (Amsterdam, 1661), frontispiece. The Bodleian Libraries, University of Oxford, 8° N 41 Med.

Lyser's harmony, but it is for precisely that reason that their cases illustrate cultural and geographical differences in understandings of coauthorship.

In July 1661, four months after the controversy began, Steno publicly presented his results in two disputations presided over by van Horne at the University of Leiden. The disputations were published a few days later by Elsevier in a short booklet, the *Disputatio anatomica de glandulis oris*.[57] They were the first illustrated disputations out of the University of Leiden.[58] Unlike in Amsterdam, academic disputations in Leiden contained original research done by the students.[59] In Steno's case this is confirmed by the fact that the disputation served as Steno's response to Blasius and that it was much longer than the traditional disputation format.[60] Bartholin responded in a diplomatic manner to both Blasius's and Steno's letters. However, behind the scenes, especially after Steno published *Disputatio anatomica*, he was clearly excited with Steno's contribution.[61] In an August letter to the Basel physician Georg Seger (1629–1678), Bartholin spoke of "our Steno, the delight of anatomy," without even mentioning Blasius.[62] In November he wrote to Joel Langelott (1617–1680), a physician from southern Denmark, informing him of the new knowledge on the glands of the mouth brought forth by "Steno, once my disciple, now living in Leiden, . . . augmenting anatomy by the great celebrity of his name."[63] Finally, in a letter to Borch, Bartholin could not contain himself: "I cannot praise enough the diligence and success of our Steno."[64] Bartholin was aware that others were also thrilled with Steno's observations. Steno was praised not just in Leiden but also in Amsterdam, where Barbette told Bartholin that "Steno does great things for us here."[65] Yet, Bartholin's praise of Steno was not just an echo of his excitement about a young fellow Dane. Bartholin knew

57. Nicolaus Steno, *Disputatio anatomica de glandulis oris* (Leiden: Johannes Elzevier, 1661).
58. Eric Jorink, "The *Myologia* by Saeghemolen and van Horne in Context: Art, Science and Religion at Leiden University, ca. 1660," in *Quatre atlas de myologie de van Horne et Sagemolen: Approche pluridisciplinaire de dessins inédits du Siècle d'or neerlandais; Actes du colloque international des 18 et 19 juin 2021*, ed. Jean-François Vincent and Isabelle Bonnard (Paris: Université Paris Cité, 2022), 39–64, esp. 55–56.
59. Evan R. Ragland, "Experimental Clinical Medicine and Drug Action in Mid-Seventeenth-Century Leiden," in "Testing Drugs and Trying Cures" [special issue], *Bulletin of the History of Medicine* 91, no. 2 (2017): 331–361, esp. 344–345.
60. A comparison between Steno's *Disputatio physica de thermis* and *Disputatio anatomica de glandulis* makes this difference in length clear.
61. This excitement only became public in 1667, when Bartholin published these letters in EMC III.
62. Bartholin to Seger, 1 August 1661, in EMC III, 131–134. Seger had asked Bartholin about Blasius's claim to priority, a point about which Bartholin remained silent; see Seger to Bartholin, 9 June 1661, in EMC III, 125–130.
63. Bartholin to Joel Langelott, 5 November 1661, in EMC III, 205–209.
64. Bartholin to Borch, 1 December 1661, in EMC III, 413–416.
65. Barbette to Bartholin, 30 July 1661, in EMC III, 194–197. On Barbette, see Daniel de Moulin, "Paul Barbette, M.D.: A Seventeenth-Century Amsterdam Author of Best-Selling Textbooks," *Bulletin of the History of Medicine* 59, no. 4 (1985): 506–514.

exactly what was at stake, and his attitude speaks to the epistemic value of observations in anatomy. In Bartholin's own words, Steno "honored the Whartonian salivary ducts" because "he had also described the exterior salivary [ducts] and added other observations on the glands of the mouth."[66]

The Historia *of the Glands*

Steno suggested that he could "willingly concede [*lubens cessissem*]" the claims for priority in discovering the parotid salivary duct, allowing Blasius to take the credit.[67] But he had many reasons not to. First, according to Steno, Blasius initially rejected his discovery. Moreover, when Blasius claimed priority for himself, the prestigious anatomists Sylvius and van Horne had already shared the discovery under Steno's name. The defense of priority thus became for Steno a matter of honesty and respect toward his professors. As he put it, "So that I am not drawn out of my conscience by modesty [*ne in conscientiam trahatur modestia*], I am forced to complain, although reluctantly, of the insult thrown upon me."[68]

There was, however, a third and deeper reason behind Steno's response to Blasius. Although both Steno and Blasius claimed to have discovered the salivary duct, their anatomical descriptions differed. Whereas Steno said that the duct carried salivary fluid from the parotid gland (another producer of saliva) into the mouth, Blasius said that the duct, "equally noticeable in the superior and inferior jaws [*maxilla*]," connected the maxillary gland with the "anterior glands of the mouth so that from there [the saliva] is expressed by the motion of the tongue"—a claim with which Steno disagreed.[69] Steno thought that his newly discovered salivary duct was entirely detached from the maxillary gland and was instead connected to the parotid gland. The young Dane thought these differences were so significant that he even suggested that this discrepancy was "a clear sign that [Blasius] had never investigated this duct [*nunquam in ductum illum inquisivisse manifesto indicio*]," neither had he "seen it" himself.[70] Thus, what for Blasius was a dispute about priority became for Steno a controversy about the duct's true location, action, and function, and ultimately about exactitude in reporting anatomical observations.

---

66. Bartholin to Langelott, 5 November 1661, in EMC III, 208.
67. Steno to Bartholin, 22 April 1661, in EMC III, 91.
68. Steno to Bartholin, 22 April 1661, in EMC III, 91. By "modesty," Steno meant remaining silent about the matter.
69. Blasius, *Medicina generalis*, 63–64.
70. Steno to Bartholin, 22 April 1661, in EMC III, 91.

The differences between Steno and Blasius's descriptions illustrate the problematic nature of making and reporting anatomical observations.[71] Steno was well aware of these difficulties. For instance, he described experimental mistakes as the "frequent tricks of Nature [*ludens sæpius natura*]" and argued that dissections had to be done slowly, because of "how little the eyes must be believed in things which are carried out quickly."[72] This care involved in making anatomical observations was also related to the epistemic significance of observations. As described previously, Steno had seriously studied medicine and anatomy through books that were structured around new observations, such as those by Borel and Pecquet.[73] More important, he saw his Dutch and Danish mentors Sylvius and Thomas Bartholin placing observations at the center of their research.[74] Thus, when writing about observations, he stated that one ought to accept "the opinion commonly accepted until another claim is proved by reliable experiments [*donec contraria sententia certis experimentis fuerit comprobata*]."[75] Convinced that observations should be in the forefront of scientific research, Steno knew that in order to convince others about the true place and function of the new duct, he had to provide observational and experimental accounts that were beyond dispute.[76] In one of his accounts, Steno even said that, after seeing the salivary ducts, he dissected many more animals, "so that I might be made more certain about them [*ut certior de illis fierem*]."[77] This was similar to Harvey's claim that he opened "many and various living animals" as epistemic support for his discovery of blood circulation.[78]

---

71. See also Domenico Bertoloni Meli, "Reliability and Generalization in Early Modern Anatomy," in *La tradizione Galileana e lo sperimentalismo naturalistico d'età moderna*, ed. Maria Monti, Biblioteca dell'edizione nazionale delle opere di Antonio Vallisneri, 8 (Florence: Leo Olschki, 2011), 1–26.

72. Steno to Bartholin, 22 April 1661, in EMC III, 89; and Steno, *Observationes anatomicæ*, 44. The latter was a comment on dissections made by Lodewijk de Bils that Steno witnessed; see chap. 3.

73. Borel, *Historiarum et observationum medico-physicarum centuriæ IV*; and Pecquet, *Experimenta nova anatomica*.

74. On Sylvius, see Ragland, "Experimental Clinical Medicine," 331–361. On Bartholin, see Thomas Bartholin, *Historiarum anatomicarum rariorum centuria* (Copenhagen, 1654–1661), which has over six hundred anatomical observations.

75. Steno, *Observationes anatomicæ*, 28.

76. Steno mentioned that Julius Casserius (1552–1616) had also seen it, although in passing and without understanding it.

77. Steno, *Observationes anatomicæ*, 6.

78. Harvey, *De motu cordis*, 20–21. See also Wear, "William Harvey and the 'Way of the Anatomists,'" esp. 239; and Bertoloni Meli, "Early Modern Experimentation in Live Animals," 209–210.

In short, Steno had to endow his description of the new salivary duct with the highest possible level of accuracy, and support that description with many dissections.

One of the first decisions that Steno made in order to make accurate claims about the salivary duct was to dissect a great variety of animals. In this way he could obtain an "accurate description [*accurata descriptione*]" of the duct, which Blasius lacked.[79] Moreover, Steno started working on other glands of the head, a research project in anatomy that began with his rebuttal of Blasius. Steno acknowledged this debt to the Blasius controversy explicitly:

> I owe much to the most famous gentleman [Blasius] because he gave me an opportunity not only of claiming my discovery but also of finding some other new things. Indeed, in the Easter holidays of this year [1661], I explored in a calf the said parotid [glands], which he challenged, and I followed the vessels where they led me. I observed, not without admiration, an elegant connection [*elegantem consensum*] of the various glands in the throat, through peculiar vessels. After having seen these things, so that I might be made more certain about them [*ut certior de illis fierem*], I dissected the head of a second calf, then I opened dogs as well, and finally I also examined the head of an ox, where many salivary ducts unexpectedly presented themselves.[80]

It was as if, by dissecting regularly, Steno were guided to his discovery by the ducts themselves.

Steno then proceeded with a careful and robust publication on the *historia anatomica* of the glands. In his first book, the *Disputatio anatomica*, Steno "described everything historically," as he later wrote.[81] This method, known as *historia anatomica*, unlike other notions of *historia* (such as *historia medica*), had recent origins in the Aristotelian empiricism of late sixteenth-century Padua.[82] Early modern anatomists followed a method best articulated by Hieronymus Fabricius ab Acquapendente (1533–1619)

---

79. Steno, *Observationes anatomicæ*, 5.

80. Steno, *Observationes anatomicæ*, 6. For this work alone, Steno also dissected a lamb, a cow, many more dogs, rabbits, and mentioned Sylvius's dissections of human cadavers at the hospital, see ibid., §16, §18, §47–48, §50, §19, correspondingly to 15, 17, 46–47, 49, 18–19.

81. Steno, *Observationes anatomicæ*, 68.

82. Pomata, "*Praxis Historialis*," 105–146, esp. 111, 114–122. *Historia anatomica* became linked to Aristotelianism owing to the rising interest in Aristotle's works on animals toward the end of the sixteenth century.

in which complete knowledge of a body part consisted in describing its structure (*historia*), action (*actio*), and purpose (*usus*).[83] These categories, which come from Galenic anatomy, parallel the Aristotelian four causes: material and formal (*historia*), efficient (*actio*), and final (*usus*) causes.[84] In the seventeenth century, the category of *historia* in anatomy widened to include the full narrative of discovery and other details.[85] For instance, when Gaspare Aselli narrated his discovery of the "lacteal vessels" in 1622, he included many details such as the date, place, and witnesses of the dissection, and even a philological discussion of the name he gave to the newly observed vessels.[86] Moreover, Harvey provided only the *historia* and *actio* of the blood circulation, but not the *usus*. Harvey believed that the *historia* (the physical description of things) was an essential step on the path toward the *usus* (what things are for).[87] Thus, for anatomists such as Aselli, Harvey, and Steno, the *historia* had such an epistemic validity and significance that it could be published as a standalone account.

Steno's *historia* of the salivary glands was a detailed account of his dissection and occupied most of the original *Disputatio anatomica*. He wanted to create in the reader a first-person impression of the dissection. To do so, he used vividly graphic narratives, such as the one he provided when he found more glands in the cheek:

> Besides the mentioned vessels, I observed peculiar ducts in them [i.e., in the cheeks] on 21 May [1661], when I dissected its transverse and thicker part a little obliquely.... Hardly indeed had it been divided when a probe [*stylus*] introduced into a tiny vessel cut loose by the original blow [*per resectum eodem ictu vasculum*] penetrated freely into the cavity of the mouth. Hence, having cut loose [*resecando*] the entire body of the cheeks from the lower jaw near the gums and slightly

---

83. Hieronymus Fabricius ab Aquapendente, *De visione; De voce; De auditu* (Venice, 1600), dedication to *De visione*. See also Peter Distelzweig, "Fabricius' Galeno-Aristotelian Teleomechanics of Muscle," in *The Life Sciences in Early Modern Philosophy*, ed. Ohad Nachtomy and Justin E. H. Smith (Oxford and New York: Oxford University Press, 2014), 65–84; and Nancy Siraisi, "*Historia, actio, utilitas*: Fabrici e le scienze della vita nel Cinquecento," in *Il teatro dei corpi: Le pitture colorate d'anatomia di Girolamo Fabrici d'Acquapendente*, ed. Maurizio Rippa Bonati and José Pardo-Tomás (Milan: Mediamed Edizioni Scientifiche Sri., 2004), 63–73.

84. Pomata, "*Praxis Historialis*," 117.

85. Pomata, "*Praxis Historialis*," 118–122.

86. Pomata, "*Praxis historialis*," 118–122. Bartholin also discussed the name "lymphatic vessels," but not philology; see Bartholin, *Vasa lymphatica*, 30–31.

87. Pomata, "*Praxis Historialis*," 117–118.

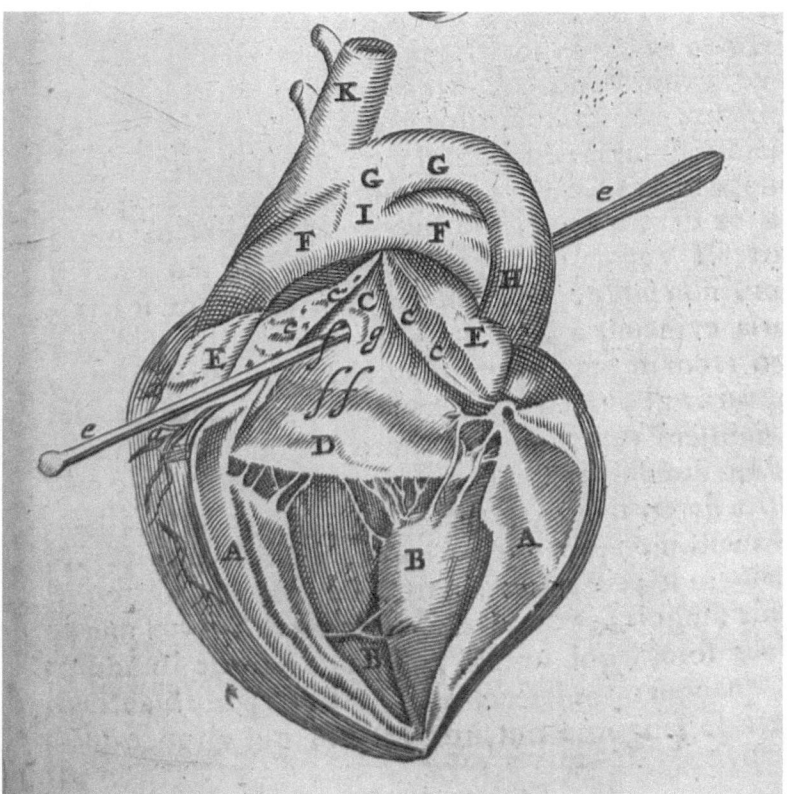

FIGURE 2.3. A probe inside the pulmonary artery of the heart in Cæcilius Folius, *Sanguinis a dextro in sinistrum cordis ventriculum ... reperta via* (Venice, 1639); shown here as reproduced in Thomas Bartholin, *Anatomia, ex Caspari Bartholini ... reformata* (The Hague, 1663). Courtesy of the Institute of the History of Medicine, Johns Hopkins University.

extending the membrane, I see on the same straight line multiple strict openings [*exigui hiatus*], through which the introduced probe easily found a passage into the gland itself.[88]

Steno strengthened his account by explicitly mentioning the instruments that he used and his sensible manipulation of the organs. He used a probe (*stylus*) of some sort to explore vessels and their depths (see fig. 2.3), a process that was sometimes so delicate he had to use an animal's "bristle

---

88. Steno, *Observationes anatomicæ*, §19, 17–18.

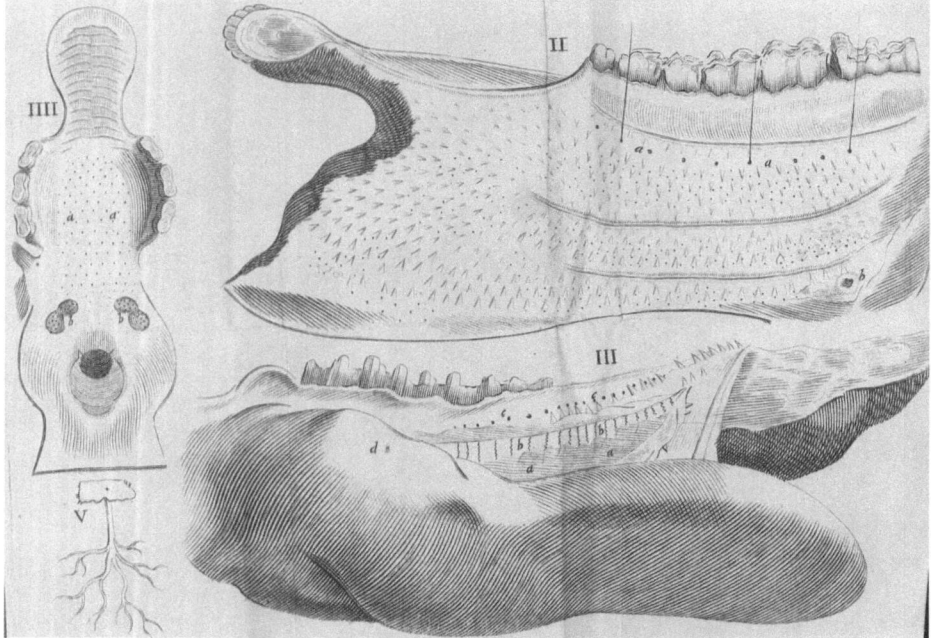

FIGURE 2.4. The palate viewed from below: (a) holes in the palate, (b) tonsils; in Steno, *Observationes anatomicæ* (Leiden, 1662). Reproduced with the permission of The President and Fellows of Magdalen College, Oxford.

[*seta*]."[89] Steno conveyed his handling of the organs in a particularly sensible way when narrating the discovery of the palatine salivary glands, at the top of the mouth:

> I noticed them for the first time on 27 May [1661] as I had already dissected almost all the head of a cow. Then indeed, as I cut out the tonsils, I saw that, by compressing the vicinity, [a] sticky humor was sifted out. Hence all that constituted the ceiling of the mouth, cut away and compressed by the fingers, displayed countless droplets of glutinous humor breaking out through very narrow holes.[90] (See fig. 2.4)

Like Aselli, Steno added images to accompany his descriptions and discussed the philological origins of the word *parotida*.[91]

---

89. Steno to Bartholin, 12 September 1661, in EMC III, 227; and Steno, *Observationes anatomicæ*, 88.
90. Steno, *Observationes anatomicæ*, §21, 20.
91. Steno, *Observationes anatomicæ*, §10, 8–9.

Steno's *historia* of the parotid gland largely surpassed the provincial aims of the dispute with Blasius. It revolutionized the anatomical understanding of the mouth glands, an understanding that remains vital to this day, and stimulated the rising field of gland studies in anatomy.[92] The first early modern publication on the anatomy of glands had been the *Adenographia sive glandularum totius corporis descriptio* (*Adenography or Description of the Glands of the Whole Body*; London, 1656), written by Wharton. When Steno arrived in Leiden, his professor Franciscus Sylvius had also developed a new theory about glands in which they were categorized as either *conglobate* or *conglomerate glands*.[93] Conglobate glands were small, oblong organs directly connected to the lymphatic vessels (today known as lymph nodes). On the other hand, conglomerate glands (today known as endocrine and exocrine glands) were large organs with no specific format that secreted bodily fluids into and out of the body, such as the salivary or pancreatic fluids. Steno learned this distinction directly from Sylvius's dissections at the Leiden hospital, where "medical practice" was taught "daily."[94]

Steno engaged seriously with Wharton's anatomical descriptions of the glands and added Sylvius's new theory to the discussion. Steno's salivary duct proceeded directly from the parotid gland, located behind the ear, into the mouth. Wharton had described the parotid gland in detail, but he did not relate it to the production of saliva because, as Steno explained, he did not see the salivary duct.[95] For Wharton, only the maxillary glands produced saliva, because he observed a pathway between them and the mouth, the *ductus whartonianus*.[96] Wharton also missed other things. According to Steno, the parotid gland described by Wharton was actually formed by two distinct glands: a small conglobate parotid gland, connected to the lymphatic system, and a larger conglomerate parotid gland, connected to the mouth via the salivary duct (see fig. 2.5).[97] Here Steno formulated a quantitative argument based on the weight of the glands (further explored in chap. 3). He then explained that the function of the conglomerate parotid gland was to produce salivary fluid, alongside the

---

92. Bertoloni Meli, *Mechanism, Experiment, Disease*, 105–129, 150–169.

93. Steno, *Observationes anatomicæ*, §§9–11, 6–8; and Bertoloni Meli, *Mechanism, Experiment and Disease*, 103.

94. Steno, *Observationes anatomicæ*, §10, 9–10.

95. Steno, *Observationes anatomicæ*, §17, 15; Wharton, *Adenographia*, 124–127.

96. Nicolaus Steno, *De musculis et glandulis observationum specimen* (Copenhagen: Matthia Godicchenius, 1664), 44; and Gray, *Anatomy of the Human Body*, 1135.

97. Steno, *Observationes anatomicæ*, §§9–11, 6–8.

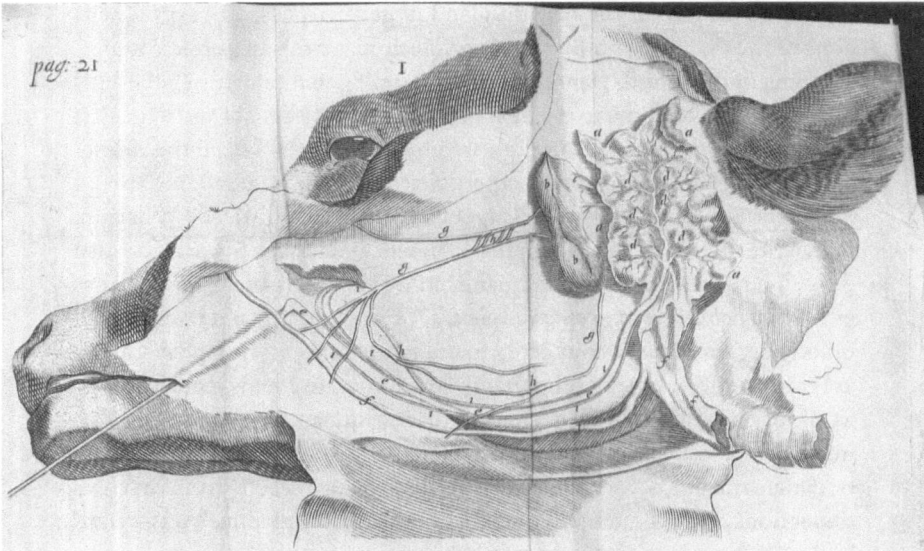

FIGURE 2.5. Parotid glands in the head of a calf. The largest, on the right, is the conglomerate gland (a), with the salivary duct exiting downwards (c); the bean shape on the left is the conglobate gland (b). From Steno, *Observationes anatomicæ* (Leiden, 1662), 21. Reproduced with the permission of The President and Fellows of Magdalen College, Oxford.

maxillary glands, which Wharton had already discovered.[98] In fact, Steno discovered other glands that also produced saliva and listed five types in total: parotid glands, maxillary glands, sublingual glands, palatine glands, and the glands of the cheek.[99] Aselli, Harvey, Pecquet, and Thomas Bartholin had discovered the lacteals, the circulation, the thoracic duct, and the lymphatics and written about them in their respective *historiæ anatomicæ*. In the same way, Steno made himself the rightful discoverer of the salivary glands by writing a *historia* of glands.

In his book, Steno said he was uncovering "knowledge of the glands [*glandularum cognitio*]" in a way that privileged the senses.[100] He stated that "reasoning deprived of the work of the senses [*sensuum ope destituta ratio*] did not find the paths carrying saliva into the mouth."[101] Steno did

98. Steno, *Observationes anatomicæ*, §17, 17.
99. See Steno, *Observationes anatomicæ*, §9, 8. On the cheek, sublingual and palatine glands, see ibid., §18, §20, §21, in 17–23.
100. Steno, *Observationes anatomicæ*, §5, in 4.
101. Steno, *Observationes anatomicæ*, vii.

not reject reasoning altogether but thought it must be based on observations. This cooperation between reasoning and observation, which became more and more important throughout Steno's works, was central to his search for certainty in anatomy.[102] He affirmed that "observation must lead to reasoning [*rationi accedat oportet observatio*] and the thing itself must be examined in all its parts as far as possible [*secundum omnes partes quantum licet*] so that its true shape, showing the thing itself, be imprinted upon the mind [*ut rem vere exhibens menti inscribatur figura*]."[103] Unless such an approach was followed, he said, one could "hardly expect certain knowledge [*vix certam cognitionem exspectare*]" of things.[104] Yet, as already mentioned, Steno also acknowledged that not all observations were reliable. His solutions to achieve observations that were reliable ranged from dissecting many animals to performing dissections slowly so that observers could follow them more easily. Moreover, as the next chapter will show, Steno complemented whatever was missing from his observations with ideas based on mathematics, chymistry, and mechanics.

Steno's description of the oral glands in print was a blow to Blasius, who in his book had barely addressed the anatomical features of the duct. A few days later Blasius wrote to Thomas Bartholin saying that "my former disciple Nicolaus Steno" had declared war upon him (*bellum indicat*).[105] Blasius supported his claim for priority not by denying Steno's account, but by adding other events to the story. According to Blasius, Steno first saw the duct "in a nebulous way" during a dissection done by Blasius before his students in the spring of 1660. Blasius concluded this dissection, he said, with "a superficial demonstration of the salivary vessels."[106] It was superficial because of the poor condition of the body parts by the end of the dissection. In addition, Blasius argued that what Steno called the "parotid gland," where the upper salivary duct began, he had called the "[upper] maxillary gland," in parallel with the lower maxillary gland, well-known to produce saliva.[107] Blasius also said that his book was not expected to have more anatomical details because it was a general textbook of medicine.[108] That was precisely the point that differed from Steno.

---

102. See also R. J. Hankinson, "Galen on the Limitations of Knowledge," in *Galen and the World of Knowledge*, ed. Christopher Gill, Tim Whitmarsh, and John Wilkins (Cambridge: Cambridge University Press, 2009), 206–242, esp. 212–213.
103. Steno, *Observationes anatomicæ*, viii.
104. Steno, *Observationes anatomicaæ*, vi.
105. Blasius to Bartholin, 16 July 1661, in EMC III, 158.
106. Blasius to Bartholin, 16 July 1661, in EMC III, 160–161.
107. Blasius to Bartholin, 16 July 1661, in EMC III, 167–168.
108. Blasius to Bartholin, 16 July 1661, in EMC III, 165.

Steno's description was different because he not only gave a more robust description of the glands, with dated observational accounts, but also placed such accounts at the center of his publication. Blasius never fully understood this. A week before writing to Bartholin, Blasius had offered an academic disputation on the subject, whose text he asked Bartholin to "compare not with Steno's *Disputatio* ... but truly with observations from dissections of animals."[109] Yet, despite this claim, Blasius did not present new anatomical observations.

"The Pen Following the Knife": The Making of a Scholar

Perhaps as a result of Barbette's mediation, requested by Thomas Bartholin, Blasius paused his letters and publications on the subject for almost a year.[110] Steno's prestige, on the other hand, continued on the rise, owing in large part to Bartholin's influence in the Republic of Letters. In January 1662 Steno republished his *Disputatio* from July 1661 as the first segment of a four-part book of anatomical observations, framing his work as part of the rising epistemic genre of anatomical *observationes*.[111] In this book, *Observationes anatomicæ quibus varia oris, oculorum, et narium vasa describuntur, novique salivæ, lacrymarum et muci fontes deteguntur* (*Anatomical Observations, in Which Several Vessels of the Mouth, Eyes, and Nose Are Described and the Sources of Saliva, Tears, and Mucus Are Uncovered*; Leiden, 1662), Steno added extensive anatomical descriptions of the glands of the eye and nose, the former of which he had begun exploring as early as November 1660.[112] The book's success was quickly confirmed by Thomas Bartholin, who mentioned this new publication to the king of Denmark as a book whose praises "would one day serve the glory of our [Anatomical] Theater."[113]

This publication showed that Steno could pursue his own research and divulge his results, not only in letters but also in the form of a book. A student's ability to produce a book was not a given, especially before

---

109. Blasius to Bartholin, 16 July 1661, in EMC III, 176.

110. Bartholin to Barbette, 28 August 1661, in EMC III, 197–200. Blasius did not seem to write on the ducts between July 1661 and May 1662. However, in January 1662, Steno mentioned a letter that Blasius had sent, though it could have been just a copy of the original letter to Bartholin; see Steno to Bartholin, 9 January 1662, in EMC III, 262–266.

111. Steno sent the book to Bartholin in this month; see Steno to Bartholin, 9 January 1662, in EMC III, 262–266.

112. Steno, *Observationes anatomicæ*, 88. By March 1661, Borch was aware of these investigations; see Borch to Bartholin, 3 March 1661, in EMC III, 360–369.

113. Bartholin to Borch, 14 February 1662, in EMC III, 426.

obtaining his medical degree. Then as now, authors had to know when to stop writing, because there is always something that can be added to the manuscript. In the letter to Bartholin that accompanied his book, Steno mentioned this challenge, "since any new observation was an instrument for the next one, [and] I could not bring this matter to an end [*ad umbilicum perducere*]."[114] This final expression explicitly referred to the completion of his book, since in classical Latin *umbilicus* means the rod in which a scroll ended.[115] The Roman poet Horace (65 BC–8 BC) used a similar expression in his *Epodes* XIV: "ad umbilicum adducere." Steno had read Horace, and this classical reference confirms that his humanist education continued to serve him well as his career progressed.[116] To produce his book, Steno paused his observations and moved from the dissection table to the writing desk. Two years later, in a letter reporting anatomical observations to a friend, Steno sighed about a similar situation: "I should have given more complete arguments and added much more if, while I had the knife in my hands, those same things, which the pen following the knife desired, had occurred to me."[117] In short, Steno's *Observationes anatomicæ* contributed to shape his scientific persona into that of a skillful dissector and a scholarly author who made careful editorial choices.

Steno's most important editorial decision was writing in the *observationes* genre, a choice he maintained throughout his career. The close relationship between the genre of *observationes* and notebooks made Steno's work simpler. The "Chaos" manuscript showed that Steno was good at taking notes, so the step toward producing a book of *observationes* was easy and convenient for him. Moreover, Steno explicitly stated that he preferred to share his results in print rather than "suppress them until they reach maturity [*dum ad maturitatem pervenirent, ea supprimere*]."[118] This genre, however, has a slightly different history from the *observationes medicæ* recently studied by Gianna Pomata.[119] Steno's

---

114. Steno to Bartholin, 9 January 1662, in EMC III, 262–266.
115. S.v. "umbilicus," in Charlton T. Lewis, *An Elementary Latin Dictionary* (New York: American Book Company, 1890), 886.
116. Horace's *Epistles* are mentioned in Steno, *Observationes anatomicæ*, 59.
117. Steno to Willem Piso, 24 April 1664, in Steno, *De musculis et glandulis*, 68–69.
118. Steno, *De musculis et glandulis*, 69.
119. For the first *observationes medicæ*, see Pomata, "Sharing Cases," 232–236. There is not yet a comprehensive history of *observationes anatomicæ*, a genre that seems to have developed in the second half of the seventeenth century with the works of Malpighi and Steno. However, there were *observationes anatomicæ* as early as the sixteenth century, e.g., Gabriele Falloppio, *Observationes anatomicæ* (Venice: Marcus Antonius Ulmus, 1561).

book of *observationes* did not follow the traditional list format used in most books of *observationes medicæ* and in Bartholin's *Historiarum anatomicarum rariorum centuria* (Copenhagen, 1654–1661). Instead, he wrote an extended account of anatomical parts, with their *historia*, *actio*, and *usus*, where he included multiple observations of the organs he was describing, such as the glands of the mouth and eyes. The book was also not organized into chapters and each paragraph is numbered, so it was a hybrid between a list of observations and the more typical anatomical *exercitatio* or *dissertatio*, like those of Harvey, Pecquet, and Bartholin.[120] Pecquet's *Experimenta nova anatomica* (Paris, 1651) was more or less a collection of essays based on observations of different kinds.[121] Since Steno's book was composed of four separate essays, he seems to have used a similar style but adapted it to the genre of *observationes*.[122] This format granted Steno freedom to publish his observations alongside other thoughts, thus textually showing his belief that observations and reasoning should go together. In short, Steno was forging his own style of anatomical writing.[123]

Despite Steno's success, no rose is without thorns. Blasius relaunched his claim to priority in the discovery of the salivary duct by sending private messages to Steno's professors in Leiden, leading Steno to ask Bartholin for help.[124] Early in May 1662, Blasius found the support of Nicolaus Hoboken (1632–1678?), an Utrecht physician, whom Blasius wanted to connect with Bartholin's correspondence network.[125] Bartholin was open to Hoboken's acquaintance but still refrained from making public claims on the dispute between Steno and Blasius. He explained that "I always declined the arbitration of this quarrel [*huius litis arbitrium semper declinavi*], because

---

120. Harvey, *De motu cordis*; Pecquet, *Experimenta nova anatomica*; and Bartholin, *Dissertatio anatomica de hepate defuncto* (Copenhagen: Christian Wering, 1662).

121. Pecquet's cover page speaks of various essays: *Experimenta nova anatomica . . . dissertatio anatomica de circulatione sanguinis . . . et chyli motu* (Paris: Sebastian & Gabriel Cramoisy, 1651).

122. Steno's *Observationes anatomicæ* had four parts: "*observationes* on the glands of the mouth"; a response to the physician Anton Deusing on the liver; "*observationes* on the glands of the eye"; and "an appendix on the vessels of the nose"; see ibid., 1–54, 55–78, 79–100, 101–108.

123. See also Domenico Bertoloni Meli, "Mechanistic Pathology and Therapy in the Medical *Assayer* of Marcello Malpighi," *Medical History* 51, no. 2 (2007): 165–180.

124. Steno to Bartholin, 21 May 1662, in EMC IV (Copenhagen, 1667), 3.

125. Nicolaus Hoboken to Bartholin, 14 May 1662, in EMC IV, 15. On Hoboken, see "Hoboken, Nicolaus," in G. A. Lindeboom, *Dutch Medical Biography: A Biographical Dictionary of Dutch Physicians and Surgeons (1475–1795)* (Amsterdam: Rodopi, 1984), cols. 871–873.

both [Steno and Blasius] are friends to me."[126] Taking this answer as a cue to become more involved, Hoboken published in November the *Novus ductus blasianus* (Utrecht, 1662), an edited volume defending Blasius's case. It contained letters exchanged between him and Blasius, Blasius's written defense to Bartholin from July 1661, and the signed testimonies of three former students of Blasius stating that Blasius had shown them the duct before Steno saw it.[127] One of the witnesses said that "the duct recently called *stenonianus*, was clearly demonstrated to me by the most illustrious gentleman G[erard] Blasius, a few months before the arrival of the false discoverer."[128] However, the testimonies were all signed in May and June 1662, more than two years after the event they claimed to have seen. Most significant, they included no new observations, nor did the rest of the book.

Steno responded in April 1663 with the visually impressive text *Apologiæ prodromus*, "in which it is demonstrated that the Blasian judge is unskilled in anatomy and a servant of his emotions," as the title went on to assert (see fig. 2.6).[129] Steno printed it in a broadsheet format, which was intended to be widely distributed.[130] The text is a masterpiece of Steno's reading and note-taking abilities and an elegant example of what has been termed seventeenth-century learned empiricism.[131] Observations, rather than being in opposition to bookish knowledge, relied strongly on the study of texts. In Steno's case, textual techniques were meant to highlight the importance of observations, as they compared what he and Blasius claimed to have seen. Steno even compared two texts of Blasius to show a lack of consistency. His text is displayed with extensive citations from Blasius's books, with abbreviations highlighted in parentheses and a two-column table comparing his and Blasius's account of the duct in a way that

---

126. Bartholin to Hoboken, 4 June 1662, in EMC IV, 18.

127. For the testimonies see Hoboken, *Novus ductus salivalis blasianus* (Utrecht: Johannis à Renswouw, 1662), 15–18. The students were Ludovicus de Keyser, Joannes Leonarts, and Gerard Croese, whose disputations they defended in Amsterdam. See van Miert, *Humanism in an Age of Science*, 383, 385, 379.

128. Hoboken, *Novus ductus salivalis blasianus*, 17.

129. Nicolaus Steno, *Apologiæ prodromus, quo demonstratur, judicem Blasianum et rei anatomicæ imperitum esse, et affectuum suorum servum* (Leiden, 1663).

130. In 1688 Boyle also published a broadsheet related to plagiarism, in Boyle, "An Advertisement About the Loss of His Writings"; see Steven Shapin, *A Social History of Truth: Civility and Science in Seventeenth-Century England* (Chicago: University of Chicago Press, 1994), 184.

131. Pomata and Siraisi, "Introduction," in Pomata and Siraisi, *Historia: Empiricism and Erudition in Early Modern Europe*, 1–38, esp. 17–31.

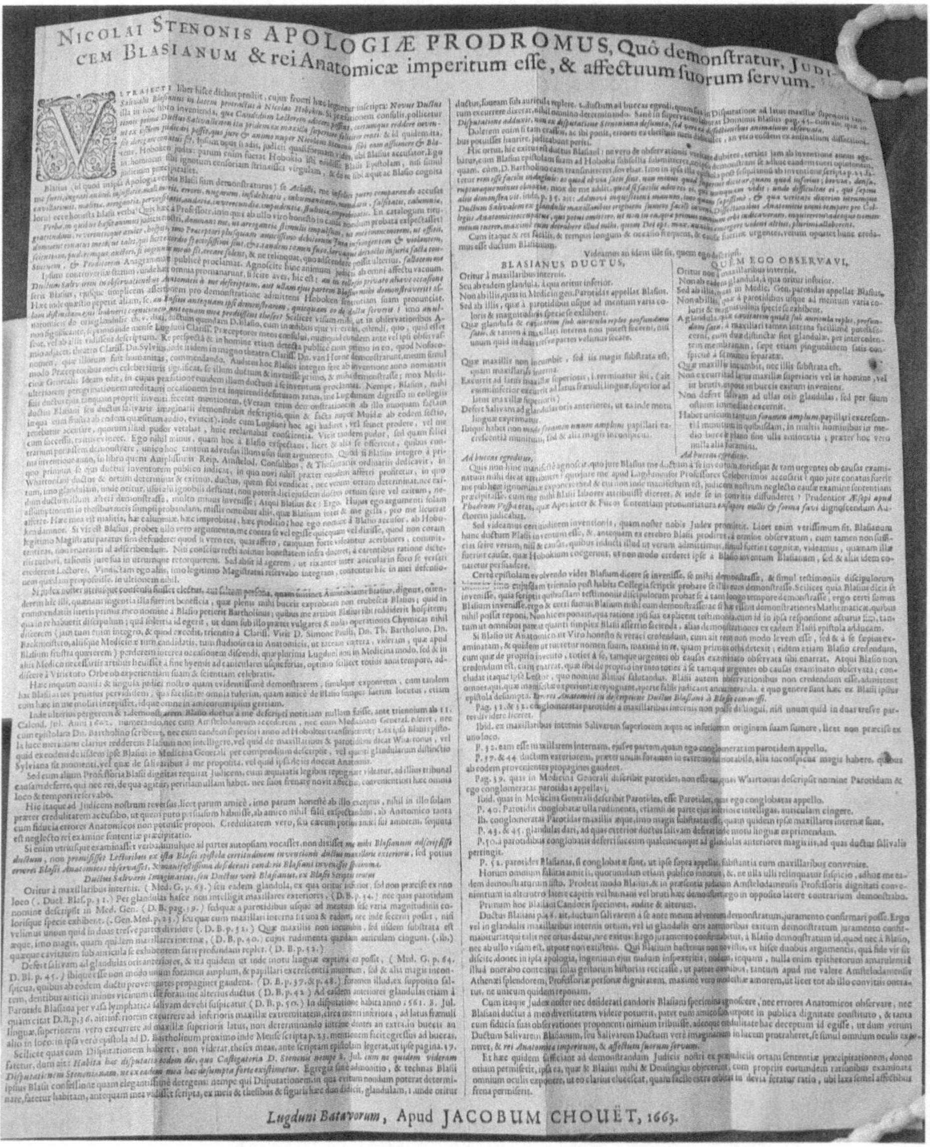

FIGURE 2.6. The broadsheet *Apologiæ prodromus* (Leiden, 1663). From the British Library Archive, General Collection 548.e.32.(2.).

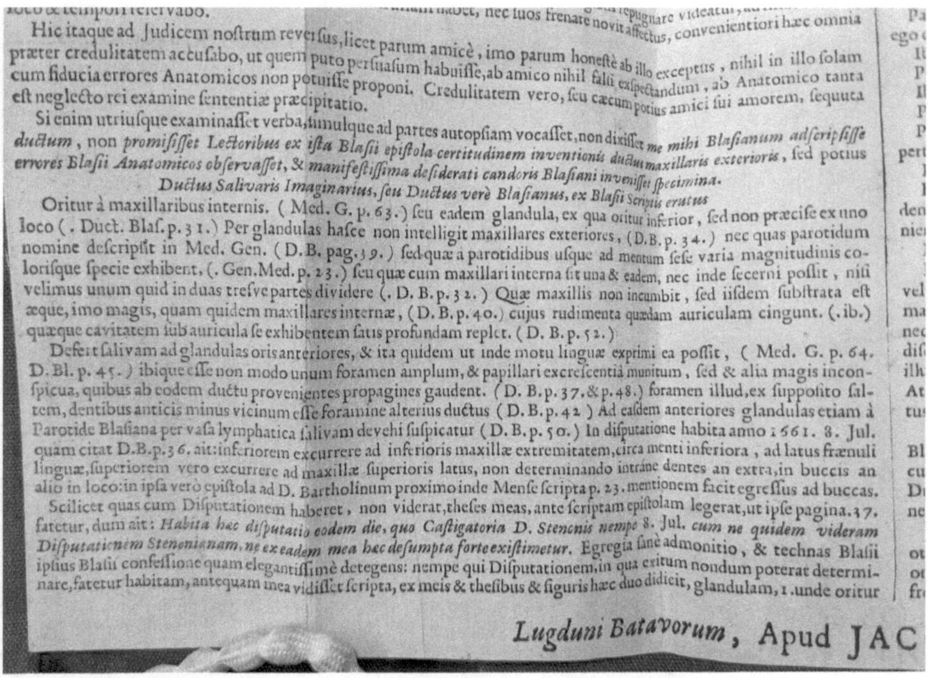

FIGURE 2.7. Paragraph with many quotations, where "D. B." refers to Hoboken's [*Novus*] *ductus blasianus* (1662) and "Med. G." to Blasius's *Medicina generalis* (1661), in Steno, *Apologiæ prodromus* (Leiden, 1663). From the British Library Archive, General Collection 548.e.32.(2.).

added force to his argument (see figs. 2.7 and 2.8). These textual comparisons would later become a hallmark of Steno's theological writings amid confessional controversies.[132]

Steno's full command of the pen, so to speak, lent itself to the intellectual destruction of Blasius and his friends. He accused Hoboken of being "ignorant in anatomy [*rei anatomicæ imperitum*]," suggested that Blasius "demonstrated something which ... does not exist," and even questioned his manliness.[133] This attitude may have contributed to ending the controversy, since Steno did not address it again, even though he had planned to do so at greater length. His text was, after all, an "introduction to an

---

132. See OTH, 1:40–48, 60–61, 66, 113–118, 147–153. See also Sobiech, *Ethos, Bioethics, and Sexual Ethics*, 51 n. 154.

133. Steno, *Apologiæ prodromus*, in OPH, 1:154, 153, 147.

> legiis Anatomicis occupatus, qui potui omittere, ut non in ea,quæ primus omnium orbi indicaveram, inquirerem: adeoque nomen meum tuerer, maxime cum detrahere illud mihi, quem Dei opt. max. auxilio emergere vident altius, plurimi allaborent.
> Cum itaque & res facilis, & tempus longum & occasio frequens, & causæ fuerint urgentes, verum oportet hunc credamus esse ductum Blasianum.
> Videamus an idem ille sit, quem ego descripsi.
>
> | BLASIANUS DUCTUS, | QUEM EGO OBSERVAVI, |
> |---|---|
> | Oritur à maxillaribus internis. | Oritur non à maxillaribus internis. |
> | Seu ab eadem glandula, à qua oritur inferior. | Non ab eadem glandula, à qua oritur inferior. |
> | Non ab illis, quas in Medicin gen. Parotidas appellat Blasius. | Sed ab illis, quas in Medic. Gen. parotidas appellat Blasius. |
> | Sed ab illis, quæ à parotidibus usque ad mentum varia coloris & magnitudinis specie se exhibent. | Non ab illis, quæ à parotidibus usque ad mentum varia coloris & magnitudinis specie se exhibent. |
> | Quæ glandula & cavitatem sub auricula replet profundam satis, & tamen à maxillari interna non potest secerni, nisi unum quid in duas tresve partes velinus secare. | A glandula, quæ cavitatem quidē sub auricula replet, profundam satis, à maxillari tamen interna facillimē potest secerni, cum duæ distinctæ sint glandulæ, per intercedentem membranam, sæpe etiam pinguedinem satis conspicuè à se mutuo separatæ. |
> | Quæ maxillis non incumbit, sed iis magis substrata est, quam maxillaris interna. | Quæ maxillis incumbit, nec illis substrata est. |
> | Excurrit ad latus maxillæ superioris, i. terminatur ibi, (ait enim: inferior excurrit ad latus frænuli linguæ, superior ad latus maxillæ superioris) | Non excurrit ad latus maxillæ superioris vel in homine, vel in brutis, utpote in buccis exitum inveniens. |
> | Defert Salivam ad glandulas oris anteriores, ut ea inde motu linguæ exprimatur. | Non defert salivam ad ullas oris glandulas, sed per suum ostium immediatè excernit. |
> | Ibique habet non *modo foramen unum amplum*, papillari excrescentia munitum, sed & alia magis inconspicua. | Habet unicum tantum *foramen amplum*, papillari excrescentia munitum in quibusdam, in multis hominibus in medio buccæ plano sine ulla eminentia; præter hoc vero nulla alia foramina. |
> | *Ad buccas egreditur.* | *Ad buccas egreditur.* |
>
> Quis non hinc manifestè agnoscit, quo jure Blasius me ductum à se inventum, totiesque & tam urgentes ob causas examinatum mihi dicat attribuere? quo jure me apud Lugdunenses Professores Celeberrimos accusarit? quo jure conatus fuerit me publicæ ignominiæ exponere? Sed & cui non inde manifestum est, judicem nostrum neglecto causæ examine sententiam præcipitasse, cum me mihi Blasii labores attribuisse diceret, & inde se in convitia diffunderet? Prudentior *Æsopi apud Phædrum Vespa* erat, quæ Apes inter & Fucos sententiam pronuntiatura *ex sapore mollis & forma favi* dignoscendum Auctorem judicabat.
> Sed videamus certitudinem inventionis, quam noster nobis Judex promittit. Licet enim verissimum sit, Blasianum hunc ductum Blasii inventum esse, &, antequam ex cerebro Blasii prodiret, à nemine observatum, cum tamen non suffi-

FIGURE 2.8. Table comparing Steno's and Blasius's claims in Steno, *Apologiæ prodromus* (Leiden, 1663). From the British Library Archive, General Collection 548.e.32.(2.).

apologia [*prodromus apologiæ*]" that he never wrote.[134] This kind of intellectual attack, however, could compromise Steno's place in the Republic of Letters, especially if he was not careful with his words. In fact, although vitriolic attacks were not uncommon in the early modern period, one of the virtues expected in the Republic of Letters, especially of its younger members, was modesty.[135] Modesty, like humility and the other virtues, was part of the ideal of civility that was central to the Republic of Letters. The true scholar was one who controlled his emotions, which is perhaps

---

134. Steno mentioned it briefly again when describing the parotid salivary duct in *De musculis et glandulis*, 30–31. Blasius maintained his stance. The expanded version of *Medicina generalis* does not mention Steno or the name *ductus stenonianus*; see Gerard Blasius, *Medicina universa* (Amsterdam: Petrus van den Berge, 1665), 303. However, almost twenty years later, Blasius mentions both his and Steno's names; see Blasius, *Anatome animalium* (Amsterdam: Johan van Someren, 1681), 253, 338.

135. Goldgar, *Impolite Learning*, 154–162.

why Steno claimed in the broadsheet's title that Blasius was unable to do so. Steno explicitly mentioned the virtue of modesty, though only in passing, amid a correspondence with Bartholin. He first wrote to Bartholin that he wanted to be modest and not respond to Blasius, but "I am forced to complain [*conqueri cogor*]."[136] His intention, he said, was to clear the names of Sylvius and van Horne, who had already announced the discovery of the new salivary duct as Steno's. Bartholin believed Steno's declaration of intentions, for he acknowledged his "modesty and composed mind [*modestiam . . . et sedatam mentem*]."[137] In the end, Steno was successful in maintaining this modest image. Throughout the controversy with Blasius there was a striking and continuous praise of Steno's lack of pretension from Borch's and Barbette's early praises in 1661, to Adolphus Vorstius's (1697–1663) in April 1663, the same month in which Steno's *apologia* was published.[138] Vorstius was then a senior professor of botany at the University of Leiden and a respected member of the medical Republic of Letters. It is therefore significant that he wrote of Steno as "a distinguished young man and one of great expectation [*juvenis egregius et summæ exspectationis*], who carries himself here very modestly."[139]

How did Steno maintain an image of modesty while responding so strongly to Blasius? The answer lies in Steno's careful writing, as if he were emotionally detached from the dispute. As he wrote, "I have always spoken of Blasius in a friendly way [*amice de Blasio semper fuerim locutus*], even when he started to undertake these things against me."[140] Indeed, he reported his desire to be modest to Bartholin. Most important, Steno's rhetoric stemmed from the confidence of one who knew that his argument hinged on observational accounts. Commenting on Blasius's work, he argued that "anatomical mistakes should not be put forward with such confidence by an anatomist [*ab anatomico tanta cum fiducia errores anatomicos non potuisse proponi*]."[141] Steno, on the other hand, contended himself with "writing down only the histories of the things that happened [*solas*

---

136. Steno to Bartholin, 22 April 1661, in EMC III, 91.

137. Bartholin to Steno, 20 May 1661, in EMC III, 96. See also Bartholin to Barbette, 28 August 1661, in EMC III, 200. Perhaps Bartholin's first letter to Steno, now lost, was about how to proceed with modesty.

138. Borch to Bartholin, 20 March 1661, in EMC III, 376; and Barbette to Bartholin, 30 July 1661, in ibid., 196.

139. Adolph Vorstius to Bartholin, 14 April 1663, in EMC IV, 405. Vorstius died in October 1663, so this was also one of his last letters. For more on Vorstius, see Lindeboom, *Dutch Medical Biography*, cols. 2088–2089.

140. Steno, *Apologiæ prodromus*, in OPH, 1:148.

141. Steno, *Apologiæ prodromus*, in OPH, 1:148.

gestorum historias recitasse] so that it appears to everybody how much I value the splendor of the Amsterdam Athenæum, the dignity of the professorial figure, and especially the love of modesty [*modestiæ amor*]."[142] Modesty became crucial to Steno's search for certainty in anatomy, especially when he wrote of his failures. In his first publication, when explaining how glands emitted a fluid into the blood, Steno admitted that the lymphatics were "so short that" that it was hard "to decide anything certain about them [*certi quid de illis si statuerem*]." To make his point clearer, he quoted Cicero, who in De natura deorum (Book I, 1) called "temerarious" those who "defend without a doubt as understood and known that which is not sufficiently explored [*quod non satis exploratè perceptum sit, et cognitum sinè ulla dubitatione defendere*]."[143] As part of his observations in a calf, Steno once also admitted that he had to use a second calf, because his "little careful hand had cut away [*in hoc subjecto parum cauta manus resecuisset*]" the original one.[144] By providing details about observations that were short of his goals, Steno thus positioned himself as an authoritative observer of the body. In a nutshell, Steno's public image of modesty was more than just a rhetorical device. It was also a crucial epistemic element of his anatomical accounts.

※

Steno's years in Leiden illustrate how he fully assimilated key trends of early modern anatomy, such as the focus on observations. As he sat down to write, his humanist education and note-taking skills shaped his book of *Observationes anatomicæ*, a genre that he continued to use in scientific publication. This book was thus foundational to Steno's career, and he rapidly acquired more intellectual maturity and social prestige. As this chapter has shown, this prestige arose in large part out of the Blasius controversy. Steno's efforts in the dispute were successful because Blasius never responded with new observational accounts, even though much ink was spilled on his side as well. Unlike Blasius's, Steno's use of rhetoric was always accompanied by more dissections. For that reason alone, Steno's replies exhibited a proficient scientific practice in anatomy. This approach was enough to satisfy Thomas Bartholin, who did not take sides publicly but quietly promoted Steno in the Republic of Letters. Techniques that Steno followed in his writing and that at first seemed to be nothing more

---

142. Steno, *Apologiæ prodromus*, in OPH, 1:153.
143. Steno, *Observationes anatomicæ*, §40, 37. See Cicero, *De natura deorum*, Book I, 1.
144. Steno, *Observationes anatomicæ*, §41, 38.

than the product of the Republic of Letters' social norms, such as the virtue of modesty, played a significant epistemic role in his publications. Rhetorical tools, rather than instruments of power, were a meaningful way to convey what Steno had really observed in the body.

In his final response, Steno made yet another revealing critique of Hoboken and Blasius. In Steno's eyes, Hoboken, although claiming to be an impartial onlooker of the events, fully sided with Blasius, "as if these were Mathematical demonstrations [*demonstrationes Mathematicæ*] to which there is no reply [*quibus nihil possit reponi*]."[145] Hoboken and Blasius lacked intellectual modesty because they did not leave room for doubt. Interestingly, Steno was at the time grappling with the improvement of certainty in anatomy, in part by applying mathematics to his studies of the body. This probably explains why he framed his criticism of Blasius in mathematical terms. As Steno was coming to discover, mathematics, mechanics, and chymistry were helpful in describing the anatomy of glands. Conveniently for Steno, the *observationes anatomicæ* was an ideally flexible genre that allowed him to gather interdisciplinary insights into an anatomy book, as the next chapter will show.

---

145. Steno, *Apologiæ prodromus*, in OPH, 1:152.

CHAPTER THREE

Dissecting with Numbers, Machines, and Mixtures

A few months after writing *Observationes anatomicæ* (Leiden, 1662), Steno wrote to Thomas Bartholin that he "had decided to lay down the anatomical knife... and to take up again the nearly cast away geometer's rod [*geometricum radium tantum non abjectum*]." However, that is not what happened. "Hardly were my fingers rid of blood, slightly besprinkled with this very pleasant powder," when the priority dispute on the salivary duct forced Steno to go back to anatomy. Nonetheless, his desire to study geometry was so strong that he saw this situation as denying him "the happiness desired for a long time" as he was forced "to return to this bloody task [*ad sanguinarium illud exercitium*]."[1] Steno had had a strong attraction to mathematics since his early days at the Copenhagen Cathedral School. He explored the field further through his multiple measurements in goldsmith workshops and by taking notes from mathematical books in the "Chaos" manuscript. But did Steno really pause his intense interest in mathematics after leaving Copenhagen? If not, how did it relate to his anatomical research? And what relationship did it have with the Cartesian mechanical ideas that were on the rise in the Netherlands?

In light of Steno's use of a mathematical model to describe muscle motion in 1667 in Florence, the city that still carried the mathematical legacy of Galileo and his disciples, one could say that Steno's mathematical interests reemerged only after his arrival in Italy in 1666.[2] This chapter shows that that was not the case. Rather than a sudden shift, Steno's turn toward mathematics was a gradual process that occurred in tandem with his anatomical work in Leiden. I argue that Steno solved the tension between his interests in mathematics and anatomy by intentionally

---

1. Steno to Thomas Bartholin, 26 August 1662, in EMC IV, 103.
2. Gustav Scherz, "Niels Stensen und Galileo Galilei," in *Saggi su Galileo Galilei*, ed. Carlo Maccagni (Florence: Barbera, 1972), 731–779, esp. 764–768.

introducing mathematics *into* his anatomical work. As Steno's reputation as a skillful and learned anatomist grew, he also grew fonder of anatomy, as his growing commitment to anatomical research suggests. Indeed, in the same sentence where he wrote of his ambivalence toward anatomy, he also could not stop himself from describing blood as a "very pleasant powder [*iucundissimus pulvis*]."[3] The truth is that Steno found anatomy attractive not only because he was good at it but also because he used knowledge from other disciplines in it, including when working on the salivary duct. When he later turned to the study of the muscles and the heart (after the glands), mathematics and mechanics gained even more prominence in his work. Steno's complaint about returning to the dispute with Blasius, therefore, was also a complaint against distractions. Most likely he wanted to focus on projects where his interest in geometry could flourish, such as the muscles. This chapter's first part describes Steno's focused interdisciplinary work in the Netherlands. It is the first historical account of his uses of mathematics, mechanics, and chymistry in his early anatomical works.

The rest of the chapter explains the various ways that Steno's eclectic interests were sustained and fostered in the Netherlands. Intellectually, rather than attaching Steno's mechanisms to the ideas of Descartes, as many have done, I demonstrate that, if anything, he viewed himself as being in opposition to them. Steno's famous objection to those who claimed that "in animals there is no soul" was above all a stroke against the rising Cartesian materialism in the Netherlands. However, I prefer to avoid historiographical labels. As a true humanist of the seventeenth century, Steno derived his thought from a diversity of sources, including from the anatomical tradition and the philosophy of Gassendi. He used them to the extent that they contributed to his search for certainty in anatomy.

Steno's ideas also did not appear in a cultural void. It is true that Thomas Bartholin did not seem interested in Steno's mathematical ambitions, since he never addressed them in their correspondence.[4] However, Steno found encouragement of his interdisciplinary pursuits in friendships that he deepened during his stay in the Netherlands, especially with his travel companion and friend Borch. This chapter thus concludes by describing the stimulating environment that Steno and other medical students

---

3. Moreover, his description of a "bloody task" does not necessarily have a negative connotation, since it refers to dissecting, which involves blood.

4. Bartholin's response does not mention mathematics; see Bartholin to Steno, 7 September 1662, in EMC IV, 113–117. Mathematics rarely appears in Bartholin's anatomical work, though he acknowledged its epistemic attractiveness; see Bartholin to Eckard Leichner, 10 March 1663, in EMC IV, 325–328.

found in the Low Countries. I show that his tours of shipyards, chymical workshops, and libraries contributed to his dissections with numbers, machines, and mixtures, as well as to his religious anxieties.

The Mathematics and Mechanics of Glands and Muscles

In *Observationes anatomicæ*, Steno described the anatomy of the salivary, lacrimal, and nasal glands by means of various observations. Most of these observations were from dissections, but he also included a broad spectrum from mathematics, mechanics, and chymistry. He was able to do so because of the flexibility granted by the new epistemic genre of *observationes* that he chose to write on. Steno used mathematical insights not only in the structural description of glands (the *historia anatomica*), but also when addressing their action (*actio*) and function (*usus*). They ranged from quantitative reasoning, typical of pure mathematics (arithmetic), to the use of analogies drawn from the physico-mathematical discipline of mechanics.

Steno used his first quantitative argument when seriously engaging the work of the English physician Thomas Wharton on the parotid gland. Wharton had written the first systematic treatment of glands in the early modern period, the *Adenographia sive glandularum totius corporis descriptio* (*Study of Glands or Description of the Glands of the Entire Body*; London, 1656). In this groundbreaking book, Wharton identified one parotid gland whose "true uses [*veri usus*]" were related to a pair of nerves and the ear, not the production of saliva.[5] In his dissections, however, Steno claimed that where Wharton saw one gland, there were two. Using categories developed by Franciscus Sylvius, his professor of medicine at the University of Leiden, Steno claimed that there was a parotid *conglobate* gland (today known as a lymph node) and a parotid *conglomerate* gland (the actual salivary gland). How did Steno support this distinction against Wharton besides describing his dissections? In his work, Wharton registered the weights of the maxillary and parotid glands in different animals (see table 1). According to his data, the maxillary salivary gland weighed almost half as much as the parotid gland in the bodies of a man and of a cow fetus (their proportions averaged approximately 0.6). This suggested that the two glands were distinct.

Since Steno had been fond of making measurements and weighing things since his early days in Danish goldsmith workshops, he decided to also weigh the glands he dissected. As I discuss in more detail below,

---

5. Wharton, *Adenographia*, 127.

TABLE 3.1. Proportion of the weights of maxillary and parotid glands as measured by Thomas Wharton and Nicolaus Steno

Dissected corpses	Maxillary gland	Parotid gland	Proportion
28-year-old man (Wharton)	9.8 g	17.6 g	0.56
Fetus of a cow (Wharton)	7.8 g	11.7 g	0.67
	proportion average for Wharton's values		0.62
Calf (Steno)	125 g	141 g	0.89

Sources: Data from Wharton, *Adenographia*, 119–120, 125; and Steno, *Observationes anatomicæ*, 10–11. The conversion of units from seventeenth-century ounces to grams is from *Nicolai Stenonis Opera philosophica*, ed. Vilhelm Maar, 2 vols. (Copenhagen: Vilhelm Tryde, 1910), 1:227–228.

the Leiden mathematician Jacob Golius (1596–1667) was present at some of Steno's gland dissections, which may have encouraged his quantitative approach. To weigh the glands, Steno separated the conglomerate from the conglobate parotid gland and weighed only the former.[6] In the end, Steno found the proportion of the weights of the maxillary and parotid conglomerate glands to be 0.9, much closer to 1, thus meaning that their weights did not differ significantly. Steno's quantitative results confirmed to him that the conglobate and conglomerate parotid glands were distinct, because the weight of the latter was similar to that of the maxillary (conglomerate) gland. Since, by definition, conglomerate glands secreted a fluid and Steno had found a salivary duct coming out of the parotid, the similar weight values of the maxillary and parotid glands also contributed to the argument that the parotid produced salivary fluid. Steno and Wharton did not calculate this precise proportion as I did in the structured format of table 3.1. However, Steno referred to it explicitly, writing that Wharton's "proportion [*proportio*] had not been precisely observed [*exacte observata*]."[7]

The differences between Wharton's and Steno's claims speak to Steno's search for clarity and rigor in anatomy. Steno pointed out that Wharton's quantitative data was not precise enough because Wharton did not specify what he had weighed. Did Wharton weigh the glands and the attached vessels, and did he remove all the blood vessels and nerves? In Steno's words, Wharton "seems to have described an abundance of matter from these glands in an undetermined quantity [*extensione non determinata*]."

6. Steno, *Observationes anatomicæ*, §12, 11.
7. Steno, *Observationes anatomicæ*, §12, 11.

Steno, on the contrary, said that he detached each gland that he weighed "from vessels and from the conglobate gland lying beside it."[8] Moreover, whereas Wharton presented the weights as a secondary detail of his anatomical description, Steno placed them at the forefront of his argument, dedicating an entire paragraph of his book to it, and commenting on its epistemic value. Ultimately, in Steno's text, the careful description of how he carried out quantitative measurements lent them greater epistemic value than they had for Wharton.

Although Steno did not use other quantitative measurements in *Observationes anatomicæ*, he resorted to other interdisciplinary approaches on the distribution and constitution of saliva.[9] As explained in the previous chapter, Steno demonstrated that saliva was produced by various salivary glands in the mouth, including the maxillary and the parotid. He did not mention the fluid production rates of each gland but commented that all the ducts transmitted "fluid equally to all parts [*omnibus æqualiter*]." The salivary fluid, he argued, reached all parts equally, so that "the upper parts moisten [the mouth] as well as the lower ones, the internal as well as the external ones."[10] When discussing the constitution of saliva, which he said required "chymical dissection [*chymicam anatomen*]," he also spoke of a quantifiable entity without actually measuring it.[11] He wrote that the "taste and smell" of saliva are "without quality [ἄποιον]" but its "sight and touch determine it less simple than water [*visus et tactus aqua minus simplicem decernunt*]."[12] Steno concluded that saliva is "not a simple but a mixed fluid [*non simplicem liquorem, sed mixtum*], and this in a singular proportion [*singulari proportione*]."[13] He discussed this proportion only in qualitative terms, in the same way as "the famous Sylvius . . . [who] thinks that in saliva there is much water, a little of volatile spirit, and very little of lixivial salt [*salus lixiviosus*] mixed with, and seasoned [*temperatus*] by a trace of oil and spirit of acid."[14] Steno was clearly following the quantitative methods of

---

8. Steno, *Observationes anatomicæ*, §12, 10–11.

9. Wharton, however, made other measurements, for example, the length of the maxillary duct in a cow; see Wharton, *Adenographia*, 131.

10. Steno, *Observationes anatomicæ*, §22, 22.

11. Steno, *Observationes anatomice*, §25, 24. Chymical dissection meant chymical separation, probably pun intended.

12. Steno, *Observationes anatomicæ*, §28, 26.

13. Steno, *Observationes anatomicæ*, §28, 27.

14. Steno, *Observationes anatomicæ*, §28, 27–28. See also Franciscus de Le Boe Sylvius, *Opera medica, hoc est, disputationum medicarum decas* (Geneva: Samuel de Tournes, 1681), 11.

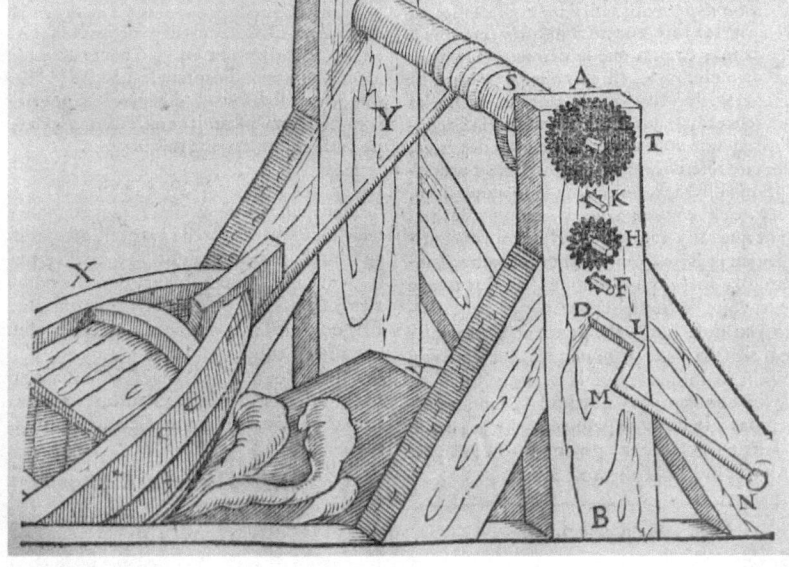

FIGURE 3.1. A machine designed to pull a boat over a flat surface with the help of a third element (grease or rolling objects) laid underneath; from Simon Stevin, *Les oeuvres mathématiques* (Leiden, 1634), 481. Reproduced with the permission of Special Collections, The Sheridan Libraries, Johns Hopkins University.

Sylvius, who also relied heavily on taste in studying the constitution of chymical substances.[15]

Steno turned his attention to the mixed mathematical discipline of mechanics in his research on the lacrimal glands. In his book, he compared the function of bodily fluids to that of lubricating fluids in the motion of machines. According to Steno, those who work on mechanics know that "to facilitate movement, the things to be moved should be smeared by some oily humor [*ut ad motum faciliorem reddendum res movendas humore unctuoso oblinerent*]." He compared the oil to a third agent that facilitates the movement, like pushing a boat over a flat surface with the help of rollers laid underneath (fig. 3.1), or "smearing the axle about which the wheel rotates with an unctuous fluid."[16] Similar to

---

15. Evan R. Ragland, "Chymistry and Taste in the Seventeenth Century: Franciscus Dele Boë Sylvius as a Chymical Physician Between Galenism and Cartesianism," *Ambix* 59, no. 1 (2012): 1–21.

16. Steno, *Observationes anatomicæ*, 85.

these mechanisms, Steno continued, animal bodies also needed fluids to help their parts move better. The salivary fluid's enhancement of the movements of the mouth and the action of fluids in the eyes were examples of that.[17]

Steno's boldest mechanical analogy at the time was comparing glands to filters or "sieves [*cribrae*]" of blood, as he wrote in a late summary of his research on glands.[18] Steno suggested that salivary glands produced saliva directly from blood and not from the nerves, as Wharton claimed:[19] "Arteries supply to the glands, besides heat, also nourishment [*nutrimentum*] and together with it the matter of saliva."[20] Wharton was open to the idea of arteries supplying some fluids, but he thought there were too few blood vessels passing through the maxillary glands to account for "the quantity of salivary matter that is excreted."[21] Steno's solution to the problem was to posit that "since saliva does not flow into the mouth with the same speed [*celeritate*] at which blood arrives, the delay of the saliva in its flowing could compensate the paucity [of blood] arriving more quickly [*celerius affluentis paucitatem compensare*]."[22]

Steno explored these ideas further in his study of lacrimal glands. At stake was the fact that Wharton and other anatomists "did not believe that so great an abundance of tears could possibly come forth from such small glands [*posse ex tam parvis glandulis tantum lacrymarum copiam prodire*]."[23] The quantity of tears that often came from the eye was so great that the Hippocratic idea that tears were produced in the brain, a large cavity, made sense.[24] Steno, however, described all eye glands as conglomerate glands, meaning they produced fluid by filtering it out of the blood. The shadowy grids in these glands' illustrations lend themselves to this interpretation (see fig. 3.2).

To address this problem of quantity of tears, Steno used a mathematical explanation: "If the magnitude of [tear]drops is compared to the time during which they are formed, no problem will appear here. For the time is not so short that as much humor could not flow in through several vessels

17. Steno, *Observationes anatomicæ*, 86.
18. Steno, *De musculis et glandulis*, 44.
19. Wharton, *Adenographia*, 134.
20. Steno, *Observationes anatomicæ*, §38, 35.
21. Wharton, *Adenographia*, 136.
22. Steno, *Observationes anatomicæ*, §37, 34.
23. Steno, *Observationes anatomicæ*, 92; and Wharton, *Adenographia*, 178.
24. Wharton, *Adenographia*, 181. See also Elizabeth Craik, "The Reception of the Hippocratic Treatise *On Glands*," in Horstmanshoff, King, and Zittel, *Blood, Sweat and Tears*, 65–82, esp. 66.

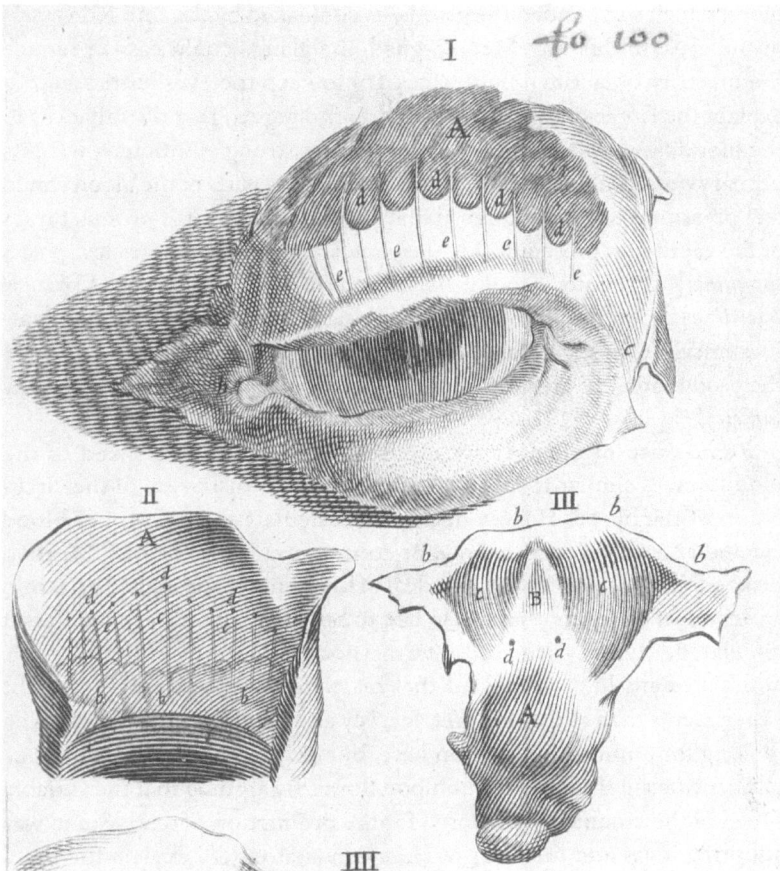

FIGURE 3.2. Illustrations of the eye glands. I. Left eye of a calf with the superior eye gland (A); II. Interior surface of the palpebra with the superior eye gland (b) and ducts (c) ending in (d); III. Lacrimal gland (A) and corresponding cartilage (B). From Nicolaus Steno, *Observationes anatomicæ* (Leiden, 1680), 100. Reproduced with the permission of The President and Fellows of Magdalen College, Oxford.

as is required to form a drop."[25] The concepts of time and flow helped him argue that the speed at which tears formed was enough to produce each teardrop. In particular, the eye glands did not have to be large because "all the humor which emanates from the eyes is [not] collected previously in the glands," but rather is produced as needed.[26] Steno associated the secretion of lacrimal fluid with "individual pulses [*singuli pulsus*]" of arterial

25. Steno, *Observationes anatomicæ*, 92.
26. Steno, *Observationes anatomicæ*, 92.

blood which surrounded the glands, as confirmed by the thin red vessels of the eyes. In this way, Steno argued, the glands could easily produce the quantity of lacrimal fluid necessary to keep the eyes lubricated. To explain the larger production of tears, Steno argued that disturbances in the blood flow—such as the ones caused by strong emotions—were its primary cause. As the blood flow increased, some parts of the blood would feel pressured to follow other paths like "the simple and porous tunics of the capillaries present inside the glands": The "blood particles [*partes sanguinis*]" that enter into the glands pour in with "greater speed [*maiori celeritate*] . . . so that the speed compensates for the transit through narrow vessels."[27] In short, the increasing speed of blood flow is "enough for the production of a great quantity of tears [*lacrymarum copiæ producendæ sufficit*]."[28]

Steno's use of quantitative arguments related to the speed of the blood flow is similar to Harvey's account of the discovery of the circulation of the blood. Harvey decided to calculate the amount of blood ejected at each forceful systole, or contraction of the heart.[29] He then realized that not only did a great deal of blood move through the heart in a very short time, but that it also had to be made somewhere—unless it circulated.[30] Harvey's quantitative method was not particularly precise, and historians have suggested that his claims involved more thought experiments than real ones.[31] Yet, Harvey and Steno were not necessarily looking for numerical precision here, but only to show the importance of quantities in thinking about blood flow.[32] By arguing that the variable speed of the circulation mattered for the production of tears, Steno was adopting ideas and methods of Harveian anatomy to explain the function of glands.

Steno blended his research on glands and muscles in *De musculis et glandulis observationum specimen* (*A Specimen of Observations on the Muscles and Glands*; Copenhagen, 1664). This book, written in Leiden, illustrates the continuity and evolution of Steno's research program.[33] From a methodolog-

---

27. Steno, *Observationes anatomicæ*, 94.
28. Steno, *Observationes anatomicæ*, 94.
29. Harvey, *De motu cordis*, chap. 9.
30. French, *William Harvey's Natural Philosophy*, 90–91.
31. French, *William Harvey's Natural Philosophy*, 92; and Cunningham, "I Follow Aristotle," 112–113.
32. Jerome J. Bylebyl, "Nutrition, Quantification and Circulation," in "Owsei Temkin at 75," [special issue,] *Bulletin of the History of Medicine* 51, no. 3 (1977): 369–385, esp. 383.
33. On the place of writing, see Nuno Castel-Branco and Troels Kardel, "Drawing Muscles with Diagrams: How a Novel Dissection Cut Inspired Nicolaus Steno's

ical perspective, he continued to rely heavily on the genre of *observationes*. The book was printed in the same year in Copenhagen and Amsterdam, but the Copenhagen edition, whose publication Steno oversaw, had a beautiful frontispiece entitling the book simply "*Observationum anatomicarum specimen* [*A Specimen of Anatomical Observations*]" (see fig. 3.3).[34] From an intellectual point of view, in the 1662 *Observationes anatomicæ* Steno used his discovery of the parotid salivary duct to address topics that were highly relevant in anatomy at the time. When explaining conglobate glands, he wrote extensively on the lymphatics; and his description of glands as filters of blood made his work speak to the circulation of the blood—the hot topic of early modern anatomy, still debated in the 1660s.[35] The jump from glands to muscles occurred because Steno associated the action of glands to the blood flow, which, he said, was controlled by the muscles around the blood vessels.[36] Therefore, starting in 1662, Steno turned his research to muscle anatomy, which, in a leap of anatomical mastery, he connected again to the heart by arguing that the heart itself was a muscle.

Steno was not the first seventeenth-century anatomist to identify the heart as a muscle. By claiming that the heart's tissue was muscular, he followed in the lineage of anatomists who had explored this matter before, especially in Leiden. In *De motu cordis* (Frankfurt, 1628), Harvey compared the contraction of the heart to that of muscles.[37] In debates about circulation against Descartes, who refused to describe the heart as a muscle, Sylvius also argued that the heart contracted like muscles.[38] But Sylvius and Harvey did not explicitly state that the heart itself was a muscle, and Harvey clearly understood it as a unique organ of the body.[39] In

---

Mathematical Myology (1667)," *Notes and Records: The Royal Society Journal of the History of Science* (2022): 8.

34. The Amsterdam edition does not have a frontispiece but has the four diagrams of the Copenhagen frontispiece inverted inside the book, suggesting that it was probably copied afterward.

35. In 1666, Michele Lipari in Messina still argued that the pulse did not depend on blood circulation; see Bertoloni Meli, *Mechanism, Experiment and Disease*, 58, 66. In 1670s France physicians also gave lectures against blood circulation; see Guerrini, *The Courtiers' Anatomists*, 207–209. See also EMC III, 308–311: "De sanguinis circulatione dissensus."

36. Steno, *Observationes anatomicæ*, 95–97. By the summer of 1662 Steno was increasingly focused on cardiological and muscle research; see Steno to Bartholin, 26 August 1662, in EMC IV, 108.

37. Harvey, *De motu cordis*, 69.

38. Ragland, "Mechanism, the Senses, and Reason," esp. 179–180.

39. Harvey, *De motu cordis*, 70; see also Bertoloni Meli, "The Collaboration Between Anatomists and Mathematicians," 697–699.

FIGURE 3.3. Frontispiece of Steno, *De musculis et glandulis* (Copenhagen, 1664), with a geometrical representation of muscle fibers in the upper left corner. Reproduced with the permission of the Wellcome Collection.

his popular anatomy textbook, Thomas Bartholin left the question open, adding that Walaeus, who also promoted Harvey's discoveries, "does not name [the heart] a muscle, but rather [says that it] is contracted in motion like a muscle [*sed constringi in motu sicut musculus*]."[40] Steno's multiple observations of muscle fibers, on the other hand, led him to affirm that

---

40. Thomas Bartholin, *Anatomia, ex Caspari Bartholini . . . reformata* (The Hague: Adriaan Vlacq, 1655), 239. The 1669 edition does not yet mention Steno's arguments.

the "*heart is truly a muscle* [*cor vere musculum esse*]."⁴¹ Steno acknowledged his predecessors, writing that Harvey did not yet have "clear knowledge [*distincta cognitio*] of the fibers" and that Sylvius was too busy to write more on the topic.⁴² Steno's arguments are significant for understanding his shift to working on fossils, so I will address them in chapter 6. What matters now is that as Steno produced cutting-edge anatomical research in the legacy of Harvey, Sylvius, and Bartholin, his mathematical yearnings continued to grow.

In *De musculis et glandulis*, Steno used mathematics to solve problems in the anatomy of muscles and commented on its epistemic roles. For instance, the intercostal muscles are hard to distinguish because they are close together and have the same function. Therefore, Steno proposed to differentiate them according to "the different angles [they make] with the ribs."⁴³ This categorization of muscles by angles carried for Steno a strong epistemic certainty. As he wrote, careful examination of "the angles formed by the back, ribs, sternum, and muscles" leads to a "demonstration perhaps not less certain than Mathematics [*demonstrationem Mathematica forte non minus certam*]."⁴⁴ Anatomists from Galen to Vesalius and beyond had often used geometrical language to describe muscle fibers, but Steno took this tradition much further.⁴⁵ Later in the same book, he wrote that muscle fibers have the specific form of "an oblique parallelogram or the figure of a rhomboid [*parallelogrammum obliquangulum, seu rhomboidea figura*]."⁴⁶ The book did not have images, but Steno included an image of these rhomboids in the top left corner of the book's frontispiece (see fig. 3.3). Steno added that "pardon should be given [*venia detur*]" to him if he gave "mathematical names to physical [and] not mathematical lines [*si Mathematicis nominibus Physicas, non Mathematicas lineas*]."⁴⁷ He may have been responding to a popular medical book that recommended the withdrawal of blood along the paths of blood vessels, not the straight lines of mathematicians.⁴⁸ But above all, Steno was addressing the epistemic problem of applying mathematics to nature, such as the fact that perfect

41. Steno, *De musculis et glandulis*, 22. Italics in the original.
42. Steno, *De musculis et glandulis*, 5. Malpighi mentioned the spiral "fleshy fibers of the heart" in 1656; see Marcello Malpighi, *Opera posthuma* (Amsterdam, 1700), 3.
43. Steno, *De musculis et glandulis*, 6.
44. Steno, *De musculis et glandulis*, 10–11.
45. Ragland, *Making Physicians*, 144–147, 153–159.
46. Steno, *De musculis et glandulis*, 15.
47. Steno, *De musculis et glandulis*, 15–16.
48. Johannes Heurnius, *Praxis medicinæ nova ratio* (Rotterdam: Arnold Leers, 1650), 518–519. See also Ragland, *Making Physicians*, 234.

geometrical lines do not exist in the material world.[49] Indeed, Steno acknowledged that these muscular lines, unlike mathematical lines, "are not deprived of all breadth [*omnis latitudinis non sunt expertes*]," so that "individual pieces of flesh represent just as many parallelograms [*totidem parallelogramma referant*]," and that two opposite tendons are not exactly two trapezoids. However, he dismissed the problem by leaving "that precision [ἀκρίβεια] to mathematicians."[50]

Finally, Steno relied on mechanical analogies to explain how muscles contract, which, he said, was "difficult to determine [*determinatu difficilis*]."[51] He concluded that muscle contraction happened because the fleshy fibers between the tendons become shorter. To explain this better—and "since an explanation through similar things [*per similia explicatio*] greatly pleases many people"—Steno mentioned a pile driver. This is a machine that pushes poles into the ground by hitting them with a large weight called a ram.[52] Dozens of men pull this ram upward and let it fall like an enormous hammer, thus driving structural posts into the ground. The ram is pulled by means of each man holding a rope linked to the ram through a main rope and a pulley (see fig. 3.4). In Steno's analogy, the ropes represent the tendons, the ram hooked to the ropes represents the mobile part, and the men are the fleshy fibers. By pulling their ropes together, said Steno, "the men indeed move the weight." Similarly, the contracting fleshy fibers, while they pull the tendon, move whatever the muscle wanted to move. Steno did not push this analogy too far, stating that it was "only a comparison [*simile hoc tantum sit*]."[53] He then used a similar mechanical analogy when explaining the motion of the diaphragm as he compared the abdomen to a pulley.[54] In short, by 1664 Steno's anatomical research was interspersed with ideas and analogies from physico-mathematics, namely mechanics and geometry.

Around the beginning of March 1664, Steno had to return home to Copenhagen. Once there, he was invited to dissect two rays, whose observations he described in a letter to Willem Piso (1611–1678), published at the

---

49. Dear, *Discipline and Experience*, 32–46.
50. Steno, *De musculis et glandulis*, 16.
51. Steno, *De musculis et glandulis*, 19.
52. Marjorie Nice Boyer, "Resistance to Technological Innovation: The History of the Pile Driver Through the 18th Century," *Technology and Culture* 26, no. 1 (1985): 56–68.
53. Steno, *De musculis et glandulis*, 20.
54. Steno, *De musculis et glandulis*, 9.

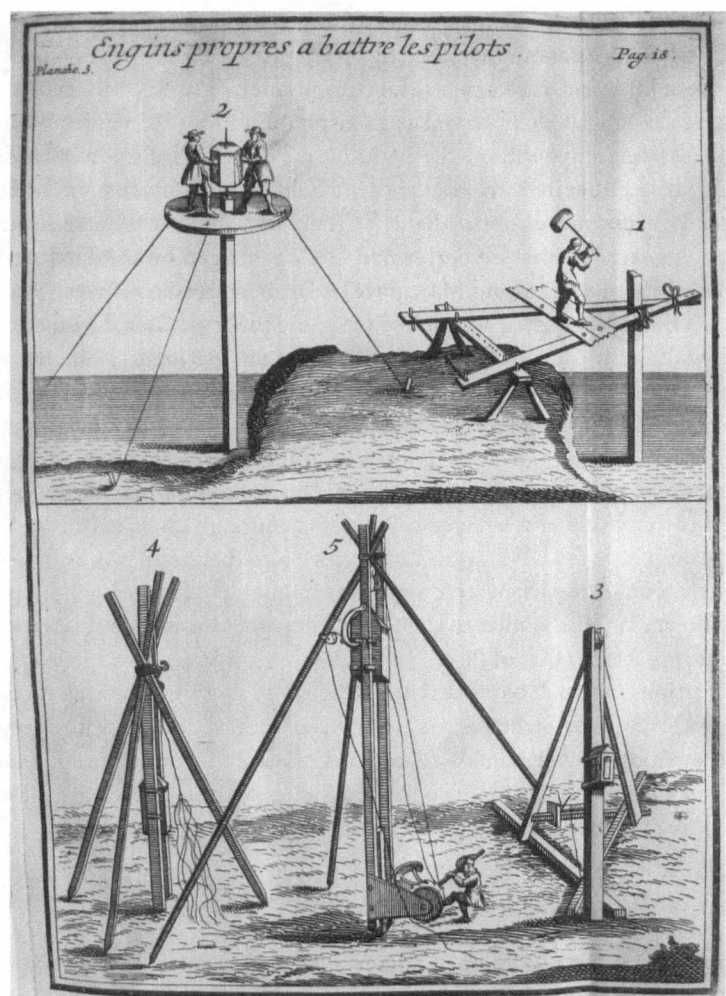

FIGURE 3.4. Seventeenth-century pile drivers. Steno's analogy uses the machine on the left (4), operated by men pulling ropes. The pulley is not visible. See Cornelis Meijer, *Traité des moyens de rendre les rivières navigables* (Amsterdam, 1696), 18. Reproduced with the permission of The Warden and Fellows of All Souls College, Oxford.

end of *De musculis et glandulis*.[55] Piso was an Amsterdam physician famous for writing, with the mathematician Georg Marcgrave (1610–1644), a popular natural history of Brazil. On 27 April 1661 Steno's friend Ole Borch met

---

55. Steno, *De musculis et glandulis*, 48–70.

Marcgrave's brother, who mentioned to him Georg's last book on "his mathematical speculations," also edited by Piso.[56] Piso and Marcgrave's work is a northern European case of a fruitful collaboration between physicians and mathematicians. It likely served as an inspiration to Steno, who continued to foster friendships with mathematicians during his travels. Piso attended Steno's dissections in the Netherlands, including some "experiments in Amsterdam [*experimentis Amstelodami*]."[57] Therefore, Steno kept him informed of two dissections that he performed in Copenhagen on 21 March 1664, especially because Piso and Marcgrave had also written about rays.[58] Steno started his observational account by weighing the rays. Then, he measured one of the rays' "length [*longitudo*]" and "width [*latitudo*]." Since these quantities were different, Steno continued, the "inequality of the diagonals [*diagonalium inaæqualitas*]" meant that the ray did not have a "square shape [*figura quadrata*]." Nor was it truly a rhomboid, since "the sides that are equal do not keep a parallel position, but rather concur toward the same angle [*quæ sibi æqualia, latera parallelum non obtineant situm, sed ad eundem angulum concurrant*]."[59] Piso and Marcgrave also measured rays and spoke of their geometrical shapes. However, the frequency of Steno's geometrical vocabulary is more similar to the text of Marcgrave, the mathematician in the group, than to that of Piso.[60] In his description Steno also spoke of the proportions in which blood arrived at the ray's heart from various parts of the body.[61] In short, when Steno left the Netherlands, he knew how to use mathematical language in his anatomical research. In Copenhagen, Steno did not even comment on the epistemic values of these approaches. It was as if using mathematics in anatomy already came naturally to him.

The (Non-Cartesian) Epistemology of Nicolaus Steno

Where did Steno's mathematical and mechanical interests come from? What relationship did they have with the mechanical descriptions of the body that followers of Descartes were spreading in the Netherlands at

---

56. OBI, 1:115, 27 April 1661. See Georg Marcgrave, "Tractatus Topographicus et Metereologicus Brasiliæ, cum Observatione Eclipsis Solaris," in Willem Piso, *De Indiæ utriusque re naturali et medica libri quatuordecim* (Amsterdam: Ludovic et Daniel Elzevir, 1658),

57. Steno, *De musculis et glandulis*, 41.

58. Willem Piso, "Historiæ rerum naturalium," in *Historia naturalis brasiliæ* (Amsterdam: Franciscus Hack, 1648), 175–176; and Piso, *De Indiæ*, 58–59.

59. Steno, *De musculis et glandulis*, 41–42.

60. Piso, "Historiæ rerum naturalium," 175–176; and Piso, *De Indiæ*, 58–59.

61. Steno, *De musculis et glandulis*, 66–67.

the time? In 1677, fifteen years after the publication of his first anatomy book, Steno confessed that he had had "a great esteem [*une très grande éstime*] for the philosophy of Descartes" but now rejected it, in large part owing to his research on muscles and the heart.[62] As a result of this statement many authors have interpreted Steno as a follower of Descartes who later abandoned Cartesianism.[63] However, as I showed in the first chapter, when Steno left Denmark he already mistrusted Cartesian medicine, an attitude that only increased in the Netherlands. Thus, how could Steno claim to have been such an admirer of Descartes? The answer lies in the specific historical context of this late claim, which Steno wrote in a letter to Leibniz. Steno had met the German philosopher in Hannover after moving there as a Catholic bishop, during a very different set of circumstances from those during his mid-twenties in Leiden. His late perception of Cartesian ideas was shaped by his new commitment to the Catholic faith as a bishop and a convert. In the 1670s Steno perceived the works of Descartes's followers as dangerous to the faith because of such materialist ideas as denying the existence of the soul and miracles. Proof of this is that Steno even requested that the Roman Inquisition place on the Index of Forbidden Books Spinoza's *Ethics*, an iconic text of this rising materialism.[64] Remarkably, Steno had been good friends with Spinoza when he lived in Leiden, as I mention below. Therefore, when he looked back at his friendships with Spinoza and "those who were praised for their knowledge of the same [Cartesian] philosophy," he saw himself as being too close to Cartesianism.[65]

Cartesian is a problematic category because many of Descartes's followers disagreed with Descartes on various points.[66] However, it was an

---

62. Steno to Gottfried Wilhelm Leibniz, 1677, in *Epistolae*, 1:366–369, esp. 367.

63. Sebastian Olden-Jørgensen, "Nicholas Steno and René Descartes: A Cartesian Perspective on Steno's Scientific Development," in Rosenberg, *The Revolution in Geology from the Renaissance to the Enlightenment*, 149–157; Miniati, *Nicholas Steno's Challenge for Truth*, 95–96; August Ziggelaar, "Commentary," in Steno, *Chaos*, 459–482, esp. 472; and Jonathan Israel, *Spinoza, Life and Legacy* (Oxford and New York: Oxford University Press, 2023), 371–372.

64. Steno to Holy Office, 4 September 1677, in Leen Spruit and Pina Totaro, *The Vatican Manuscript of Spinoza's "Ethica"* (Leiden: Brill, 2011), 68. On materialism, see Israel, *Spinoza*, 3, 9, 17–18.

65. Steno to Leibniz, 1677, in *Epistolae*, 1:366–369, esp. 367. See also Eric Jorink, "'Outside God, There Is Nothing': Swammerdam, Spinoza, and the Janus-Face of the Early Dutch Enlightenment," in *The Early Dutch Enlightenment in the Dutch Republic, 1650–1750*, ed. Wiep van Bunge (Leiden: Brill, 2003) 81–107.

66. Tad Schmaltz, *Early Modern Cartesianisms: Dutch and French Constructions* (Oxford and New York: Oxford University Press, 2017).

actors' category, and one which Steno used to distance himself from those he labeled Cartesians, including when doing anatomy in Leiden. Steno respected Descartes, certainly more than he did his followers. Otherwise it would be hard to explain his friendships with Spinoza and other so-called radical Cartesians in Amsterdam.[67] He also tempered his attacks on Descartes with some praise.[68] In a comment to Thomas Bartholin, Steno devastated Descartes' newly published ideas on the brain writing that the more brains he dissected, the less he found a correspondence between that "most ingenious structure [*ingeniosissima fabrica*] of the animal brain devised by the most noble Descartes, otherwise very suitable for explaining animal actions [*admodum alias conveniens actionibus animalibus explicandis*]."[69] However, such praise, when not ironic, does not necessarily mean that he approved of Descartes's ideas.

The truth is, contrary to what Steno suggested later in his life as a bishop, the young anatomist Nicolaus Steno did not think of himself as Cartesian. The writings of Steno's friends, colleagues, and mentors at the University of Leiden and his own anatomical publications point away from any such attachment. First of all, rejecting Cartesian claims about the body had been a common trope among Leiden anatomists since the 1640s. According to Evan Ragland, Leiden anatomists regularly refuted Descartes's errors in anatomy on the basis of dissections and observations.[70] In his anatomy textbook, Thomas Bartholin explicitly rejected the description of the pineal gland provided by Descartes and "his followers."[71] Even Italian anatomists who often used mechanical analogies, such as Malpighi and Borelli, also thought little, if anything, of Cartesian ideas about the brain

---

67. See Jorink, "*Modus politicus vivendi*." Spinoza's comments on anatomy are closer to Steno's ideas than Descartes's; see Raphaële Andrault, "Spinoza's Missing Physiology," *Perspectives on Science* 27, no. 2 (2019): 214–243, esp. 219–228.

68. Raphaële Andrault, "Anatomy, Mechanism and Anthropology: Nicolas Steno's Reading of *L'Homme*," in *Descartes' Treatise of Man and Its Reception*, ed. Delphine Antoine-Mahut and Stephen Gaukroger (Cham: Springer, 2016), 175–192, esp. 182–185.

69. Steno to Bartholin, 5 March 1663, in EMC IV, 348–359, esp. 358. As a bishop, Steno also praised Descartes's philosophy, because "all humans have certain aspects that you praise and others that you criticize." See Nicolaus Steno, *Defensio et plenior elucidatio epistolæ de propria conversione* (Hannover, 1680), in OTH, 1:388.

70. Evan Ragland, "Mechanism, the Senses, and Reason," 180.

71. Thomas Bartholin, *Anatomia ex Caspari Bartholini parentis institutionibus . . . reformata* (Leiden, 1651), 336–337. See also Raphaële Andrault, "Human Brain and Human Mind: The *Discourse on the Anatomy of the Brain* and Its Philosophical Reception," in Andrault and Lærke, *Steno and the Philosophers*, 87–112, esp. 91–92; and Andersen, *Thomas Bartholin: Laegen & anatomen*, 52–62.

and the eyes.[72] As a skillful anatomist, Steno was not an exception to this trend. When discussing the origin of tears in *Observationes anatomicæ*, Steno rejected the ideas of the "most clever [*ingeniosissimus*] Descartes."[73] He also rejected other opinions by Felix Plater (1536–1614) and Conrad Schneider (1614–1680), because they "do not agree with each other," a decision that speaks to his ongoing commitment to accurate knowledge.[74] Most importantly, he continued, "experience [*experientia*] widely shows other ways more convenient to the normal way of nature [*ordinario naturae modo*]" of forming tears, such as lacrimal glands drawing tears from the blood flow.[75]

This rejection of Cartesian anatomy resonates with what Steno encountered in medical circles across the Netherlands. Borch's travel diary is particularly illustrative of the opposition to Cartesian ideas that Steno saw and heard in those years, since he and Borch were often together. In April 1661 Borch attended a medical disputation about asthma by a certain Hidding, a Swede, who Borch said "opposed a Cartesian physician, among others."[76] In September 1661 Sylvius told Borch that "Descartes was very hostile [*valde fuisse obliquum*] to those who contradicted him." The French philosopher had accused Sylvius of not understanding mechanics "when in a dissection of a rabbit everything did not conform to Descartes's principles [*non respondebant omnia Cartesii principiis*]." However, Sylvius concluded, "by mechanics [*per mechanicam*] he understood nothing other than the fabrications of his own philosophy [*quam suæ philosophiæ commenta*]."[77] Two years later, when dissecting a swan with Steno, Borch noted that the fluid stored in the bird's crop, a little sac near the esophagus used to store food, was not "ferment from the stomach [*fermentum ventriculi*] . . . as the Cartesians wanted [*idem volunt Cartesiani*]."[78] Johannes van Horne was also not a fan of Cartesian anatomy of the heart because,

---

72. Bertoloni Meli, *Mechanism, Experiment, Disease*, 84–88; and Maria Conforti, "'Se fusse meno cartesiano lo stimarei molto': Anti-Cartesian Motifs in Italian Medicine," in *Descartes and Medicine: Problems, Responses and Survival of a Cartesian Discipline*, ed. Fabrizio Baldassarri (Turnhout: Brepols, 2023), 437–449.

73. Steno, *Observationes anatomicæ*, 91.

74. Steno, *Observationes anatomicæ*, 91.

75. Steno, *Observationes anatomicæ*, 91.

76. OBI, 1:82, 10 April 1661.

77. OBI, 1:216, 14 September 1661. I thank Evan Ragland for the reference.

78. OBI, 2: 282–283, 25 February 1663. Steno mentions this dissection in Steno to Bartholin, 5 March 1663, in EMC IV, 357. The young Dane Matthias Jacobsen was there too, and mentions it in a letter to Bartholin, 3 March 1663, in ECM IV, 340–344.

in van Horne's words, "experience hardly proved it [*vix haec experientia probabit*]."[79]

In the summer of 1662, Descartes's *De homine* (*On Man*; Leiden, 1662) was published in Latin by the Dutch physician Florentius Schuyl (1619–1669). That summer, shortly after the book came out, Borch visited Schuyl at his home in 's-Hertogenbosch as part of a longer tour of the southeastern Low Countries. Steno did not accompany Borch on this visit but was likely informed of it, since Steno also read *De homine* early on.[80] One of the memories that Borch had of that visit was seeing "various excerpts from *Antoniana Margarita* of Gómez Pereira" at Schuyl's home.[81] This reference to Gómez Pereira (1500–1567), a Spanish physician who claimed that animals did not have souls, was an implicit attack on Descartes's originality. Indeed, a year before, Borch had written to Thomas Bartholin that Pereira had claimed animals did not have souls "a hundred years before" Descartes.[82] Soon after *De homine* was published, Steno wrote to Bartholin that the book had beautiful images of the brain. Yet, in a display of literary mastery, he also wrote that the images had definitely originated in an "ingenious brain [*ex ingenioso cerebro*], but whether such images could be seen in any brain [*in ullo cerebro*]" was doubtful.[83] Steno's criticism may be the source of a comment made two months later by Johannes de Raey (1622–1702), a professor of natural philosophy at the University of Leiden, who told Borch that Schuyl had misunderstood things related to the brain cavities.[84] It was in fact Schuyl who had drawn the images, not Descartes.[85]

The environment described in Borch's diary does not mean that every university scholar whom Steno met in Leiden was anti-Cartesian. Schuyl, who translated *De homine*, became professor of medicine at the University

79. OBI, 2:254, 23 December 1662.
80. Borch traveled outside of Leiden from 16 August to 9 September 1662; see OBI, 2:179–208. It appears Steno did not go, since he signed a letter to Bartholin from Leiden on 26 August 1662, in EMC IV, 1–10. On Schuyl, see Tad M. Schmaltz, "The Early Dutch Reception of *L'Homme*," in *Descartes' Treatise on Man and Its Reception*, ed. Stephen Gaukroger and Delphine Antoine-Mahut (Cham: Springer, 2016), 71–90, esp. 82–89.
81. OBI, 2:179–180, 16 August 1662.
82. Borch to Bartholin, 3 March 1661, in EMC III, 368–369. On Gómez Pereira, see José Manuel García Valverde and Peter Maxwell-Stuart, *Gómez Pereira's "Antoniana Margarita": A Work on Natural Philosophy, Medicine and Theology* (Leiden: Brill, 2019), esp. vii–xii, 55–60.
83. Steno to Bartholin, 26 August 1662, in EMC IV, 113.
84. OBI, 2:217, 22 October 1662.
85. René Descartes, *De homine* (Leiden: Petrus Leffen & Franciscus Moyardus, 1662), title page: "figuris et latinate donatus a Florentio Schuyl."

of Leiden in 1664, diversifying the influence of Cartesianism there.[86] De Raey also taught Cartesian thought in his natural philosophy course.[87] More significant, Jan Swammerdam, whose close friendship with Steno will be explored in the following chapter, seemed to be a committed Cartesian.[88] Swammerdam agreed with Steno's praise of observations but, unlike his Danish friend, highlighted Descartes's epistemic emphasis on them.[89] His anatomy of respiration was also in line with Cartesian physiology, since he believed, along with Descartes, that the function of respiration was to cool down the blood in the heart.[90] Swammerdam's attitude speaks to the complex nature of Cartesian epistemologies and their convoluted reception at the time.[91]

Despite the different Cartesianisms in circulation, Steno's epistemology is best understood in contrast to that of Descartes himself.[92] What can humans know about the world? How can such knowledge be reached? In *Discours de la méthode* (Leiden, 1637), Descartes stated that his goal in the natural sciences was "to reach certainty and reject shifting ground," and that he "succeeded reasonably well [at it] ... by clear and certain reasoning."[93] Steno, on the contrary, began his *Observationes anatomicæ* with the remark that "there are some people who convinced themselves that ... it is not necessary that everything be subjected to the

---

86. Wiep van Bunge, "The Early Dutch Reception of Cartesianism," in Nadler, Schmaltz, and Antoine-Mahut, *The Oxford Handbook of Descartes and Cartesianism*, 417–433, esp. 429.

87. Antonella del Prete, "Teaching Cartesian Philosophy in Leiden: Adriaan Heereboord (1613–1661) and Johannes De Raey (1622–1702)," in *Descartes in the Classroom: Teaching Cartesian Philosophy in the Early Modern Age*, ed. David Cellamare and Mattia Mantovani, Medieval and Early Modern Philosophy and Science 35 (Leiden: Brill, 2023), 60–78.

88. There is no extensive study of Swammerdam's epistemology. For a good start, see Charlotte Sleigh, "Jan Swammerdam's Frogs," *Notes and Records: The Royal Society Journal of the History of Science* 66 (2012): 373–392.

89. Jan Swammerdam, *Bybel der Natuure, of Historie der Insecten/Biblia naturæ, sive Historia insectorum* (Leiden: Isaak Severinus, Boudewyn van der Aa, and Pieter van der Aa, 1738), 2:868–873.

90. Jan Swammerdam, *Tractatus physico-anatomico-medicus de respiratione usuque pulmonum* (Leiden, 1667), 8, 10–11; and René Descartes, *A Discourse on the Method*, trans. Ian Maclean (Oxford and New York: Oxford University Press, 2006), part V, esp. 44.

91. Evan R. Ragland, "Between Certain Metaphysics and the Senses: Cataloguing and Evaluating Cartesian Empiricisms," *Journal of Early Modern Studies* 3, no. 2 (2014): 119–139.

92. Schmaltz, *Early Modern Cartesianisms*, 1–14.

93. Descartes, *A Discourse on the Method*, 25.

external senses [*nec opus, ut externis omnia sensibus subiiciantur*], they think that reason alone [*sola ratio*] supplies the rest."[94] For Steno, "reasoning deprived of the work of the senses [*ratio sensuum ope destituta*]" was worthless in anatomy.[95] Steno's comments are strikingly similar to those of his professor Sylvius, who repeatedly affirmed that medicine should refer "to Experiments, or to Reasonings deduced from foregoing experiments."[96] Yet, as already mentioned, the trope that Cartesians do not rely on observations was not true, since Descartes and his followers did carry out experiments.[97] In disputes with Harvey about the motion of the heart, Descartes even supported his claims with dissections. But the difference was not in the making of observations per se, but rather on the greater epistemic emphasis that Harvey, Sylvius, and Steno placed on them.[98]

Observations in Practice: The Lymphatics Controversy

The role of observations in Steno's epistemology emerged with special detail and clarity when he entered a controversy on the lymphatics that was shaking the medical Republic of Letters. By the 1660s the lymphatics were being thoroughly discussed across Europe, with medical practitioners commenting on them in places as varied as France, the Low Countries, and England.[99] Steno's intervention dealt with a specific dispute between Thomas Bartholin and the medical entrepreneur Lodewijk de Bils.[100] Steno addressed three themes related to this dispute: the direction of lymphatic flow, the existence of lymphatic valves, and the nature of the lymphatic fluid, also known as lymph. In all these cases, he used

---

94. Steno, *Observationes anatomicæ*, vi.

95. Steno, *Observationes anatomicæ*, vii.

96. Franciscus Sylvius, "Epistola apologetica," translated in Evan R. Ragland, "The Contested *Ingenia* of Early Modern Anatomy: Continuities and Conflicts in Medical Training at Leiden University, 1592–1678," in *Ingenuity in the Making: Matter and Technique in Early Modern Europe*, ed. Richard J. Oosterhoff, José Ramón Marcaida, and Alexander Marr (Pittsburgh: University of Pittsburgh Press, 2021), 112–130, 124.

97. For the case of Descartes, see Jed Z. Buchwald, "Descartes's Experimental Journey Past the Prism and Through the Invisible World to the Rainbow," *Annals of Science* 65, no. 1 (2008): 1–46.

98. Ragland, "Mechanism, the Senses, and Reason," 177.

99. For Guy Patin and Jean Pecquet in France, see EMC I, 545–547; for Paul Barbette and van Horne in the Netherlands and Thomas Wharton and Walter Charleton in England, see Bartholin, *Spicilegia bina*, 91–92, 100–101.

100. On de Bils, see Harold J. Cook, *Matters of Exchange: Commerce, Medicine, and Science in the Dutch Golden Age* (New Haven, CT: Yale University Press, 2007), 268–276.

dissections with distinct epistemic strengths to make his points, further supporting Thomas Bartholin's original claims. De Bils, on the other hand, although he dissected often, gave priority to reasoning over observations. Interestingly, the sides on the de Bils controversy seem to be separated mostly along Cartesian lines.[101]

Thomas Bartholin, who first discovered lymphatic vessels other than those in the mesentery, argued that lymph flowed from different organs of the body into the more central thoracic duct. Bartholin built upon Pecquet's discovery that the chyle, that is, lymph originating in the lacteals, sidestepped the liver and moved instead to the thoracic duct, as explained in chapter 1. The problem is that Bartholin never saw lymph flowing in undisturbed states. He based his argument on experiments in which he placed ligatures on lymphatics and noticed which side of the vessels inflated.[102] Harvey also used this method in his studies of the circulation of the blood.[103] De Bils, on the contrary, argued that lymph flowed in opposite directions, from the thoracic duct toward the rest of the body. De Bils lived in Rotterdam, not far from Leiden, so Steno made two visits to see him.[104] In one, Steno asked de Bils why he did not use ligatures to find the direction of the lymphatic flow. De Bils replied that ligatures should not be used with the lymphatics because there was "no circular motion," as there was in the blood.[105] Thus, de Bils used reasoning to reject observations with ligatures. However, in the particular case of the lacteals, de Bils claimed to have seen lymph moving from the lacteals to the liver, counteracting the opinion of Bartholin and Pecquet. In his other visit to de Bils, Steno watched him dissect a dog "to show the movement of lymph into the liver." Unfortunately,

---

101. Pietro Omodeo, "Lodewijk de Bils' and Tobias Andreae's Cartesian Bodies: Embalment Experiments, Medical Controversies and Mechanical Philosophy," *Early Science and Medicine* 22, no. 4 (2017): 301–332. On one side there was Steno, Bartholin, Sylvius, and van Horne; on the other, de Bils, Anton Deusing, and Anton Everhardi. This subject still needs to be further researched.

102. Bartholin, *Vasa lymphatica*, 13–21, 46.

103. Bertoloni Meli, "Early Modern Experimentation in Live Animals," 206–207.

104. Steno visited de Bils with Henricus à Moinichen in the fall of 1660 and with Ole Borch in March 1661. See Thomas Bartholin, *Responsio de experimentis bilsinianis* (Copenhagen: Petrus Haubold, 1661), 18–19; and OBI, 1:38–40.

105. Steno, *Observationes anatomicæ*, §42, 41–42. De Bils's response was more sophisticated but was still based on reasoning a priori. De Bils explained that the lymphatics "attracted" the lymph, so ligatures would break that attraction and change the original flow. It should be mentioned that de Bils's method of dissecting without spilling blood was done by holding the blood in a section of the body by means of ligatures; see Steno to Bartholin, 22 April 1661, in EMC III, 86–95, esp. 93–94.

Steno claimed that he did not see anything clearly because the dissection "was done at such a speed that neither mine nor the eyes of those present could discern such movement."[106] That is, even if observations mattered for Steno, not every new observation counted because of the handicap of the senses or the recklessness of who performed it.[107] As Steno put it, "We know how little the eyes must be believed in things which are carried out quickly [*quam parum oculis credendum in rebus quæ magna cum celeritate peraguntur*], even though we observe them at clear sight."[108]

Similarly, de Bils denied the existence of valves in the lymphatics. Thomas Bartholin, on the contrary, did "not doubt" the existence of such valves.[109] But again, Bartholin never saw the valves. He only saw that air blown through an extremity of the lymphatics did not flow to the other end. Moreover, since he observed a valve at the intersection point between the lymphatics and the jugular vein, he assumed all the lymphatics had valves, too. Given the weak observational evidence Bartholin presented, it was reasonable for de Bils to deny their existence, since no one had ever seen them.[110] For all these reasons, Steno decided to perform more experiments himself. He used not just ligatures but also probes to feel the resistance to the probe's motion created by the lymphatic valves.[111] Steno's experiments convinced him of their existence, even though he never saw them either—valves were only seen a few years later by Steno's colleagues and friends Frederik Ruysch (1638–1731) and Jan Swammerdam, who embalmed the lymphatic vessels.[112] Steno's experiments, however, contrasted starkly with de Bils's dissection of the lacteals, because Steno's experiment

---

106. Steno, *Observationes anatomicæ*, §44, 44.

107. Steno, too, faced his own poor treatment of vessels; see Steno to Bartholin, 22 April 1661, in EMC III, 89.

108. Steno, *Observationes anatomicæ*, §44, 44.

109. Bartholin, *Vasa lymphatica*, 36.

110. De Bils accepted the existence of a valve in the thoracic duct; see Steno, *Observationes anatomicæ*, §46, 45–46.

111. Steno, *Observationes anatomicæ*, §46–§48, §52–§54, 45–47, 51–54. Steno also showed the existence of valves by feeling obstacles when forcing fluids and probes up the vessels, in the direction contrary to the flow.

112. See Dániel Margócsy, "Advertising Cadavers in the Republic of Letters: Anatomical Publications in the Early Modern Netherlands," *British Journal for the History of Science* 42, no. 2 (2009): 187–210. For Swammerdam's possible priority in discovering the valves, see Johan Nordström, "Swammerdamiana: Excerpts from the Travel Journal of Olaus Borrichius and Two Letters from Swammerdam to Thévenot," *Lychnos* 14 (1954–1955): 21–65, esp. 41–50.

with probes could easily be repeated by moving the probe back and forth. He therefore counted them as observational evidence.

Finally, de Bils and his supporters claimed that lymph was just the chyle, that is, a "dew-carrying juice [*succum roriferum*]" digested from food, carried into the lymphatics in the mesentery, and then transported to the whole body, including the glands and liver.[113] According to de Bils's theory, the glands secreted the white chyle because fluids in the lymphatics were often transparent or whitish, similar to the color of glandular fluids such as saliva, milk, and semen. Steno, however, thought impossible that lymph correspond to "all the aqueous humors in the body," including glandular secretions, and at the same time be nutrition for the whole body, a task that he still attributed to the chyle.[114] On the contrary, Steno argued, the chyle and these other fluids were all transported together within the blood flow. Glandular secretions, he argued, were chymically mixed in the blood flow and filtered out by the glands.[115]

To strengthen the empirical basis of his claim, Steno had to search outside of anatomy. His interdisciplinary focus thus turned from mechanics and mathematics to chymistry. He explained that blood looked red despite carrying various non-red substances by arguing that superficial colors did not reveal anything about the constitution of the blood: "Who, I ask, among those who admit the synthesis and analysis of things [*qui σύγκρισιν rerum et διάκρισιν admittunt*], who attach some value to chymistry, ignore how unreliable it is to put faith in colors?"[116] Steno then presented what he considered to be "a very evident example": that "beneath the brightest, indeed manifest red [*sub fulgentissimo rubore, et quidem perspicuo*] which appears if you sprinkle spirit of niter onto butter of antimony, such a white matter lies hidden."[117] Steno's rejection of the epistemic value of colors is unsurprising, since he reached similar conclusions when taking chymical notes in the "Chaos" manuscript.[118] What is remarkable is how easily he now applied it to the study of the body. In short, Steno did not shy away from supporting his case with

---

113. Steno explains de Bils's theory in *Observationes anatomicæ*, §42, 41–42.
114. Steno, *Observationes anatomicæ*, §42, 41.
115. Steno, *Observationes anatomicæ*, §§32–33, 29–30.
116. Steno, *Observationes anatomicæ*, §33, 30.
117. Steno, *Observationes anatomicæ*, §33, 30.
118. See Domenico Bertoloni Meli, "The Color of Blood: Between Sensory Experience and Epistemic Significance," in Daston and Lunbeck, *Histories of Scientific Observation*, 117–134; Frank, *Harvey and the Oxford Physiologists*, 183–188; and Newman, *Atoms and Alchemy*, 80, 120–123.

indirect observations, such as the use of probes in valves and chymical experiments.

Interestingly, Steno advised great caution when there was no observational evidence whatsoever. He said that the "great power of the mind [*ingenii magna vis*]" reveals itself when inferring new images from "previously observed ones" and when devising various causes for the same thing. "However, where the matter itself is silent, whatever it [i.e., the mind] says may indeed be possible [*quicquid illud loquitur, posse quidem id ita esse*], as the strongest arguments demonstrate, but it does not stand to assert that it is so [*esse autem asseverare non sustinet*]."[119] In practice, although Steno once saw, "in the presence of the famous Borch," several lymphatics connecting the conglobate glands between themselves, he wrote that "it is uncertain that these lymphatics have this origin nowhere [*nude autem lymphatica hæc originem habeant, incertum*]."[120] In 1664 he wrote again that "the first origin of the Bartholinian lymphatic vessels [*prima lymphaticorum Bartholinianorum origo*] remains uncertain [*incertior*] because it allows several probable modes of explanation [*plures admittit explicandi modos probabiles*]."[121]

Steno's explicit refusal to make definite claims when lacking observations remained a strong mark of his scholarship. It contrasts with the epistemology of Descartes, who claimed in *Discourse on the Method* that "by clear and certain reasoning, I found none [i.e., no propositions] so doubtful that I could not draw some reasonably certain conclusion from it."[122] In fact, Descartes's physiology relied so much on a priori ideas (like propositions) that amid his dispute with Harvey, Descartes claimed that if his explanation of the heart's motion was shown to be false, "then the rest of my philosophy is entirely worthless."[123] At the same time, by 1664 Steno became confident of making claims about things for which he lacked observations. But he caveated such claims as "inferred only by the number of probabilities [*probabilium verò tantum numero inferendas*]."[124] Probabilistic claims would become especially important for his accounts of the history of the Earth, whose past was also inaccessible to the senses.

119. Steno, *Observationes anatomicæ*, preface, vii.
120. Steno, *Observationes anatomicæ*, §41, 40.
121. Steno, *De musculis et glandulis*, 40–41.
122. Descartes, *A Discourse on the Method*, 25.
123. René Descartes to Marin Mersenne, 9 February 1639, as quoted in Ragland, "Mechanism, the Senses, and Reason," 175.
124. Steno, *De musculis et glandulis*, 41.

Galen, God, and Gassendi

If not necessarily in Descartes, where did Steno find inspiration for his observational and mechanical epistemology?[125] The first place is anatomy itself. Anatomists knew that there was already a tradition of using mechanical analogies in anatomy since Galen and Erasistratus (c. 304–c. 250 BC).[126] Thomas Wharton, author of the first early modern book on glands and no Cartesian himself, used a mechanical analogy to compare a muscle in the mouth to a pulley.[127] There were, however, differences in the way anatomists adopted mechanical analogies. Galen thought that, although useful, mechanical analogies fell short of the full anatomical reality.[128] He said that unlike machines that needed to be operated, natural things, such as living bodies and planets, moved by themselves because they were endowed by the gods with special powers.[129] Steno agreed that animal and human bodies were superior to machines but went beyond Galen and reduced the difference between men and machines to the "humor which is supplied and where it is supplied," which reveal "a skill far greater [*maius artificium*]."[130] Galen also promoted an epistemology grounded in observations and reasoning, where primacy was given to the senses.[131]

For Steno and other anatomists, mechanisms pointed to, and not away from, God because of their intrinsic final causes, that is, the end for which they existed. Indeed, despite superficial similarities, Steno's mechanical analogies contrasted greatly with contemporary efforts to rid the body of an immaterial soul. Descartes famously stated that in studying nature one should "never derive any explanations from the purposes which God or nature may have had in view . . . and we shall entirely banish from our

---

125. See Andrault, "Human Brain and Human Mind," 100–104, where Andrault argues that Steno's empirical epistemology could still come from Descartes's influence but acknowledges other possible origins.

126. Ragland, "Mechanism, the Senses, and Reason," 183–184.

127. Wharton, *Adenographia*, 131. On Wharton as non-Cartesian, see Wharton, *Adenographia*, 154; and Wharton to Mrs. Church, 15 May 1673, in Thomas Wharton, *Thomas Wharton's Adenographia*, trans. Stephen Freer, with a Historical Introduction by Andrew Cunningham (Oxford and New York: Oxford University Press, 1996), 311.

128. Berryman, "Galen and the Mechanical Philosophy," esp. 242–248; and Bertoloni Meli, *Mechanism*, 11–16.

129. See Galen, *De usu partium* IV, 156–7 and III, 168 (Kühn notation), as explained in Berryman, "Galen and the Mechanical Philosophy," 243–244, 247–248.

130. Steno, *Observationes anatomicæ*, 86.

131. Hankinson, "Galen on the Limitations of Knowledge," 214–222.

philosophy the search for final causes."¹³² This was a prior methodological decision, since Descartes still thought that there were final causes, and they played a greater role in his work than he articulated.¹³³ However, this view of the world led a new generation of scholars to abandon this teleological outlook, ultimately rejecting God's role in the world and the immateriality of soul. That was indeed the path followed by authors whom Steno associated with Cartesianism such as Regius and Spinoza.¹³⁴ In March 1661 de Raey said to a philosophy class at the University of Leiden, probably attended by Steno, that Regius "had been for a while Socinian and now seemed to be an atheist [*videri esse Atheum*]."¹³⁵ In his journal, Borch also registered another rejection of Cartesianism when, in late May 1661, he was visited by his "friend Hoyer," about whom little else is known. Hoyer spoke about the omnipotent role of God in the world and warned him about atheists in Amsterdam, "many of whom were Cartesians [*ex iis plures esse Cartesianos*]," especially a "certain shameless Jew," most likely Spinoza.¹³⁶

Since the beginning of his anatomical research, Steno explicitly rejected this materialist path. In the very beginning of *Observationes anatomicæ*, he wrote that "it should be lamented that men of great reputation [*dolendum, fuisse inter magni nominis viros*]" spoke of things done in animals as mere superficialities, since "nothing is more repugnant to the mind and divine purpose [*menti consilioque divino nihil magis repugnet*]."¹³⁷ In one of his first letters to Thomas Bartholin, in September 1661, he repeated this criticism while at the same time targeting the certitude that Cartesians claimed to have:

> I do not torture these animals in such long torment without horror. Cartesians boast much about the certitude of their philosophy [*de philosophiæ certitudine multum gloriantur Cartesiani*]: I wish they could

---

132. René Descartes, *Principles of Philosophy*, in *The Philosophical Writings of Descartes*, trans. John Cottingham, Robert Stoothoff, and Dugald Murdoch, 2 vols. (Cambridge: Cambridge University Press, 1985), 2:202.

133. Peter Distelzweig, "The Use of *Usus* and the Function of *Functio*: Teleology and Its Limits in Descartes's Physiology," *Journal of the History of Philosophy* 53, no. 3 (2015): 377–399.

134. Schmaltz, *Early Modern Cartesianisms*, 1–2, 40, 239–260, esp. 255–260; and Steven Nadler, "Spinoza, Descartes, and the 'Stupid Cartesians,'" in Nadler, Schmaltz, and Antoine-Mahut, *The Oxford Handbook of Descartes and Cartesianism*, 659–677.

135. OBI, 1:43, 19 March 1661.

136. OBI, 1:128, 27 May 1661.

137. Steno, *Observationes anatomicæ*, 2.

convince me as certainly as they have convinced themselves that in animals there is no soul [*nullam esse brutis animam*] and that it makes no difference whether you touch, cut, or burn the nerves of a living animal or the cords of an automaton that is moved in action [*automati, quod actu movetur, chordas*].[138]

Historians of anatomy know this quote well because it speaks to Steno's discomfort with cutting open live animals, something which he continued to practice, nonetheless.[139] But perhaps more significant, these comments testify to Steno's aversion toward a materialistic mechanical philosophy that denied a place for the animal soul, like Descartes and most Cartesians did.[140]

As a good anatomist, Steno knew that each anatomical part had its own purpose (*usus*). Therefore, his mechanical analogies were imbued with final causes that pointed toward a divine creator.[141] More specifically, for Steno and other anatomists such as Malpighi, not just the whole organism but indeed each bodily part pointed toward teleology and therefore to God's creative power.[142] As Steno wrote:

> all their [the animals'] limbs and the parts contained in the limbs almost say themselves that they were made with foresight, because

---

138. Steno to Bartholin, 12 September 1661, in EMC III, 228.

139. See also Raphaële Andrault, "Tirer sur la corde: Douleur et vivisection animale à l'âge classique," in *L'homme et la brute au XVIIe siècle: Une éthique animale à l'âge classique*, ed. Marine Bedon and Jacques-Louis Lantoine (Lyon: ENS Éditions, 2022), 83–104.

140. Descartes's dualism of *res extensa* and *res cogitans*, together with his claim that only humans have reason (see Descartes, *A Discourse on the Method*, 47–48) means that animals were only material, i.e., *res extensa*. See also Schmaltz, "The Early Dutch Reception of *L'Homme*," 83.

141. Final causes pointed to a divine creator insofar as non-intelligent things have purpose, such as a wooden chair's purpose of being suitable for someone to sit on, as decided by a carpenter. Therefore, "some intelligent being exists by whom all natural things are directed to their end; and this being we call God." See Thomas Aquinas, *Summa theologiae*, I, q. 2, a. 3, as translated by Fathers of the English Dominican Province (see www.newadvent.org/summa/1002.htm). Gassendi's argument for the existence of God was a version of this proof from final causes; see Osler, *Divine Will and the Mechanical Philosophy*, 161.

142. On Malpighi, see Bertoloni Meli, *Mechanism*, 134–138. For a deeper discussion, see Raphaële Andrault, "Divine Organs? Leibniz's 'Hymn to Galen' and the Best of All Possible Bodies," in *Galen and the Early Moderns*, ed. Matteo Favaretti and Emanuela Scribano (Cham: Springer, 2022), 135–153.

nothing seems to be so small as not to be destined to its use [*cum tam minutum nihil videatur, quod suo non destinatum sit usui*], and nothing to be so low that it does not teach the wisdom of the Creator [*tam abjectum, quod Conditoris sapientiam non doceat*].[143]

Steno also spoke of an architect who falls into error when, in his project, he "leaves even the very smallest space unused or not serving any clear purpose."[144] God, on the contrary, imbued everything with meaning and "even in the smallest things lies hidden the greatest, indeed the most admirable skill."[145] Quoting Socrates in Greek, Steno added, "The structure of animals is the work of a wise and life-giving craftsman."[146] As historians of philosophy say, Steno adopted operational rather than ontological mechanisms, unlike Descartes.[147]

Steno's epistemology and mechanical philosophy strongly resonates with that of Gassendi, another so-called mechanical philosopher whom Steno studied in Copenhagen. Gassendi did not believe in making absolute claims about the world, choosing instead a probabilistic epistemology, to which Steno was attracted, especially when observations were lacking.[148] Moreover, Gassendi's mechanical philosophy included final causes, which Margaret Osler describes as "divine intentions reflected in the design of the creation."[149] Like Steno (and most seventeenth-century scholars), Gassendi also attributed the existence of a non-rational soul to animals, following the threefold Aristotelian division of vegetative, animal and rational.[150] Interestingly, even when Steno condemned Epicureanism, he followed Gassendi's rejection of non-Christian aspects of Epicurus's philosophy. First, Steno said that the only way one could deny the role of the senses and final causes was if "our Providence had presented itself as Epicurean," that is, if it had no order.[151] Then he rejected Blasius's description of the salivary duct, saying that

---

143. Steno, *Observationes anatomicæ*, 1.
144. Steno, *Observationes anatomicæ*, 2. Galen also speaks of an architect in relation to nature's designs; see Berryman, "Galen and the Mechanical Philosophy," 247.
145. Steno, *Observationes anatomicæ*, 2.
146. Steno, *Observationes anatomicæ*, 1.
147. Denis Des Chene, "Mechanisms of Life in the Seventeenth Century: Borelli, Perrault, Régis," *Studies in History and Philosophy Science Part C: Studies in History and Philosophy of Biological and Biomedical Sciences* 36, no. 2 (2005): 245–260.
148. Osler, *Divine Will and the Mechanical Philosophy*, 106–113.
149. Osler, *Divine Will and the Mechanical Philosophy*, 49.
150. Osler, *Divine Will and the Mechanical Philosophy*, 64–66. In this categorization, only the rational soul was eternal.
151. Steno, *Observationes anatomicæ*, 2.

it exists only "in the inhabitants of the moon or [in] those crawling in the spaces between the worlds of Epicurus"—an allusion to Epicurus's plurality of worlds.[152] Both points were aspects of Epicureanism that Gassendi also refused.[153] Therefore, if anything, Steno's mechanical philosophy was much more inspired by Gassendi than Descartes. But labels are not always useful when it comes to describing seventeenth-century mechanical philosophies, which often varied from scholar to scholar. Indeed, as I will now show, Steno's research methods drew from a much greater variety of sources.

The Mechanical and Chymical Worlds of the Netherlands

In his early anatomical works, Steno used methods and ideas directly associated with early modern pure and mixed mathematics. Such methods and ideas included quantification measurements, geometrical models, mechanical analogies, and the axiomatic structure of mathematical treatises.[154] Anatomists since Galen had had recourse to knowledge and methods that were available to them in these mathematical disciplines.[155] The main difference between the ancients and seventeenth-century anatomists is their different historical contexts. Seventeenth-century physico-mathematics used not only mathematical methods, but also new experimentation techniques and analogies with devices that were popular at the time among mathematicians to explain natural phenomena.[156] Those who studied and practiced mechanics—the *mechanici*, as Steno called them—relied on mathematics as their main tool to describe the natural world.[157] Their aim was to achieve higher levels of certainty in their description of the natural world.[158] Kircher's *Magnes sive de arte magnetica*

---

152. Steno, *De musculis et glandulis*, 30.
153. Osler, *Divine Will and the Mechanical Philosophy*, 45, 56–57. For Epicurean ideas that Gassendi rejected, see Olivier René Bloch, *La philosophie de Gassendi: Nominalisme, matérialisme et métaphysique*, Archives internationales d'histoire des idées/International Archives of the History of Ideas 38 (The Hague: Martinus Nijhoff, 1971), 300.
154. Andersen and Bos, "Pure Mathematics," 696–723, esp. 702. Simple quantification like counting does not necessarily entail interest in mathematical methods except when used to make stronger epistemic claims, as Steno did.
155. Berryman, "Galen and the Mechanical Philosophy"; and Bertoloni Meli, *Mechanism*, 11–16.
156. Dear, *Discipline and Experience*, 32–62, 151–179; and Domenico Bertoloni Meli, "Machines of the Body in the Seventeenth Century," in Distelzweig, Goldberg, and Ragland, *Early Modern Medicine and Natural Philosophy*, 91–116.
157. Steno, *Observationes anatomicæ*, 85.
158. Dear, *Discipline and Experience*, 32–44.

(Cologne, 1643) was a physico-mathematical treatise on magnetism in which experiments were presented as theorems.[159] Steno carefully studied this book as a university student in Copenhagen, including chapters where Kircher illustrated magnetic attraction with hydrostatic devices.[160] Therefore, Steno's use of mechanical analogies speaks to the interests that anatomists had in physico-mathematics.

However, unlike other anatomists, Steno was interested in mathematics even before starting his studies in anatomy. As I explained in the first chapter, Steno acquired these interests while studying at the Cathedral School of Copenhagen under its headmaster, Georgius Hilarius. Hilarius's influence was so significant that Steno dedicated his first book, the *Observationes anatomicæ*, to him, calling him a "mathematician and man of letters."[161] When he left Copenhagen, Steno continued to search for mentors in mathematics: another of the six professors to whom Steno dedicated his book was the University of Leiden mathematics professor Jacob Golius.[162] Golius seems to have grown close to Steno, because he attended Steno's dissections with pleasure and "did not disdain to watch when I [i.e., Steno] prepared the salivary and lacrimal ducts in a calf."[163] Indeed, Golius may have been the one who encouraged Steno to measure and compare the weights of maxillary and salivary glands. It was precisely while dissecting a calf that Steno performed "the examination [*examen*] of the parotids and a comparison [*collatio*] with the lower of the maxillary glands."[164] Moreover, their relationship extended beyond mathematics. After several of Steno's Danish friends fell ill in the summer of 1661, Golius informed Steno of a fever epidemic in Amsterdam.[165]

When Steno met Golius, the Dutch mathematician was already a senior professor and a symbol of the lively tradition of mathematics in seventeenth-century Leiden. Together with the previous chair of mathematics at the University of Leiden, Willebrord Snellius, Golius contributed to make mathematics a learned and humanist discipline in the

159. Dear, "Mixed Mathematics," 156; and Dear, *Discipline and Experience*, 172–179.
160. Gal. 291, fol. 39r (22 March), in Steno, *Chaos*, 122; and Kircher, *Magnes sive de arte magnetica*, 527–529.
161. Steno, *Observationes anatomicæ*, 80.
162. Steno also dedicated it to four professors of medicine and anatomy: Sylvius and van Horne in Leiden, Bartholin, and Simon Paulli in Copenhagen. It is significant that Steno *did not* dedicate the book to Erasmus Bartholin (1625–1698), professor of mathematics at the University of Copenhagen and an admirer of Descartes.
163. Steno, *Observationes anatomicæ*, 59.
164. Steno to Bartholin, 22 April 1661, in ECM III, 92.
165. Steno to Bartholin, 12 September 1661, in ECM III, 230.

Netherlands. They worked on publishing ancient Greek mathematics, such as recovering the lost chapters of Apollonius's *Conics*, of which Golius had a rare Arabic manuscript copy. When Steno arrived in Leiden, Golius had been following the efforts of Borelli to translate another rare Arabic copy of that text in Florence. Golius may have shared what he knew with Steno, which would help explain why Steno visited Borelli upon his arrival in Florence.[166] Golius was also well connected with other intellectuals in the Netherlands, such as Constantin Huygens (1596–1687), the father of Christiaan Huygens, and Descartes. It was Golius who introduced his mathematics student Frans van Schooten (1615–1660) to Descartes. Van Schooten then produced the mathematical diagrams of some of Descartes's works and a very popular second edition of Descartes's *Geometria* (Leiden, 1659–1661).[167] The mathematical environment that Steno found at the University of Leiden was thus a thriving one, including one in which the name of Descartes was highly praised, in contrast to his anatomical context.

Mathematicians also used seventeenth-century physico-mathematics, and especially mechanics, for the understanding and command it gave over phenomena whose operations were marvelous and unseen or secret. These operations might be using a lever to hoist weights that were impossible to lift otherwise or mastering the details of a hidden mechanism inside a machine.[168] Simon Stevin (1548–1620), a leading mathematician whose works Snellius translated from Dutch into Latin, described the power of mechanics in his motto, stating that "a wonder is really no wonder."[169] Steno thought highly of the work of "the most talented Stevin," whom he mentioned in the preface to his text on lacrimal glands.[170] Unsurprisingly, Steno spoke of the body in a way that conveyed a wonder that mechanics could surpass. He wrote that "the [body] parts are arranged so that the liquid concealed in the vicinity [of glands], as in a storeroom, is secreted more sparingly or more copiously depending on the more or less intensive

---

166. For more details, see chap. 5.

167. Jantien Dopper, "A Life of Learning in Leiden: The Mathematician Frans van Schooten (1615–1660)" (Ph.D. diss., Utrecht University, 2014), 21–64.

168. Pamela O. Long, *Openness, Secrecy, Authorship: Technical Arts and the Culture of Knowledge from Antiquity to the Renaissance* (Baltimore, MD: Johns Hopkins University Press, 2004), 14–15, 172, 245–47.

169. Simon Stevin, *De Beghinselen der Weeghconst* (Leiden: Christoffel Plantijn, 1586), title page, as translated in Ragland, *Making Physicians*, 69. See also Dirk Struik, *The Land of Stevin and Huygens: A Sketch of Science and Technology in the Dutch Republic During the Golden Century* (London: D. Reidel, 1981), 52–60, esp. 58.

170. Steno, *Observationes anatomicæ*, 82.

usage, without our noticing it."¹⁷¹ Steno also thought that the human body operated even "more skillfully or, I should say, more divinely" than machines.¹⁷² Jesuit scholars such as Kircher and Schott, about whom Steno read in his final year in Copenhagen, also used physico-mathematics to unmask the hidden phenomena of nature.¹⁷³ In short, through mechanical analogies, Steno made the invisible operations of the human body intelligible.

The culture of craftsmanship that Steno experienced daily in Amsterdam and Leiden also fostered his interest in mechanics. For example, before comparing glandular fluids to lubrication oils, he mentioned that examples of lubrication "are numerous here and there in workshops."¹⁷⁴ Steno was likely thinking of lens-grinding workshops, which were particularly common in the Low Countries.¹⁷⁵ Golius was familiar with these workshops and spoke to his Danish students about the making of convex lenses.¹⁷⁶ On another occasion Steno reported to Bartholin observations made with other Danes with convex lenses.¹⁷⁷ In the spring of 1663 Steno also used a microscope to replicate observations from Malpighi's recently published *De pulmonibus observationes anatomicæ* (Bologna, 1661), becoming one of the first to do so.¹⁷⁸ Malpighi's book was composed of two epistles addressed to Borelli, who, Malpighi wrote, had "a geometric eye for the contemplation of anatomical things [*in anatomicis contemplatione geometricum oculum*], which, with a single stroke [*ictu solo*], could discern the truth or falsehood of the things gathered from these observations."¹⁷⁹ This was yet another example of a respected anatomist praising the knowledge and skills of mathematicians in a place miles away from northern Europe but to which Steno would also travel. Most likely, reading Malpighi's

---

171. Steno, *Observationes anatomicæ*, 86.
172. Steno, *Observationes anatomicæ*, 86.
173. Waddell, *Jesuit Science and the End of Nature's Secrets*, 5–15, 161–186.
174. Steno, *Observationes anatomicæ*, 85.
175. Edward Ruestow, *The Microscope in the Dutch Republic: The Shaping of Discovery* (Cambridge: Cambridge University Press, 1996), esp. 19–32.
176. OBI, 1:227, 31 September 1661.
177. Steno to Bartholin, 21 May 1662, in EMC IV, 2. Thomas Walgesten (1627–1681), a Danish lens grinder and natural philosopher, also lived with Steno and Borch in these years, as mentioned in Borch and Braem's journals.
178. Steno to Bartholin, 5 March 1663, in EMC IV, 348–349. Steno confirmed some of Malpighi's observations but not all of them. Borelli, too, had a hard time replicating them; see Bertoloni Meli, *Mechanism, Experiment, Disease*, 45.
179. Marcello Malpighi, *De pulmonibus observationes anatomicæ* (Bologna: Giov. Baptista Ferroni, 1661), 3.

praises of the epistemic powers of mathematics reinforced Steno's commitment to it. Around this time Steno also met Spinoza, whose "only occupation," according to Borch, was "the manufacture of telescopes and microscopes."[180] In his later writings on religion, Steno said that Spinoza visited him daily "to see the anatomical investigations of the brain that I carried out on several animals."[181] Spinoza was much more Cartesian than Steno and was not an anatomist. Yet, the two shared several interests, including a fascination with the axiomatic structure of mathematical works, as their later works show.[182]

Steno was also intrigued by the machinery associated with navigation and the shipbuilding culture of Amsterdam during the so-called Dutch Golden Age. Stevin too had this interest, since he wrote extensively on mathematical navigation.[183] In late July 1662 Borch, Steno, and some other Danes visited Zaandam, a town close to Amsterdam with many windmills and a large shipyard, where they saw at least "forty finished ships" and several mills.[184] It was here or in a similar place that Steno saw the pile driver, to which he compared to the action of muscle fibers. Fittingly, a French book on machines from a few decades later describes the specific machine mentioned by Steno as typical of the Netherlands.[185] Perhaps thinking of these environments, Steno once compared the function of the "oily fluid" on the skin of fishes with the "grease and other fats" with which the surfaces of boats were covered to help them cut through the sea.[186] That month they also met Nicolaus Witsen (1641–1717), who was then a student at the Amsterdam Athenæum and gave them a tour of the city's

---

180. OBI, 1:214, 10 October 1661.

181. Steno to Holy Office, 4 September 1677, in Rome, Archivum Congregationis Doctrinæ Fidei, SO, Censuræ librorum, 1680–1682, Folia extravagantia, n. 2, as translated in Spruit and Totaro, *The Vatican Manuscript of Spinoza's "Ethica,"* 10. See also Wim Klever, "Steno's Statements on Spinoza and Spinozism," *Studia Spinozana* 6 (1990): 303–313.

182. Baruch Spinoza, *Renati Descartes principiorum philosophiæ . . . more geometrico demonstrate* (Amsterdam: Johannes Riewerts, 1663); and Steno, *Elementorum myologiæ specimen*.

183. Struik, *The Land of Stevin and Huygens*, 41.

184. OBI, 2:168, 27 July 1662. Unlike Borch, Braem mentions Steno in his journal; see Copenhagen, Royal Library of Denmark, Thott 1926 kvart, as mentioned in Scherz's biography in Kardel and Maquet, *Nicolaus Steno: Biography and Original Papers*, 94.

185. Cornelis Meijer, *Traité des moyens de rendre les rivières navigables avec plusieurs desseins de jetée* (Amsterdam: Pierre Mortier, 1696), 16–17. See also Boyer, "Resistance to Technological Innovation," 63–65.

186. Steno, *Elementorum myologiae specimen*, 72–73.

arsenal.[187] Witsen was part of the Amsterdam bourgeoisie and became mayor of the city in 1682.[188] Steno and Witsen remained in contact, since Witsen quoted information provided to him by "the famous and highly learned man Nicolaus Steno" in his 1671 book on shipbuilding.[189] Steno later translated parts of this book from Dutch into Italian for Viviani, the mathematician whom Steno befriended in Florence.[190]

In Leiden, just as in Copenhagen, Steno's workshop life continued to undergird his interests in chymistry, which he increasingly used in anatomical research. There is no doubt that Sylvius, Steno's professor of medicine in Leiden, influenced his chymical arguments: Steno explicitly referred to him when addressing the chymical constitution of saliva.[191] But he also experienced the chymical culture of the Netherlands in other contexts, thanks especially to his friendship with Borch. In his journal Borch wrote of a vivisection performed in Leiden with Steno in which they injected chymical substances, such as oil of tartar, spirit of vinegar, or water of alum, into the circulatory system of a dog.[192] Together with Borch, in Amsterdam Steno also visited the chymist Johannes Glauber (1604–1670), the young Theodor Kerckring (1638–1693), and the Italian alchemist Giuseppe Borri (1627–1695); and in Leiden, a certain Baron Sonnenthal, who had a large laboratory there.[193] Borri had studied with the Jesuits in Rome, but his alleged mystical experiences and unusual theological ideas forced him out of Italy and brought him to Amsterdam.[194] Kerckring, on

---

187. For Witsen at the Athenæum, see van Miert, *Humanism in an Age of Science*, 262, 391. For Borch's meeting with Witsen, see OBI, 2:169, 28 July 1662.

188. W. Veder, "Witsen (Mr. Nicolaas)," in *Nieuw Nederlandsch Biographisch Woordenboek*, 10 vols. (Leiden: A. W. Sijthoff, 1911–1937), 4:cols. 1473–1479.

189. Nicolaus Witsen, *Aeloude en Hedendaegsche Scheeps-Bouw en Bestier* (Amsterdam: Commelijn & Broer and Jan Appelaer, 1671), 15.

190. Steno dictated the translation to Viviani; see Florence, BNCF, Gal. 251, fols. 1r–12v.

191. Steno, *Observationes anatomicæ*, §25, 24. On Sylvius's chymistry, see Ragland, "Chymistry and Taste in the Seventeenth Century."

192. OBI, 2:109–110, 16 May 1662.

193. Steno accompanied Borch to visit Glauber's furnaces; see OBI, 1:13, 20 December 1660, and 1:7, 1:17, 1:184, 28 July 1661. Borch also talked with Borri many times, though the sources are not as clear with regard to Steno's presence. For Borch's visit to Kerckring, see OBI, 2:109, 25 April 1662. Steno would later speak of his acquaintance with Kerckring, so he probably visited too. On Sonnenthal, see OBI, 2:230, 10 November 1662. It is not clear that Steno was there, but he was aware of it, because Borch speaks of Sonnenthal elsewhere in his diary, so it was likely a common place to visit in Leiden.

194. Hans-Reinhard Koch and Konrad R. Koch, "Borri, the Prophet, on the 'Restitutio Humorum' and on Lens Aspiration in the 17th Century/Der Prophet Borri

the other hand, would make important contributions on anatomy and chymistry, including the publication of a translation with commentary on a popular alchemical work by Basil Valentine.[195] Kerckring enrolled as a medical student at the University of Leiden in 1659 and wrote many years later of meeting Steno in a "community of studies in the Netherlands [*communia studiorum in Hollandia*]."[196] Kerckring and Steno would reunite again around 1680 in Hamburg, after they had both, and independently, converted to Catholicism.[197]

Steno also encountered an interest in chymistry in the circles of Catharina Questiers (1631–1669), once described as a "Roman Catholic playwright" from Amsterdam.[198] Borch and Steno became friends with Questiers, especially after the former met her in the summer of 1661. In a letter to Thomas Bartholin, Borch mentioned seeing many rare things at her house, and noted that she painted, sculpted, wrote poetry, and knew chymistry. She claimed to be able to distill "powerful waters from all kinds of plants without fire, ice, and snow in any time of the year."[199] Steno may have met Questiers before Borch, when he was living in Amsterdam with the Blasius brothers.[200] In fact, Questiers dedicated a poem to Steno in a book of Dutch poetry published in 1662, recently discovered by Eric Jorink. The poem, which had been written in Steno's now-lost *album amicorum*, reads: "When science is joined by virtue, / Time does not affect a man's life. / His name will stand, even if he is stricken down, / Minerva has written [his name] in steel."[201] Steno's relationship with Questiers is one of various intellectual friendships with women that Steno had throughout his life. Swammerdam was also particularly close to Questiers and, like Steno, would be deeply influenced by another woman on religious matters.

The environments that Steno visited in the Low Countries not only supported his interdisciplinary interests, but also speak to the plural religious experiences that he had around that time. Besides Questiers,

---

über die 'Restitutio Humorum' und die Linsen-Aspiration im 17. Jahrhundert," *Sudhoffs Archiv* 101, no. 2 (2017): 160–183.

195. "Kerckring, Theodorus," in Lindeboom, *Dutch Medical Biography*, cols. 1030–1032; on Kerckring's chymistry, see J. R. Partington, *A History of Chemistry*, 4 vols. (London: Macmillan, 1961), 2:208.

196. Theodor Kerckring to Steno, 24 April 1680, in Florence, ASF, MdP, filza 2662, fol. 25r–v.

197. Angeli, *Niels Stensen*, 286–287.

198. Jorink, "Modus politicus vivendi," 19.

199. Borch to Bartholin, 25 March 1662, in EMC III, 434.

200. Jorink, "Modus politicus vivendi," 19, 30.

201. Jorink, "Modus politicus vivendi," 31.

Borch mentions visiting a French woman in Amsterdam who gave them some pharmaceuticals and books on moral theology from Pierre de Moulin (1568–1658).[202] In scientific circles, the Danes had conversations on religious topics and witnessed different religious experiences. In August 1661 Borch and Steno visited the philosopher Petrus Serrarius (1600–1669), who spoke to them about quieting the soul.[203] In a visit to the home of Borri in Amsterdam, Borch found him and others in a "back room during a Quaker worship ritual . . . with taciturn faces waiting for the Holy Spirit."[204] Catholics were a minority in the Netherlands, but that did not stop Borch and Steno from hearing a lot about Jesuits and about Catholicism more generally. In a conversation with David Stuart (c. 1630–?), a recently appointed professor of philosophy in Leiden, Borch learned of Sorbière's conversion to Catholicism a few years earlier.[205] Steno was familiar with Sorbière and his ideas on the use of mathematics in anatomy, which he had copied into the "Chaos" manuscript years before. Steno and Borch also used Sorbière's published itinerary as a guide for their travels in Flanders.[206] And, at a dinner in Rotterdam, before traveling to Flanders in the summer of 1663, the Danes met a certain Mr. Hubert, who had been a Dutch diplomat in Denmark. Hubert told them that the Catholic religion was the least tolerated in the Netherlands, because "Catholics did not want to be subjects of the Magistrates, but rather of the Pope."[207] In Leiden, Steno also worked with the young anatomist Regnier de Graaf (1641–1673), a Roman Catholic, who in his first publication on the pancreatic juice spoke of Steno as "my most entire [*integerrimo*] friend."[208]

---

202. OBI, 1:234, 13 October 1661. This was probably not Antoinette Bourignon (1616–1680), who only arrived in Amsterdam from Flanders in 1667; see Joyce Irwin, "Anna Maria van Schurman and Antoinette Bourignon: Contrasting Examples of Seventeenth-Century Pietism," *Church History* 60, no. 3 (1991): 301–315.

203. OBI, vol. 1:187, 9 August 1661; see also OBI, 2:48–49, 67–68. On Serrarius, Steno, and Borch, see Jorink, "*Modus politicus vivendi*," 29–30. Borch also mentioned interactions with the theologian Jan Commenius (1592–1670) in OBI, 1:185, 2:48–49.

204. OBI, 2:67–68. Once Borri told Borch that "all of Christ's miracles were done naturally [*omnia Christi miracula naturaliter facta fuisse*]"; see OBI, 1:133, 24 May 1661.

205. See OBI, 1:200, 19 August 1661. On another occasion, Stuart told Borch about atheists who lived morally upright lives; see OBI, 2:74. On Sorbière's conversion, see Samuel Sorbière, *Discours du sieur de Sorbiere, sur sa conversion à l'Église Catholique* (Paris: Antoine Vitré, 1654).

206. Samuel Sorbière, *Relations, lettres et discours* (Paris: François Clousier, 1660). For Borch's notes on it, see OBI, 1:121–124, 12 May 1661.

207. OBI, 3:9, 19 May 1663.

208. Reinier de Graaf, *Disputatio medica de natura et usu succi pancreatici* (Leiden: Hack, 1664), as published in Johan Vesling, *Syntagma anatomicum*, 2nd ed.

Finally, Steno had the occasion to meet some Jesuits in person in the Low Countries and see with his own eyes the fruits of their mathematical work, which included architecture.[209] In May 1663, traveling through Flanders with Borch and other friends, Steno visited the Jesuit college in Antwerp, whose four-chamber library greatly impressed Borch.[210] When they visited the main Jesuit church in the city, the Dane Corfitz Braem (1639–1688), another member of their group, noted in his private journal that the church's architects were the Jesuit mathematician François d'Aguillon (1567–1617) and the painter Peter Paul Rubens (1577–1640).[211] Steno also met in Ghent the famous mathematician Grégoire de Saint-Vincent (1584–1667), an eighty-year-old Jesuit who became famous for his works on the quadrature of the circle and mechanics and who might have attracted Steno's interest.[212] These encounters with Jesuit mathematicians were significant for Steno's future. First, Steno was then working on the book in which he distinguished intercostal muscles with geometric angles, compared muscles to pulleys, and suggested that muscle fibers resembled parallelograms.[213] It is possible that he may have exchanged ideas with Jesuits, including Grégoire. Second, these meetings were Steno's first interactions with the intellectual life and architectural beauty of early modern Catholicism, which laid the groundwork for his later conversion. If Steno had had an interest in meeting Kircher during his travels, it was probably reinforced after meeting these Flemish Jesuits. As Steno traveled southward, he encountered other Catholic cultures in Paris and Florence. But the welcome, beauty, and intellectual livelihood that he first encountered in Flanders would be especially felt in what ancients and moderns alike

---

(Amsterdam, 1666), 409. In this book de Graaf emphasized Steno's discovery of the parotid salivary duct.

209. In a conversation about China with Ole Borch (and probably Steno), the mathematics professor Jacob Golius said that Jesuit missionaries were "very experienced in astronomy and philosophy"; see OBI, 2:60, 7 February 1662.

210. OBI, 3:10, 21 May 1663.

211. See Scherz's biography in Kardel and Maquet, *Nicolaus Steno: Biography and Original Papers*, 109. For Braem's journal, see Copenhagen, Royal Library of Denmark, Thott 1926 kvart. Rubens also contributed with images for d'Aguillon's famous book on optics.

212. OBI, 2, 26 May 1663. See also Borch to Bartholin, 10 August 1663, in EMC IV, 516–540. See also Geert Vanpaemel, "Jesuit Science in the Spanish Netherlands," in *Jesuit Science and the Republic of Letters*, ed. Mordechai Feingold (Cambridge, MA: MIT Press, 2003), 389–432, esp. 391–397, 405–406, and 418–420.

213. Steno's first conclusions on the muscles were reported in April 1663; see Steno to Bartholin, 30 April 1663, in EMC IV, 414–421.

called the Eternal City.[214] As I show in the book's last two chapters, Steno's first trip to Rome had a significant intellectual and cultural impact on his religious conversion. It is unlikely that while still in the Netherlands Steno was seriously thinking about becoming Catholic. However, the religious and cultural experiences that he had there remained in his memory, even if only as signs of the intellectual confusion he came to associate with the Netherlands.[215]

✹

In 1664 Steno returned hastily to Copenhagen to say goodbye to his mother, who was dying.[216] As soon as he arrived, he wrote the aforementioned "letter on the anatomy of a ray" to Piso, with accounts of dissections he performed in Denmark.[217] The letter was published at the end of Steno's most recent book, the *De musculis et glandulis observationum specimen* (1664), published in Copenhagen and Amsterdam. This was Steno's final book on the glands, in which he summarized his previous research on the eye and salivary glands and added more observations on the sweat and mammary glands.[218] But the greatest novelty, mentioned at the very beginning of the book, was Steno's observation of the structure of the muscles, which defined his future anatomical research.

"Anatomy, our goddess [*nostra Diva, Anatome*]," as he wrote, was at the center of Steno's life.[219] Even in dire personal circumstances he did not stop dissecting. Through his publications on the glands and intense controversies with Blasius and de Bils, Steno had become a well-established anatomist. He also gradually increased the use of mathematics in his research and further

---

214. Stephanie Malia Hom, "Consuming the View: Tourism, Rome, and the Topos of the Eternal City," *Annali d'italianistica* 28 (2010): 91–116.

215. Steno, *De propria conversione epistola*, in OTH, 1:126.

216. Steno's mother died in June 1664; see Scherz's biography in Kardel and Maquet, *Nicolaus Steno: Biography and Original Papers*, 122. The idea (suggested by Scherz and Miniati) that Steno was looking for a university chair in Copenhagen that was taken by Matthias Jakobsen is hard to support. Indeed, in 1663 there were already talks of giving that chair to different scholars (such as Jakobsen or Paulli), and Steno was not under consideration. See Bartholin to Johannes Scheck, 4 March 1663, in EMC IV, 119–122. Steno also says his research on the heart was interrupted "by the deaths of my relatives"; see Nicolaus Steno, *De solido intra solidum naturaliter contento dissertationis prodromus* (Florence: Stella, 1669), 7.

217. Steno, *De musculis et glandulis*, 48.

218. Steno, *De musculis et glandulis*, 42, 45. Research on sweat had been at the center of Santorio Santorio's quantitative experiments, which Steno studied in Copenhagen.

219. Steno, *De musculis et glanduli*, 30.

developed his mechanical approach to anatomy. Mechanics illustrated his ideas well and helped him speak more clearly about the body—and to marvel at it as God's creation. Free of classes and exams, with no family commitments and a large inheritance, Steno was open to go wherever he wanted, geographically and intellectually.[220] But he did not want do so alone, probably because he had experienced the benefits of friendships with such intellectuals as the mathematician Gollius, the shipyard master Witsen, and the chymist and author Questiers. In Amsterdam and Leiden Steno had also become close friends with Swammerdam, who shared Steno's commitment to certainty and interdisciplinary research. Thus, when Steno left Copenhagen again, his path led him back to Swammerdam, who was then in Paris.

---

220. Steno, *Epistolae*, 1:11 n. 2. After his mother died, Steno received an inheritance of 300 Rigsdaler. This inheritance was part of the Royal Court's debt to Steno's stepfather (which then went to his mother, and later to him). In his diary entry for 19 August 1661, Borch mentions that a monthly quote for a good place to stay in Paris was 12 Rigsdaler [*imperiales*], which suggests that 300 was a good budget for a year and a half.

CHAPTER FOUR

In the Cradle of the Académie des Sciences

Early modern Paris was different from anything Nicolaus Steno had seen in northern Europe and the Low Countries. Paris was home not only to the largest urban population of Europe, but also to some of Steno's scientific heroes.[1] As a young student in Copenhagen, Steno had studied the writings of several philosophers who had worked in Paris such as Descartes and Gassendi, in addition to Pecquet and Borel. Steno stayed in Paris almost a year, from around November 1664 to September 1665. As it turned out, Paris welcomed him warmly. Steno's social skills and anatomical expertise, as displayed in conversations and dissections, resonated well with the learned society of Paris, rich in informal academies or salons.[2] The social aspects of these academies, such as good manners, social status, and the influence of the host, have often been described as framing the kind of research that was practiced therein.[3] However, Steno's dissections and the way he made a name for himself in Paris show that the reverse is also true—that science also shaped these groups. As this chapter shows, salons ran the risk of disappearing without the specific scientific practices that took place in their meetings. Thus, the academies that Steno frequented were centered not only on their host, but also on the activities of their participants. In this case, these activities were experiments at the intersection of anatomy with physico-mathematics.

1. Colin Jones, *Paris: The Biography of a City* (New York: Penguin Books, 2006), 132–172.
2. I use the terms "salons" and "informal academies" interchangeably, according to Antoine Lilti, *The World of Salons: Sociability and Worldliness in Eighteenth-Century Paris* (Oxford and New York: Oxford University Press, 2005), 1–11, 28.
3. Shapin, *A Social History of Truth*; and Biagioli, *Galileo, Courtier*. For the specific case of French salons, see Dena Goodman, *The Republic of Letters: A Cultural History of the French Enlightenment* (Ithaca, NY: Cornell University Press, 1994).

Steno's focused interdisciplinary research intensified in Paris in intensity owing to his successful use of knowledge from other disciplines in anatomy while in Leiden. However, this rise in intensity was also the result of the collaboration between Steno and Swammerdam, which flourished in these months, and which forged a friendship that endured for years. The two friends applied their interdisciplinary methods to the study of respiration, the circulation of the blood (beyond the work of Harvey), and the motion of muscles. Steno developed a sophisticated mechanical analogy involving the muscles of respiration to account for how blood returns to the heart—a fascinating result that has so far escaped the attention of most historians. Steno was also working then on a book about the brain, the draft of which he wrote as he dissected animal brains before various audiences—a testimony to how the pen and the knife continued to go hand in hand for him. Steno's *Discours sur l'anatomie du cerveau* (Paris, 1669) became a standard reference in brain anatomy in the following decades because of its critical review of works by Thomas Willis (1621–1675) and Descartes. Due to its prominence in the history of neuroscience, this book has been thoroughly studied by a number of historians, so it will not be exhaustively addressed in this chapter.[4] It should be said, however, that Steno brought his mechanical analogies to another level in his brain studies. In his book, and probably orally to his audiences, he bluntly stated that "the brain being a machine [*le cerveau étant une machine*]" it should be studied by "dismantling all its parts piece by piece [*de démonter pièce a pièce tous ses ressortes*] and consider what they might do together and in a group."[5] Ultimately, however, Steno concluded that little was still known about the brain, a sign that the quest for knowledge that was certain remained central to his research.

Steno's sojourn in Paris coincided with what Roger Hahn has called the prehistory of the Académie Royale des Sciences, founded by King Louis XIV in the fall of 1666, exactly a year after Steno left Paris.[6] It also coincided with the prehistory of the Académie de Physique de Caen, the other scientific academy established under royal patronage in seventeenth-century

---

4. Gustav Scherz, ed., *Steno and Brain Research in the Seventeenth Century*, Analecta Medico-Historica 3 (Oxford: Pergamon Press, 1968); Bertoloni Meli, *Mechanism, Experiment, Disease*, 76–88; and Niels Stensen, *Discours sur l'anatomie du cerveau*, ed. Raphaële Andrault (Paris: Éditions Classiques Garnier, 2009).
  5. Nicolaus Steno, *Discours sur l'anatomie du cerveau* (Paris: Robert de Ninville, 1669), 32–33.
  6. Roger Hahn, *The Anatomy of a Scientific Institution: The Paris Academy of Sciences, 1666–1803* (Berkeley and Los Angeles: University of California Press, 1971), 4.

France.⁷ One of its founders, the physician André Graindorge (1616–1676), visited Paris in the final months of Steno's stay. As this chapter shows, those months alongside Steno were transformative for Graindorge and, consequently, for the research program of the Académie de Physique, begun only a year later. Despite Graindorge's being much older, Steno became like a mentor to the Frenchman. On the other hand, given that Graindorge was a convert to Catholicism, his religious conversion may have inspired Steno's own three years later, though little else is known about Graindorge's religion.⁸

Graindorge was only one of a number of scholars whom Steno attracted with his commitment to collaborative research, physico-mathematical experimental methods, and search for accuracy. In his research in Paris on muscles and the brain, Steno again detached himself from Cartesianism, which further contributed to his popularity among the founding members of the Académie Royale des Sciences, making him almost one of them. The praise that French intellectuals lavished on Steno confirms that his influence on the scientific culture of Paris and Caen was more significant than is usually assumed. Jean Chapelain (1595–1674), an influent courtier and member of the Académie Française, wrote in the spring of 1665 that Steno "definitely surpasses all the ancients and moderns in this discipline [of anatomy] and, since he is not yet thirty years old, we may expect from him many sure novelties about the human body and great secrets for the perfection of medicine."⁹

Monsieur Stenon, a Sçavant Danois in Paris

When Nicolaus Steno arrived in Paris, he found an intense intellectual life mediated by events where philosophers, mathematicians, physicians, and courtiers gathered regularly. At least since the 1640s, scholars interested in the natural sciences had been meeting in Paris at the convent of the famous Minim friar Marin Mersenne.¹⁰ Mersenne was a natural philosopher

---

7. For the history of this academy, see David Lux, *Patronage and Royal Science in Seventeenth-Century France: The Académie de Physique in Caen* (Ithaca, NY: Cornell University Press, 1989).

8. On Graindorge, see Lux, *Patronage and Royal Science*, 17–22; on his conversion, see ibid., 10 n. 6.

9. Jean Chapelain to Pierre Daniel Huet, 6 April 1665, in *Lettres de Jean Chapelain, de l'Académie française*, 2 vols., ed. Philippe Tamizey de Larroque (Paris: Imprimerie Nationale, 1880–1883), 2:393 n. 3.

10. Guerrini, *The Courtiers' Anatomists*, 41–49, esp. 43–44. On Mersenne's correspondence, see Grosslight, "Small Skills, Big Networks."

and mathematician who contributed much to the spread of physico-mathematics in early modern Europe. He was one of the leading scholars propounding the ideas of Galileo and Descartes through his vast correspondence network.[11] The scholars attending Mersenne's meetings included the philosopher and priest Gassendi, the mathematicians Pascal and Roberval, and the physicians Pecquet and Pierre Bourdelot (1610–1685).[12] Pecquet probably learned about and saw at Mersenne's meetings the recent experiments on the vacuum performed by Pascal, Roberval, and Auzout and integrated them in his explanation of the motion of chyle in the thoracic duct. Most of these new experimental results on the vacuum were first published in 1651, in the first edition of Pecquet's *Experimenta nova anatomica*, in a section called "Physico-Mathematical Experiments on the Vacuum."[13] Sorbière, another philosopher in Mersenne's circle, wrote that Pecquet, albeit an anatomist, was "abundantly imbued with geometrical precepts [*geometricis imbutus abunde præciptis*], and most skilled in mechanical inventions."[14] As explained in chapter 1, Steno studied Pecquet's book in depth and was fully aware of the fruitful interactions taking place in Paris between the life sciences and physico-mathematics.

Similar to early modern salons, Mersenne's scientific gatherings greatly depended on their host, who was the center and origin of his group.[15] Mersenne sometimes called his group of scholars the "Academia Parisiensis," but his academy and those that Steno joined in Paris were not like the Académie Royale des Sciences, for they lacked a royal charter and did not produce a report from every meeting.[16] Borch, who attended similar meetings in Paris with Steno, simply called them a "gathering of honorable people [*honoratorum*]" or of "the learned [*eruditorum*]," or even an "academy

---

11. Peter Dear, *Mersenne and the Learning of the Schools* (Ithaca, NY: Cornell University Press, 1988), esp. 4, 117–170.

12. See Guerrini, *The Courtiers' Anatomists*, 43–47, esp. 46; and René Taton, *Les origines de l'Académie royale des sciences* (Paris: Palais de la découverte, 1966), 16–17.

13. See Bertoloni Meli, "The Collaboration Between Anatomists and Mathematicians," 672 n. 8.

14. Sorbière to Pecquet, in Pecquet, *Experimenta nova anatomica*, 166. On Sorbiére and Gassendi, see Guerrini, "Experiments, Causation, and the Uses of Vivisection," esp. 239–241; and Guy Patin, *Correspondance complète et autres écrits de Guy Patin*, ed. Loïc Capron (Paris: Bibliothèque interuniversitaire de santé, 2018), n. 26 in letter to Charles Spon, 22 March 1648, www.biusante.parisdescartes.fr/patin/?do=pg&let=0152&cln=26.

15. Lilti, *The World of Salons*, 10, 15–39.

16. Grosslight, "Small Skills, Big Networks," 344, 353. Mersenne only mentioned the "academia parisiensis" in four letters of his vast correspondence; see Grosslight, "Small Skills, Big Networks," 366 n. 76. On the lack of sources on Mersenne's academy, see Guerrini, *The Courtiers' Anatomists*, 82.

of curious people [*academia curiosorum*]."[17] The latter two terms speak to the diversity of themes discussed there, as one would expect of learned humanists.[18] After Mersenne's death in 1648, the meetings continued in the house of other hosts, especially Henri Louis de Montmor (c. 1600–1679), a close friend of Gassendi.[19] Montmor invited more scholars to his meetings, such as the intellectual Pierre Daniel Huet (1630–1721), the mathematician Christiaan Huygens (1629–1695), the author and courtier Jean Chapelain, and the former diplomat Melchisedec Thévenot (1620–1692).[20] However, this diversity of guests led to a polarization of ideas, and by the early 1660s Montmor's home had become less hospitable to some of his meetings' participants.[21] Sorbière thus began searching for a more generous host— "kings and wealthy sovereigns," he wrote, who could support "a physical academy where there would be constant experimentation."[22] Eventually Jean-Baptiste Colbert (1619–1683), Louis XIV's chief minister, asked a few courtiers with intellectual ambitions to identify suitable candidates to form an academy hosted by the king himself.[23] Chapelain's skills as a courtier and his long membership in the literary Académie Française fitted him to the task, and when Montmor's meetings ceased in 1663, Chapelain began to look elsewhere for candidates.[24]

It was during this time, between the end of the Montmor meetings and the formation of the Académie des Sciences, that Borch and Steno arrived in Paris. By then two other scholars—Bourdelot and Thévenot— had taken the reins of Montmor's academy and invited intellectuals to

---

17. See OBI, 4:173, 18 November 1664; 4:39, 24 July 1664; and 4:61, 6 August 1664.

18. Kenny, *The Uses of Curiosity*, 183–193.

19. Gassendi lived his final years in Montmor's house, and Montmor wrote the introduction to Gassendi's complete works; see Suzanne Delorme, "Montmor, Henri," in *Complete Dictionary of Scientific Biography* (Charles Scribner's Sons, 2008), 9:497–499.

20. Guerrini, *The Courtiers' Anatomists*, 80–84; and Taton, *Les origines de l'Académie*, 21–27.

21. Harcourt Brown, *Scientific Organizations in Seventeenth Century France (1620–1680)* (Baltimore, MD: Williams & Wilkins, 1934), 117–134; and Delorme, "Montmor, Henri," 498.

22. Hahn, *The Anatomy of a Scientific Institution*, 7; and Brown, *Scientific Organizations*, 124–128.

23. See Taton, *Les origines de l'Académie*, 29–31.

24. Brown, *Scientific Organizations*, 117–134. Charles Perrault (1628–1703) also gave advice to Colbert on this matter; see Guerrini, *The Courtiers' Anatomists*, 84, 88. Perrault's memoirs, however, do not mention Steno; see Charles Perrault, *Mémoires de ma vie*, ed. Paul Bonnefon (Paris: Librairie Renouard, 1909).

their individual houses in Paris.²⁵ Borch faithfully attended both academies while in Paris, from August 1663 to May 1665. The meetings occurred at least every other week. Sometimes Borch went to Bourdelot's home on one day and to Thévenot's the next.²⁶ Steno also attended both meetings when he was in Paris, according to Borch's diary and other sources. In late November 1664, a few weeks after Steno's arrival, the mathematician Pierre Petit (1598–1677) wrote to Huygens that at Thévenot's house in Paris they did "many things, especially about anatomy because Monsieur Stenonius is here."²⁷ In fact, Steno stayed at Thévenot's home for most of his Parisian sojourn, alongside Swammerdam and the mathematician Bernard Frénicle de Bessy (1605–1675).²⁸ As for Bourdelot's meetings, we can deduce Steno's presence from the topics discussed at them, especially Steno's novel claim that the heart was a muscle, as laid out in his most recent book, *De musculis et glandulis specimen* (Copenhagen, 1664). Borch wrote that in a December 1664 meeting, all those present saw the sternum of a dog and the back of a viper moving "as if true systole and diastole [*vera systole et diastole*]," which was due to nothing but "the motion of the intercostal muscles." Borch went on to say that if the motion of muscles looked like the motion of the heart, then "neither the systole and diastole of the heart are anything but the motion of the muscles of the heart."²⁹ This analogical reasoning was the method used by Steno to claim in his book that the heart was a muscle, which means Steno was probably the one exposing these ideas to Bourdelot's assembly. Steno remained in contact with both Thévenot and Bourdelot in the following years, although not much more is known about the correspondence with the latter.³⁰

25. Guerrini, *The Courtiers' Anatomists*, 85–88; and Brown, *Scientific Organizations*, 231–236.

26. See, e.g., OBI, 3:439, 8–9 June 1664, and 3:464, 24–25 July 1664; and OBI, 4:274, 2–3 March 1665.

27. Pierre Petit to Christiaan Huygens, 28 November 1664, in Christiaan Huygens, *Oeuvres complètes*, 22 vols. (The Hague: M. Nijhoff, 1888–1950), 4:266–271, esp. 270–271. Because this letter is wrongly dated in Huygens's *Oeuvres complètes* (corrected in the errata), it has been thought that Steno was in Paris before 1664; see Matthew Cobb, *Generation: The Seventeenth-Century Scientists Who Unraveled the Secrets of Sex, Life, and Growth* (London: Bloomsbury, 2006), 46.

28. On Frenicle, see Melchisedec Thévenot, *Biblioteca thevenotiana, sive catalogus impressorum et manuscriptorum librorum bibliothecae viri clarissimi D. Melchisedecis Thevenot* (Paris, 1694), ii. On Swammerdam, see Jan Swammerdam, *Miraculum naturæ, sive uteri muliebris fabrica* (Leiden: Severinus Matthæi, 1672), 41.

29. OBI, 4:192, 29 December 1664.

30. On Steno corresponding with Bourdelot, see John Stratchey to John Locke, 30 August 1666, in *The Correspondence of John Locke*, ed. E. S. De Beer (Oxford and

Steno is surprisingly absent from Borch's journal in Paris in a way that speaks to the two men's evolving relationship, as Steno went from student at the Cathedral School to fellow traveler and researcher. In the journal there are a few references to Steno at Thévenot's place, but Borch does not even mention his presence at Bourdelot's.[31] One reason for this could be that Steno was always with Borch, as in Leiden, and therefore there was no need for Borch to mention him.[32] This meant that Borch simply looked at Steno as another participant in these salons and thus not worthy of reference. This attitude, however, contrasts with the writings of those who met Steno for the first time in Paris and who could not help but notice him. Indeed, Steno intentionally enhanced his academic image at the time of his arrival in France. Unlike Borch and Swammerdam, who had not yet graduated, Steno received a medical degree from the University of Leiden shortly after arriving in Paris, in December 1664. Steno actively sought this degree because he requested special permission to receive the M.D. in absentia, without having to travel back to Leiden from Copenhagen.[33] This academic move contributed to Steno's social integration in a community of highly prestigious scholars. With a medical degree he was no longer merely a young student but a peer who could speak to them on a common academic level.

Dissections were at the center of Steno's reputation in Paris as a new scientific publication that started around those months in Paris makes clear. Indeed, Steno's time in Paris coincided with the appearance of the first learned journal of Europe, the *Journal des sçavans*, published for the first time in January 1665 in Paris.[34] The *Journal* was produced under the patronage of Colbert at the recommendation of Chapelain, meaning that it was a product of the intellectual circles that Steno frequented. When Borch visited the *Journal*'s editor, Denis de Sallo (1626–1669), in December 1664, the latter said that his task was to "publish critiques of every book from all Europe, which will be printed in this year and the following, and to publish something similar every week."[35]

New York: Oxford University Press, 1976–1989), 1:290–292 (Letter 208); and Henry Justel to Henry Oldenburg, December 1667, in *The Correspondence of Henry Oldenburg*, ed. Rupert Hall and Marie Boas Hall (Madison: University of Wisconsin Press, 1965–1986),4:29–31. On corresponding with Thévenot, see *Epistolae*, 1:178, 381–382.

31. For Steno at Bourdelot, see Jean Gallois, *Conversations académiques, tirées de l'Académie de Monsieur l'Abbé Bourdelot* (Paris: Claude Barbin, 1674), 102.

32. See OBI, 1:xxxiii–xxxix.

33. *Epistolae*, 1:183.

34. Guerrini, *The Courtiers' Anatomists*, 105–107.

35. OBI, 4:184–185, 5 December 1664, 184–185.

One of the first issues of the *Journal*, from March 1665, includes a long review of Steno's *De musculis et glandulis*, a book which "gives to the public this project of anatomical observations [*ce projet de ses observations Anatomiques*]."[36] From the review, readers learn that Steno, "this Danish *savant*[,] is currently in Paris, where he performs dissections every day in the presence of many curious people; and he did them in the School of Medicine, where he is admired by everyone thanks to his new discoveries."[37] This praise is particularly relevant because the review also cast doubt on Steno's claim that "the heart does not differ in any way from other muscles," a claim that "will reverse that which is most well established [*ce qu'il y a de plus constant*] in medicine."[38] Yet, the author of the review, possibly de Sallo himself, insisted that Steno "has something peculiar, because he renders most of these things so tangible that one is compelled to remain convinced [of them] [*il rend la plus-part de ces choses si sensibles, qu'on est obligé d'en demeurer convaincus*] and to be surprised that such things escaped all anatomists who preceded him."[39] De Sallo wrote as if he or someone close to him had witnessed Steno's dissections in person. And Steno seemed to have a powerful way of winning others to his side.

Steno's success with his dissections speaks to the dynamics of social approval in the scientific salons of seventeenth-century Paris. Academies such as Mersenne's and Thévenot's thrived by means of sharing results in person, not by the writings of its participants.[40] In this environment, the focus of intellectual production moved away from the Republic of Letters to in-person scientific collaboration and discussion. In the words of Dena Goodman, the "traditional epistolary relations were now enriched by direct verbal ones."[41] A similar phenomenon occurred a few years later in Paris with Wilhelm Homberg (1653–1715), whose reputation derived

---

36. Anon., "Review of Steno's *De musculis et glandulis*," *Journal des sçavans* (23 March 1665): 139–142, at 139.

37. Anon., "Review of Steno's *De musculis et glandulis*," 141.

38. Anon., "Review of Steno's *De musculis et glandulis*," 139–140.

39. Anon., "Review of Steno's *De musculis et glandulis*," 141.

40. Pecquet writes of Auzout as if he were family and that he was "present" at Roberval's experiments. He also speaks of Pascal as "witness" of other experiments; see *Experimenta nova anatomica*, 9, 51, 55. Scientific correspondence networks relied more on in-person interactions than previously thought; see David S. Lux and Harold J. Cook, "Closed Circles or Open Networks? Communicating at a Distance During the Scientific Revolution" *History of Science* 36, no. 2 (1998): 179–211.

41. Goodman, *The Republic of Letters*, 90. Goodman writes of the eighteenth century, but these dynamics seemed already to exist in the seventeenth.

in large part from his direct personal contacts and oral reports.[42] In the case of Steno, most of what is known about his Parisian sojourn comes from private, unpublished accounts of those who witnessed his dissections and spoke with him, such as Borch, de Salo, and others, and not from Steno's own publications. It is telling that the book Steno wrote in Paris, the *Discours sur l'anatomie du cerveau*, has a strong oral character because it is based on dissections that Steno performed in Paris, possibly at Thévenot's salon.[43] The rapid spread of Steno's renown due to his dissections also shows the value of seeing a dissection in person and not simply through texts. This social value of witnessing dissections in person would also prove useful to Steno's rapid integration into the Medici court in Florence and to the impressions that he made on the few members of the Royal Society of London that he met in Montpellier in the fall of 1665.

As the review in the *Journal des sçavans* mentioned, Steno performed dissections daily, so they were not limited to the meetings of Bourdelot and Thévenot. Indeed, these two salons were only a sample of the French scientific culture at the time. Some of the dissections happened at the School of Medicine, as the *Journal* pointed out. Others took place at the meetings organized by "the surgeon Morel," whom Steno and Borch visited on 7 November 1664.[44] There are not many sources available on Morel, but Borch describes him as "surgeon of the Duchess of Orléans."[45] In addition, an eighteenth-century historical dictionary of medicine describes a Claude Morel (d. 1703) as the personal physician of Marguerite of Lorraine (1615–1672), wife of the Duke of Orléans.[46] Morel was a member, and at some point the president, of the confraternity of surgeons of Saint-Côme, whose members have been described as "learned surgeons."[47] This

---

42. Lawrence Principe, *The Transmutations of Chymistry: Wilhelm Homberg and the Académie Royale des Sciences* (Chicago: University of Chicago Press, 2020), 62.

43. On the oral character of the *Discours*, see Raphaële Andrault, "Introduction," in Stensen, *Discours sur l'anatomie du cerveau*, ed. Andrault, 7–74, esp. 16.

44. OBI, 4:163, 7 November 1664.

45. OBI, 3:337, 2 April 1664.

46. N. F. J. Eloy, *Dictionnaire historique de la médecine ancienne et moderne*, 2 vols. (Paris, 1756), 2:341. See also Paris, Bibliothèque numérique Medica, "Liste funèbre des chirurgiens de Paris, qui sont morts depuis l'année 1315 . . . ," 194–195, www.biusante.parisdescartes.fr/histmed/medica/page?ms02118&p=201.

47. Eloy, *Dictionnarie historique*, 2:341. See also Gustav Scherz, "Introduction," in *Nicolaus Steno's Lecture on the Anatomy of the Brain*, ed. Gustav Scherz (Copenhagen: Nyt Nordisk Forlag Arnold Busck, 1965), 61–103, esp. 64. On the surgeons of Saint-Côme, see Laurence Brockliss and Colin Jones, *The Medical World of Early Modern France* (Oxford and New York: Oxford University Press, 1997), 219–225.

organization had its own anatomical amphitheater, where Morel's meetings could have taken place.[48] Borch attended these meetings several times before Steno arrived in France.[49] Once, in May 1664, he saw at Morel's a brain dissection "according to the Cartesian method."[50] And a month later, Borch himself "showed the lymphatic vessels in a hanged female body at Mr. Morel's."[51] Morel's anatomical interests and his hands-on experience as a surgeon made him an interesting friend to Steno, and indeed the two worked together again with Francesco Redi a few years later in Italy.[52] Morel's group confirms the role of nonacademic environments in attracting scholars interested in the new sciences, and especially in fostering anatomical research centered on dissections.[53] Since their days in Denmark and the Netherlands, Steno and Borch continued to frequent both academic and nonacademic environments, confirming the overlapping of these cultures at the time.

The Académie de Physique de Caen

Steno's dissections had an impact that extended beyond Paris. After the founding of the Académie des Sciences in 1666, other scientific academies also received royal patronage across the country.[54] One of them was the Académie de Physique de Caen, founded by Huet and Graindorge in 1666, which gained status as a royal academy in 1667. Huet had attended Montmor's academy in the 1650s and knew how an informal academy of natural sciences operated. Graindorge, on the other hand, learned about it in his visit to Paris in 1665, where he attended the academies of Bourdelot and Thévenot.[55] What Graindorge saw at Thévenot's meetings, especially

48. Guerrini, *The Courtiers' Anatomists*, 29.
49. Borch mentions visiting Morel several times; see OBI, 3:378, 29 April 1664, and below.
50. OBI, 3:379, 1 May 1664.
51. OBI, 3:448, 13 June 1664.
52. Francesco Redi, *Esperienze intorno alla generazione degl'insetti* (Florence: Stella, 1668), 74. In later correspondence with Redi, Morel mentions the duchess of Orléans; thus, Redi's "Maurel" is the same Morel mentioned by Borch. See Morel to Redi, January (?) 1674, in Florence, BML, Redi 214, fols. 161r–162v. Claude Morel's name sometimes appeared as "Jean Charles Morel"; see Eloy, *Dictionnarie historique*, 2:341.
53. Sandra Cavallo, *Artisans of the Body in Early Modern Italy: Identities, Families and Masculinities* (Manchester: Manchester University Press, 2010); and Bertoloni Meli, *Mechanism, Experiment, Disease*, 6–12.
54. Lux, *Patronage and Royal Science*, 1–7.
55. See Graindorge to Huet, 1 July 1665, in Léon Tolmer, "Vingt-deux lettres inédites d'André Graindorge à P.-D. Huet," *Mémoires de l'Académie nationale* 10 (1942): 288.

the dissections of Steno, made his Parisian sojourn an intensive formative period for the foundation of the Académie de Physique.[56]

Graindorge's letters to Huet in Caen give a clear picture of Steno's work in Paris.[57] In his first letter, from early May, Graindorge wrote about "the group of Mr. Thévenot," where he "had the satisfaction of seeing the dissections of Mr. Steno, who is very skillful and performs with good grace [*et qui s'en acquitte de bonne grâce*]."[58] Ten days later Graindorge could not contain his enthusiasm about Steno's dissections, writing that "this Mr. Steno makes a furor [*ce Mr. Sténon fait rage*]" because he "dissects every day," and "there is no butterfly or fly that escapes his work."[59] Graindorge carefully observed everything Steno did, and his letters give an account of almost all of Steno's research projects at the time. They saw "in the head of a calf, certain glands," the "exterior salivary duct"; "after dinner, the eye of a horse"; "something very considerable for the lymphatics in the neck"; "dissections of the brain"; "the fibers of the heart"; and the "constitution of muscles, which clearly shows the mechanics used by nature for the motion of muscles."[60] "To tell the truth," wrote Graindorge to Huet, "we are but apprentices before him [*nous ne sommes que des apprentis auprès de lui*]."[61]

Steno became Graindorge's de facto mentor, despite the fact that Graindorge was twenty-two years older than Steno. Graindorge noted the extraordinary availability of Steno, who had "an inconceivable patience and, through practice, has acquired a skill above the ordinary [*par routine il a acquis une adresse au-dessus du commun*]."[62] In Paris as in Leiden, Steno's manners served both a social and a scientific purpose.[63] As part of his search for certainty in anatomy, Steno wanted to make his dissections more accessible to observers. In Graindorge's words, "without putting either the eye, or the scissors, or a small instrument anywhere other than his hand, which he always kept exposed to those there [*sans mettre ni l'oeil,*

---

56. Lux, *Patronage and Royal Science*, 29–56, esp. 29–31.

57. Copenhagen, Royal Library of Denmark, NKS 4660 kvart. They are also transcribed in Tolmer, "Vingt-deux lettres," 245–337.

58. Graindorge to Huet, 9 May 1665, in Tolmer, "Vingt-deux lettres," 268–269. Borch also mentions meeting Graindorge at Thévenot's on 3 May 1665; see OBI, 4:317, 3 May 1665.

59. Graindorge to Huet, 19 May 1665, in Tolmer, "Vingt-deux lettres," 269–272.

60. Graindorge to Huet, 9 May, 19 May, 30 May, and 16 September, in Tolmer, "Vingt-deux lettres," 269–270, 273, 322.

61. Graindorge to Huet, 19 May 1665, in Tolmer, "Vingt-deux lettres," 270.

62. Graindorge to Huet, 19 May 1665, in Tolmer, "Vingt-deux lettres," 270."

63. On manners, see Mario Biagioli, "Etiquette, Interdependence, and Sociability in Seventeenth-Century Science," *Critical Inquiry* 22, no. 2 (1996): 193–238.

ni les ciseaux, ni un petit instrument autre part que dans sa main qu'il tint toujours exposée à la compagnie], he showed us all that can be observed in the constitution of the eye."[64] Once Graindorge asked Steno to "show me a heart, which he promised with singular kindness."[65] In July Steno left Paris with Swammerdam to stay at Thévenot's summer house in Issy for a month. But Steno answered Graindorge's questions even from a distance. In a letter dated 8 July 1665 and sent from Paris to Huet in Caen, Graindorge said that Steno, still in Issy, had told him of an experiment on the lymphatics of a dog, to which Graindorge asked "if he had found anything on the milk ducts [sic, likely chyle ducts]," and Steno replied that he had not.[66] Interestingly, Graindorge also commented that without Thévenot, who "was in the countryside with his Mr. Steno, ... the conversation [in Paris] took place in general terms."[67] This confirms that without a host, scientific salons could easily lose their direction, even if only for a meeting. But it also points to the role that Steno and his lively dissections played in such meetings. Perhaps not accidentally, a few months after Steno and Swammerdam left Paris, Thévenot stopped gathering scholars at his place despite being back in Paris.[68]

Steno taught Graindorge not only the contents of anatomy, but also various techniques such as those needed to see "the thread of the [muscle] fibers of the heart very distinctly": "You have to boil it, remove the flesh and follow the course of these fibers. This cannot be explained. He has shown us in the same way in the tongue a furious extension of fibers [*un furieux échappement de fibres*] from top to bottom, sideways and long; but it takes good practice."[69] This description beautifully illustrates Steno's hands-on approach to anatomical dissections, which were an extension of his artisanal skills. Indeed, just as a workshop master teaches his

---

64. Graindorge to Huet, 30 May 1665, in Tolmer, "Vingt-deux lettres," 272–273.
65. Graindorge to Huet, 19 May 1665, in Tolmer, "Vingt-deux lettres," 270.
66. Graindorge to Huet, 8 July 1665, in Tolmer, "Vingt-deux lettres," 291.
67. Graindorge to Huet, 1 July 1665, in Tolmer, "Vingt-deux lettres," 288–290.
68. Steno left Paris in September 1665. By December 1665 Chapelain was complaining about Thévenot's absence from Paris; see Chapelain to Steno, 8 December 1665, in *Lettres de Jean Chapelain*, 2:424. Graindorge reports the academy's last session took place in March 1666; see Lux, *Patronage and Royal Science*, 52 n. 61. The start of the Académie royale des sciences is another possible reason for the end of Thévenot's academy, but Thévenot would not become a member of the Académie until decades later.
69. Graindorge to Huet, 5 August 1665, in Tolmer, "Vingt-deux lettres," 305. Although not mentioned, it was also necessary to dry the fibers, just as Malpighi did with the lungs; see Bertoloni Meli, *Mechanism, Experiment, Disease*, 45.

apprentice, these dissection skills were best learned by direct observation and not by writing, as Graindorge told Huet.

Back in Caen, Graindorge knew that he "wanted to undergo the effort of performing everything I have seen."[70] Months after Steno's departure from Paris, when discussing the role of anatomical dissections in the Académie de Physique's research program, Graindorge reminded Huet of what he had learned with Steno, recalling the latter's "patience and dexterity."[71] In Caen, Graindorge continued to speak of his friend and teacher as a model to imitate, saying that a large number of "ladies [were] eager to see a heart unraveled in the style of Mr. Steno."[72] Dissections had been part of the original scientific activities led by Huet and Graindorge since 1662, but they became a major focus only after 1666, at the time of the academy's founding.[73] The timing of Graindorge's encounter with Steno in Paris—right before the beginning of regular dissections in Caen—is therefore revealing of Steno's influence in the Académie de Physique's foundational research program.

But perhaps the most important thing that Steno taught Graindorge was the need to search carefully for certainty in the natural sciences. According to the historian David Lux, doubts "planted at Thévenot" led Graindorge to think "that no one had yet arrived at anything close" to the truth about nature.[74] This awareness that much knowledge was still lacking shaped the Académie's research program. Graindorge had seen with his own eyes Steno's groundbreaking discoveries, such as the salivary glands and the muscle-fiber structure of the heart. He also saw how Steno successfully persuaded his audiences by relying on new dissection techniques, which, when done in public, were often accompanied by a discourse on the need for reliable knowledge in anatomy. Therefore, as its main leader, Graindorge shaped the research of the Académie de Physique de Caen so that it was about finding new things (or verifying the validity of old ones), based on experiments and collaboration.[75] As for collaboration, Graindorge learned not only from working directly with Steno, but also from seeing the fruitful collaboration between Steno and Swammerdam.

---

70. Graindorge to Huet, 30 May 1665, in Tolmer, "Vingt-deux lettres," 273.
71. Graindorge to Huet, 20 January 1666, in Florence, BML, Ashburnham 1866, 551, as translated in Lux, *Patronage and Royal Science*, 46.
72. Graindorge to Huet, 14 May 1666, in Florence, BML, Ashburnham 1866, 556, as translated in Lux, *Patronage and Royal Science*, 68.
73. Lux, *Patronage and Royal Science*, 27.
74. Lux, *Patronage and Royal Science*, 50.
75. Lux, *Patronage and Royal Science*, 50–51, 66–75.

Breathing Physico-Mathematics: The Friendship Between Steno and Swammerdam

In Paris, Steno was deeply immersed in new and collaborative research on the muscles with Jan Swammerdam. Swammerdam is widely known in the history of science for his natural history of insects and his extensive uses of the microscope. However, his contributions to early modern science also included making experiments borrowed from the disciplines of physico-mathematics and chymistry. Steno had met Swammerdam when he arrived in Amsterdam in the spring of 1660, at the meetings organized by Blasius. Possibly because of their similar circumstances and interests, the two young researchers became close friends.[76] Performing dissections together, Steno and Swammerdam reached new insights about the anatomy of respiration. The problem of respiration and its relation to the heart was a hot topic at the time because it was one of the research questions left open by William Harvey.[77] As Swammerdam later put it, "Respiration is as certain as its nature is difficult and obscure to explain [*quam certum est esse Respirationem, tam abstrusa et difficilis explicatu est illius Natura*]."[78] In a letter dated March 1663 from Leiden, after commenting on Malpighi's groundbreaking observations of pulmonary alveoli, Steno told Thomas Bartholin what "the very intelligent young man and very diligent in anatomical exercises, Mr. Swammerdam, showed to me, [when] propelling air in the heart by way of the lungs [*aerem per medios pulmones in cor propellens*]."[79] Swammerdam's experiment consisted in "simulating respiration [*motum respirationis æmulum*]" in a dead animal. He saw that bubbles of air descended "into the heart through branches of the pulmonary vein."[80]

Swammerdam's observation of air entering the heart seemed to confirm the ancient theories of respiration in which air entered and cooled the heart through the lungs. This theory had also been adopted by Harvey's

---

76. Jorink, "*Modus politicus vivendi*."

77. Frank, *Harvey and the Oxford Physiologists*, 1–20, esp. 20; and Leonard G. Wilson, "The Transformation of Ancient Concepts of Respiration in the Seventeenth Century" *Isis* 51, no. 2 (1960): 161–172, esp. 162–164.

78. Jan Swammerdam, *Tractatus physico-anatomico-medicus de respiratione usuque pulmonum* (Leiden: Daniel, Abraham, and Adrian à Gaasbeeck, 1667), 1.

79. Steno to Bartholin, 5 March 1663, in EMC IV, 352.

80. Steno to Bartholin, 5 March 1663, in EMC IV, 351. Sylvius also saw Swammerdam do this vivisection on 15 January 1663; see Franciscus de Le Boe Sylvius, *Disputationum medicarum pars prima* (Amsterdam: Johannes van den Bergh, 1663), Disp. Septima, Additamentum, esp. 126–127.

professor Fabricius d'Acquapendente and by Descartes but ran counter to Harvey and Malpighi's new physiology of the lungs, the latter published in 1661.[81] For the latter two, the air did not mix with blood nor did it reach the heart; rather, it provided only the necessary pressure in the lungs to affect the blood flow.[82]

Malpighi, Steno, and Swammerdam were not the only ones interested in these topics. Across the English Channel, a few members of the Royal Society of London also investigated this question by carrying out experiments on the vacuum. Swammerdam's work on respiration, although leading him to traditional conclusions, also relied on these novel research methods, such as using physico-mathematical experiments with siphons and bladders. This is why he entitled his book a *Tractatus physico-anatomico-medicus de respiratione* (*Physico-Anatomico-Medical Treatise on Respiration*; Leiden, 1667). Swammerdam portrayed his research visually in the book's frontispiece, with illustrations of more than ten experiments, each accompanied by the page number where they are discussed in the book (fig. 4.1).[83] Swammerdam's book was published sometime after March 1667, and it arrived shortly afterward in England, in time to be reviewed in the *Philosophical Transactions* in October 1667.[84] It was also around this time that the physician Richard Lower (1631–1691) and Robert Hooke (1635–1703) developed new experiments to artificially inflate the lungs of a dog.[85] After this and other experiments, Lower understood that fresh air alone, not the motion of

---

81. Wilson, "The Transformation of Ancient Concepts of Respiration," 162; and Malpighi, *De pulmonibus*. Harvey developed this view only after the publication of his work on the circulation of the blood; see Frank, *Harvey and the Oxford Physiologists*, 38–42.

82. Wilson, "The Transformation of Ancient Concepts of Respiration," 163, 166; and Bertoloni Meli, *Mechanism, Experiment, Disease*, 46–47.

83. Most references to this frontispiece focus on the hermaphrodite snails at the bottom; see Eric Jorink, "Cartesian Sex: Dutch Anatomists on Genitalia, Lust, and Intercourse (ca. 1660–1680)," in *Les libertains néerlandais/The Dutch Libertines*, ed. Nicole Gengoux, Pierre Girard, and Mogens Lærke, Libertinage et philosophie à l'époque classique (xvie–xviiie siècle) 19 (Paris: Classiques Garnier, 2022), 249–290, esp. 259.

84. Swammerdam's book was printed either in or after March 1667 because it includes a letter to Thévenot signed that month; see Swammerdam, *Tractatus physico-anatomico-medicus de respiratione*, viii. For the review, see Anon., "An Account of some books ... II. Joh. Swammerdam, M.D. Amsterodamentis *De respiratione et usu pulmonum*," *Philosophical Transactions* 2, no. 28 (1667): 534–535.

85. On 2 May 1667 Lower told the Royal Society that he had been filling the lungs of animals; see Frank, *Harvey and the Oxford Physiologists*, 197.

FIGURE 4.1. Frontispiece of Jan Swammerdam, *Tractatus physico-anatomico-medicus de respiratione* (Leiden, 1667). Courtesy of Balliol College, University of Oxford (ref. 905 f 5 (2)).

the lungs, was essential to life.[86] This result portrayed the motion of the lungs as less important than anatomists had thought until then for the circulation of the blood. Because Swammerdam insisted on the ancient role of respiration as a cooling of the heart, his work on respiration remains understudied.[87] But regardless of theoretical differences, Swammerdam and Lower alike demonstrated the methodological benefits of using contrived mechanical experiments in physiological research.

Steno also had something new to add about respiration and its effect on the circulation of the blood. Harvey had written about the lungs propelling blood out of them from the pulmonary artery into the heart, an idea originally proposed by Galen.[88] Lower, on the contrary, came to disregard this motion as nonessential. Yet, when Swammerdam showed Steno his first observations on the lungs, Steno noticed during a vivisection that the two venae cavae, and not the pulmonary vein, pushed the blood forward toward the heart "every time the animal distended its thorax in inspiration."[89] Steno concluded that venous blood moved upward in the vena cava toward the heart due to an inflation movement linked to the "action of the muscles" of the thorax. This had two important implications. First, this phenomenon could shed light on the "function of respiration in the circulation of blood."[90] Steno's focus was not on the veins of pulmonary circulation mentioned by Harvey, but on the superior and inferior vena cava, which received venous blood from all the body (fig. 4.2). Second, this observation presented Steno with an opportunity to expand Pecquet's use of the elasticity of the air in his explanation of the motion of blood toward the heart. Steno said that since the "fear of vacuum seems entirely defeated," it "seems truer" to explain the motion of blood through "pulsing alone."[91] Steno was following the lead of Pecquet, who said that the opinion of the fear of vacuum "is refuted by means of hydraulic machines" [*hydraulicis exploditur machinis*].[92]

---

86. Frank, *Harvey and the Oxford Physiologists*, 193–205; and Bertoloni Meli, *Mechanism, Experiment, Disease*, 52–54.

87. Other reasons may be that Swammerdam did not write in English and that his famous works on insects and microscopy overshadowed it. See, however, Jorink, "'Outside God, There Is Nothing,'" 92.

88. Wilson, "The Transformation of Ancient Concepts of Respiration," 163.

89. Steno to Thomas Bartholin, 5 March 1663, in EMC IV, 351–352.

90. Steno to Thomas Bartholin, 5 March 1663, in EMC IV, 353.

91. Steno to Thomas Bartholin, 5 March 1663, in EMC IV, 352. See also Pecquet, *Experimenta nova anatomica*, 25.

92. Pecquet, *Experimenta nova anatomica*, 25, 62–66. Pecquet included the opinion of fear of the vacuum within the broader "opinion of attraction."

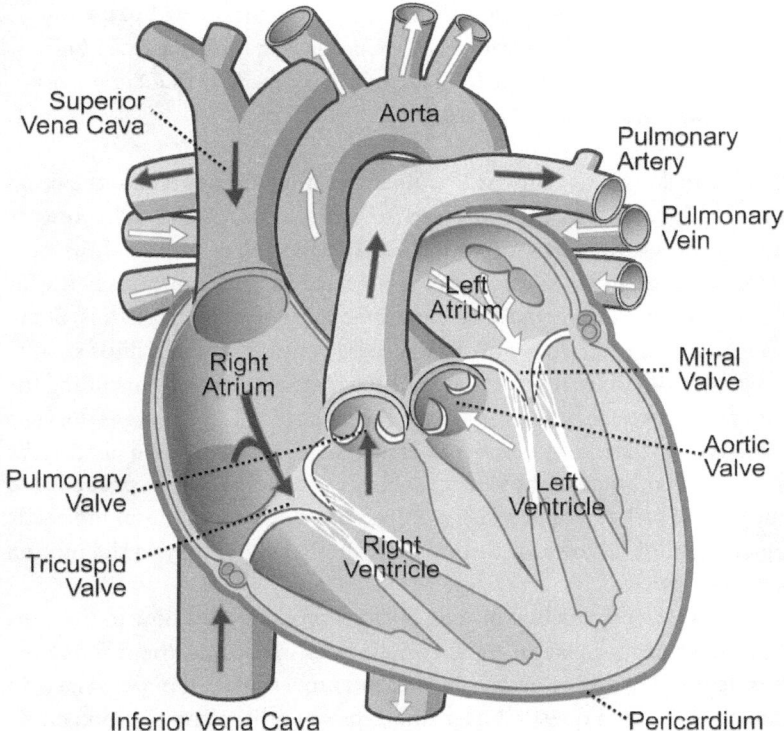

FIGURE 4.2. Diagram of the human heart. While Harvey associated the motion of the lungs with the flow of blood into the pulmonary vein, Steno associated it with the superior and inferior vena cava. © Wikimedia Commons.

Steno was aware that no one had previously linked the motion of the lungs to the blood flow in the vena cava, as Bartholin reminded him.[93] Therefore, he developed a purely mechanical argument in support of his idea:[94]

> I think it is beyond controversy that when air suffers a force by something which pushes it, it is either condensed in the place where it is by the same force, or it is propelled to another place. Mechanics, however, teaches that a great force is necessary to condense it [*ad condensandum autem magnam requiri vim mechanica docet*], and experience shows that condensation never occurs unless all the surrounding bodies are strong

---

93. Bartholin to Steno, 7 April 1663, in EMC IV, 360.
94. Steno to Bartholin, 30 April 1663, in EMC IV, 414–421.

enough to resist the compressed air. But if even a small part of the surroundings is unable to resist it, air propelled by a thrusting cause finds there a crack through which it escapes [*illico à trudente causa propellius aer, qua elabatur, invenit rimam*].[95]

Steno's analogy is as follows: During inspiration, air pours into the lungs so that "the ribs, which move away from each other, and the diaphragm, moving outward, exert a force on air."[96] This motion of the ribs shows that they "yield to air by as much as the external surface compresses the air [*quantum exterior superficies aerem premit, tantum ei cedant*]." Thus, Steno concluded, "the air does not condense but pushes where it finds smaller resistance"—that is, it pushes everything inside the thorax, including the lungs and a section of the vena cava. The force of the air squeezes the vena cava so that "the blood which is outside the thorax, in order to distend the yielding membrane of the vena cava inside the thorax, is thus propelled to the place which does not resist it."[97] In short, the force of the air indirectly pushes the blood toward the heart, thus returning agency to the motion of the lungs.

Steno never published his conclusions on the blood flow in the vena cava, perhaps because of Bartholin's lack of interest in them.[98] But the problem of explaining exactly how blood moves upward in the vena cava had been an open question ever since Harvey's discovery of blood circulation.[99] At stake is the problem of how such a large amount of venous blood returns to the heart, especially in humans, where most of the blood is below the heart (fig. 4.3). Neither the pumping of the heart nor the so-called siphon effect is enough to counter the force of gravity in the inferior vena cava of humans and other animals that stand upright.[100] Pecquet sug-

95. Steno to Bartholin, 30 April 1663, in EMC IV, 419.
96. Steno to Bartholin, 30 April 1663, in EMC IV, 419–420.
97. Steno to Bartholin, 30 April 1663, in EMC IV, 420.
98. Bartholin recommended that Steno look more closely at the heart, with no reference to respiration again; Bartholin to Steno, 25 July 1663, in EMC IV, 421–427.
99. Steno's comments are like those made almost a hundred years later by Albrecht von Haller (1708–1777); see Lord Cohen of Birkenhead, "On the Motion of Blood in the Veins," *British Medical Journal* 3, no. 5774 (1971): 551–557, esp. 553. Haller quotes Steno from Thomas Bartholin's *Epistolarum medicinalium centuriæ*; see Haller, *Dissertationem inauguralem medicam et anatomicam de motu sanguinis per cor* (Gottingen, 1737), 407–408.
100. Modern articles attribute priority in noticing this problem to Franciscus Donders, *Physiologie des Menschen* (Leipzig, 1856), but Pecquet noticed it first in *Experimenta nova anatomica*, 45, 47. For a modern explanation, see Roger S. Seymour

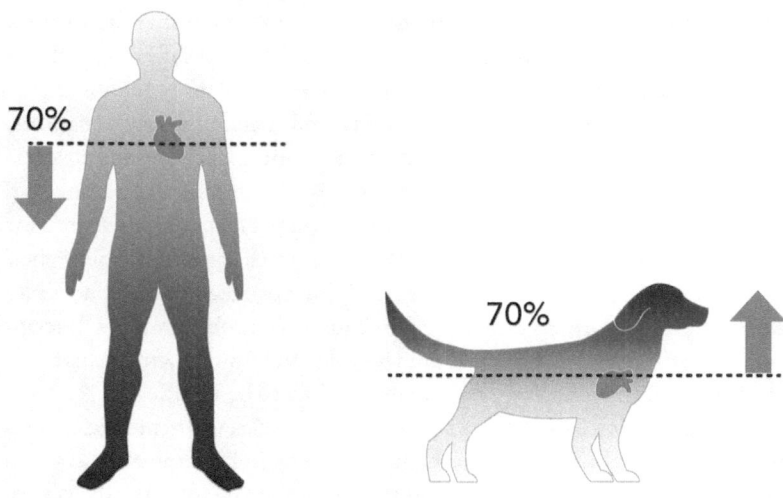

FIGURE 4.3. Gravity-dependent distribution of blood in a human and a dog. Designed by Luís Ribeiro. Based on Furst, *The Heart and Circulation*, 2nd ed. (Cham: Springer, 2020), 324.

gested that this motion was related to the contraction of the walls of the veins, but he did not explain it further, since his focus was on the motion of chyle.[101] Steno thus expanded upon Pecquet with new observations of the blood flow in the vena cava and a sophisticated mechanical argument. Specifically, he concluded that this flow depends on the muscles of respiration (the diaphragm and intercostal muscles), which is still one of the mechanisms proposed today.[102] But more historically interesting is the fact that, then as now, studies of the action of respiration on the circulation of blood relied heavily upon a fruitful interaction between anatomy, physics, and mathematics.[103]

---

and Kjell Johansen, "Blood Flow Uphill and Downhill: Does a Siphon Facilitate Circulation Above the Heart?," *Comparative Biochemistry and Physiology Part A: Physiology* 88, no. 2 (1987): 167–170.

101. Pecquet, *Experimenta nova anatomica*, 75. For a comparison between the explanations given by Pecquet, Steno, and Borelli, see Castel-Branco, "Physico-Mathematics and the Life Sciences."

102. Branko Furst, *The Heart and Circulation: An Integrative Model*, 2nd ed. (Cham: Springer, 2020), 199–202.

103. For a modern case, see the joint article by physicians, physicists, and mathematicians on the role of respiration in the circulation of the blood: Gerrit J. Noordergraaf, Johnny T. Ottesen, Wil J. P. M. Kortsmit, Wil H. A. Schilders, Gert J.

The collaboration between Steno and Swammerdam reached its peak when the two friends met in Paris. They had followed different paths in their research on respiration: Swammerdam invested more in new experiments and instruments, while Steno favored mechanical explanations. Both, however, fully relied on analogical reasoning between their experiments and the body. Moreover, Steno's new research on the anatomy of muscles was motivated by his interest in respiration and by his search for the causes behind it. As Steno wrote in *De musculis et glandulis*, published a few months before his arrival in Paris, "The absence of a most accurate examination [*accuratius examen*] of the muscles has obscured the description of respiration."[104] The problem was that "the doubts which surround the theory of the muscles" were many and had to be clarified.[105]

Swammerdam also became interested in muscles while in Leiden, and the two friends continued to study them in Paris. In December 1664, when vivisecting frogs before Borch, Swammerdam showed that "the nerves [of a frog], rubbed by a knife [*cultro affrictos*], showed convulsive motions for a long time."[106] Steno had already observed this phenomenon back in Leiden while dissecting the muscles of a dog.[107] Swammerdam enigmatically concluded that "the contractive motion of the muscles came from the outside."[108] Modern writers have been quick to assume that Swammerdam was rejecting the traditional theory that animal spirits were needed for muscular motion.[109] However, this was not Swammerdam's conclusion at the time, although it would be years later. In fact, in a report made in October 1665 in Amsterdam, Swammerdam wrote about his experiments with the muscles of frogs in Paris but refrained from commenting on the nature of their motion.[110]

Scheffer, and A. Noordegraaf, "The Donders Model of the Circulation in Normo- and Pathophysiology," *Cardiovascular Engineering* 6, no. 2 (2006): 53–72.

104. Steno, *De musculis et glandulis*, 9.
105. Steno, *De musculis et glandulis*, 11.
106. OBI, 4:186, 8 December 1664.
107. Steno, *Elementorum myologiae specimen*, 59.
108. OBI, 4:186, 8 December 1664.
109. Raphaële Andrault, "The Chain of Motions and the Chain of Thoughts: The Diachronic Mechanism of Spinoza's Friends," in *Mechanism, Life and Mind in Modern Natural Philosophy*, ed. Antonio Clericuzio, Paolo Pecere, and Charles Wolfe (Cham: Springer, 2022), 119–135; and Matthew Cobb, "Exorcizing the Animal Spirits: Jan Swammerdam on Nerve Function," *Nature Reviews: Neuroscience* 3, no. 5 (2002): 395–400, esp. 397.
110. [Gerard Blasius,] *Observationes anatomicae selectiores collegii privati Amstelodamensis* (Amsterdam: Caspar Commelin, 1667), 29–30.

According to traditional anatomical concepts, the contraction of the muscles, which coordinated animals' movements, was caused by animal spirits.[111] The nerves transported these spirits from the brain to the muscles, which then became inflated, thus explaining why the muscles seem full when contracted. These ideas, originally proposed by Hippocrates and Galen, had been revisited and accepted in the seventeenth century by Descartes.[112] In *De homine*, which was published in its original French right before Steno's arrival in Paris, Descartes said that "animal spirits . . . enter the cavities of the brain," whence "they proceed to the nerves" and thus "are able to change the shapes of the muscles . . . and in this way to move all the members."[113] Descartes went so far as to compare this system with the "fountains in the gardens of our kings," in which the animal spirits were "the water which drives" the garden's "mechanisms."[114] In Steno's Paris, almost everyone adopted this or a similar explanation, as can be seen in a December 1664 meeting of the Bourdelot academy. Borch reported seeing the "motion of intercostal muscles" in a dead dog owing to "the presence of [animal] spirits."[115] Months later, in June 1665, Graindorge mentioned the same idea when writing about a chicken that was able to survive without a brain for "several months." Graindorge, who accepted Galen's theory, said that the only possible explanation was that the "animal spirits may be in storage, because they are made while sleeping, which, in order that they are not dissipated, are stored in some reservoir for the operations of an entire day."[116] Strikingly, even Swammerdam accepted this reasoning. In *Tractatus de respiratione*, Swammerdam attributed the contraction of the muscles to the "continuous influx, or perhaps forceful entry, of animal spirits through the nerves."[117] But, just like Steno, Swammerdam also spoke of "the most difficult doctrine of the contraction of muscles" in the very title of that chapter.[118]

Meanwhile, Steno grew increasingly skeptical of animal spirits. A few anatomists since Harvey had questioned whether they really existed, but

---

111. Siegel, *Galen's System of Physiology and Medicine*, 183–195.
112. On Hippocrates, see C. U. M. Smith, Eugenio Frixione, Stanley Finger, and William Clower, *The Animal Spirit Doctrine and the Origins of Neurophysiology* (Oxford and New York: Oxford University Press, 2012), 15–16.
113. René Descartes, *Treatise on Man*, ed. and trans. Thomas Steele Hall (Cambridge, MA: Harvard University Press, 1972), 21.
114. Descartes, *Treatise on Man*, 21–22.
115. OBI, 4:192, 29 December 1664.
116. Graindorge to Huet, 3 June 1665, in Tolmer, "Vingt-deux lettres," 277.
117. Swammerdam, *Tractatus physico-anatomico-medicus de respiratione*, 61, 66.
118. Swammerdam, *Tractatus physico-anatomico-medicus de respiratione*, 59.

none as bluntly as Steno.[119] When studying the glands, even Steno spoke of muscle motion in terms of animal spirits.[120] The trigger for his skepticism was related to his studies of muscles and the heart. In *De musculis et glandulis*, Steno addressed the current knowledge about muscle contraction, though he refrained from reaching a definite conclusion: "How this contraction occurs is difficult to determine. Many people attribute it to a filling of the fibers, some to their emptying, and some to both. I would be temerarious if I arbitrated between them [*temerarius essem, si hic meum interponerem arbitrium*]. Therefore, I publicly proclaim [*apertè pronuntio*] that the causes and ways of action are not obvious."[121] Steno was reluctant to use animal spirits as an explanation for motion. After demonstrating that the heart was just a muscle, he concluded that the heart was not "the producer of some spirits, such as vital spirits."[122] A year later, in his famous discourse on the brain, Steno spoke at length about the problematic existence of animal spirits:[123]

> One sees even less certainty on the subject of animal spirits [*on voit encore moins de certitude, sur le sujet des esprits animaux*]. Is it the blood? Could it be a particular substance separated from the chyle in the glands of the mesentery? Might the serosities not be their sources? There are some who compare them to the spirit of wine, and one can wonder whether it might not be the very matter of light? In the end, the dissections we usually use cannot clarify our understanding on any of these doubts [*ne nous peuvent éclaircir l'esprit, sur aucun de ces doutes*].[124]

Unlike Swammerdam, Steno intentionally cast doubt on the traditional theory of animal spirits, even though he was not yet able to provide an alternative.

---

119. Harvey questioned whether vital spirits were separate from blood; see Frank, *Harvey and the Oxford Physiologists*, 15–16, 38–44. In the 1660s de Graaf also noticed that he could not observe "effects consonant with the idea that the nerves contained a nervous fluid." See Evan R. Ragland, "Experimenting with Chymical Bodies: Reinier de Graaf's Investigations of the Pancreas," *Early Science and Medicine* 13, no. 6 (2008): 615–664, esp. 635.

120. Steno, *Observationes anatomicæ*, 36–37, 97–98.

121. Steno, *De musculis et glandulis*, 19.

122. Steno, *De musculis et glandulis*, 27.

123. Scherz argues that the actual dissection of the brain took place around February 1665; see Scherz, "Introduction," 69–70. Chapelain's letter to Huet mentions a brain dissection before April 1665.

124. Steno, *Discours sur l'anatomie du cerveau*, 6–7.

Swammerdam and Steno discussed these matters extensively in Paris. After their time in France, Steno went on to Italy and Swammerdam returned to the Netherlands; each quoted the other when writing on muscles.[125] Steno and Swammerdam stressed the intellectual link with Thévenot in the books that they published immediately after leaving Paris in 1667. Steno included the responses to objections against his research on muscles in the form of a letter to Thévenot at the end of his book,[126] and Swammerdam dedicated his *Tractatus* on respiration to Thévenot.[127] By hosting Steno and Swammerdam, Thévenot provided a platform where they could develop their ideas at will, improve their collaboration, and strengthen their friendship.

But why did Swammerdam not immediately follow Steno's lead in dismissing animal spirits? This is a relevant question considering that Swammerdam also acknowledged the epistemic value of observations and had observed the contraction of muscles without the influx of any fluid or spirit after rubbing them. One reason could be Swammerdam's attachment to Cartesian ideas. In the *Tractatus de respiratione*, he explained respiration in terms of the cooling of the heart, which was also Descartes's theory.[128] Indeed, Swammerdam seems to have had a much more positive attitude toward Cartesianism than Steno, although further research is needed to confirm this.[129] However, several years after the two friends parted ways, Swammerdam developed a series of experiments on the muscles of frogs that finally convinced him that animal spirits were not needed for the motion of muscles.[130] Swammerdam set forth his conclusions in an essay published posthumously in his *Bybel der Natuure/Biblia naturæ* (Leiden, 1738). Interestingly, he acknowledged Steno's priority on the matter, but not without some caveats. He wrote that "although the most outstanding anatomist Steno uncovered very many curious things in this regard, in the middle of the way, however, he also hesitated [*medio tamen in curriculo is quoque haesitavit*]."[131]

---

125. Swammerdam, *Tractatus physico-anatomico-medicus de respiratione*, 60–62; and Steno, *Elementorum myologiæ specimen*, 59.

126. Steno, *Elementorum myologiæ specimen*, 48.

127. Swammerdam, *Tractatus physico-anatomico-medicus de respiratione*, i.

128. Swammerdam, *Tractatus physico-anatomico-medicus de respiratione*, 8, 10–11; and Descartes, *A Discourse on the Method*, 44.

129. R. P. W. Visser, "Theorie en Praktijk van Swammerdams Wetenschappelijke Methode in Zijn Entomologie," *Tijdschrift voor de Geschiedenis der Geneeskunde, Natuurwetenschappen, Wiskunde en Techniek* 4 (1981): 63–72, esp. 66. See also Sleigh, "Jan Swammerdam's Frogs."

130. See Andrault, "The Chain of Motions"; and Cobb, "Exorcizing the Animal Spirits."

131. Swammerdam, *Bybel der Natuure/Biblia naturæ*, 2:836.

Despite their different conclusions at the time, Steno and Swammerdam remained close friends. In 1672 Swammerdam fondly remembered Steno in print, who had been to him "as a friend [*amice*]," and then, after Paris, "where they lived in the same house [of Thévenot]," Steno was "like family [*familiariter*]."[132] In fact, Herman Boerhaave (1668–1738), who edited and published Swammerdam's most famous book, the *Bybel der Natuure/Biblia naturæ*, stressed that Swammerdam "developed a friendship with Nicolaus Steno, which he maintained from then on until he died."[133] Later both Swammerdam and Steno would develop an intense interest in religion, to the point of altering their personal and professional lives because of it, albeit in different ways. But in Paris their friendship was centered on their common interest in muscles and respiration and in the search for new research methods. This kind of scientific collaboration, which frequently turned to physico-mathematical concepts or experiments, was exactly what French scholars such as Sorbière and Chapelain had been looking for.

The Benefits of Being Anti-Cartesian

Almost three years after the foundation of the Académie des Sciences, Chapelain wrote about its origins to his friend François Bernier (1625–1688). The story matches the events already narrated in this chapter. Chapelain wrote that "Mr. de Monmor established a meeting on natural philosophy at his place [which] attracted many learned men. It lasted for four or five years with great reputation."[134] Despite its success, however, Montmor's group "dispersed, and the doctrine of Mr. Descartes, which they were trying to establish there, weakened more than half of it [*en fust affoiblie de plus de moitié*]." Thévenot then gathered "in his home the debris of the previous assembly [*le débris de cette assemblée*] and, for more than a year, admirable dissections were made there by a Dane named Steno."[135] The story then continues with a reference to the academy of Bourdelot and the royal foundation of the Académie with the help of Colbert. It is remarkable, however, that among all the scholars that attended Thévenot's meetings, Chapelain mentions only Steno. It could have been because

---

132. Swammerdam, *Miraculum naturæ*, 41.
133. Boerhaave, "Vita D. Joannis Swammerdammii," in Swammerdam, *Bybel der Natuure/Biblia naturæ*, 1:fol. 2r.
134. Chapelain to François Bernier, 16 February 1669, in *Lettres de Jean Chapelain*, 2:622.
135. Chapelain to Bernier, 16 February 1669, in *Lettres de Jean Chapelain*, 2:622.

Bernier was a physician and so was especially interested in anatomy, but there were other physicians in these meetings, not to mention more famous mathematicians and natural philosophers. One thing is certain: in Chapelain's eyes at least, Steno was a central character in the prehistory of the Académie.

For Chapelain, Steno's crowning achievement was his explicit rejection of Cartesian anatomy. Early in 1665 Steno dissected a human brain before a few audiences, possibly because of his interest in the origins of animal motion. He began the dissection with a shocking statement: *"Messieurs, rather than promising to please your curiosity about the anatomy of the brain, I make here a sincere and public confession: that I know nothing about it [je vous fais icy une confession sincere et publique, que je n'y connois rien]."*[136] The main goal of the whole essay and dissection was to show that "the substance of the brain is poorly known," as well as "the true manner of dissecting it."[137] This objective was in line with Steno's research program, driven by a search for certainty in anatomy. Steno commented on several opinions about the brain, including those from Thomas Willis's recent *Cerebri anatome* (London, 1664). But one of his main intentions was to show the "great difference between the machine which Mr. Descartes has imagined [*la machine que Monsieur des Cartes s'est imaginée*] and that which we see when we make the anatomy of human bodies."[138] Reminiscent of the days when he refuted Blasius's claims on the salivary duct, Steno ended his discourse by comparing seven passages from Descartes's *L'homme* with what he saw in dissections.[139] In the end, Steno destroyed Descartes's brain anatomy. Nonetheless, this level of devastation was possible because Steno had read Descartes's work very carefully, as Raphaële Andrault has convincingly argued.[140] The fact that Steno took Descartes seriously speaks to the intellectual prestige Descartes had in some of Steno's circles, and especially in Swammerdam's mind.

Chapelain was thrilled with Steno's uncovering of Descartes's mistakes. In April 1665 he wrote that "Mr. Steno, a Dane, has done in this [anatomical] art the most beautiful proofs that one has seen until now, such that he has forced the Cartesians, those ever so obstinate dogmatists, to fall out of agreement with the error of their patriarch in regard to the gland of the brain and its use, in the presence of the most honest men in the

---

136. Steno, *Discours sur l'anatomie du cerveau*, 1–2.
137. Steno, *Discours sur l'anatomie du cerveau*, 7.
138. Steno, *Discours sur l'anatomie du cerveau*, 15.
139. Steno, *Discours sur l'anatomie du cerveau*, 58–60.
140. Andrault, "Human Brain and Human Mind," 89–97.

city, in which however he placed all the operations of the rational soul."[141] The recipient of this letter was Huet, who was also no endorser of Cartesianism, to say the least.[142] After Louis XIV issued a series of prohibitions against the public teaching of Cartesian philosophy, starting in 1671, Huet published two books explicitly against Cartesianism.[143] As Sophie Roux has explained, in the early 1660s a few leading scholars in Paris, such as Chapelain and Sorbière, grew disillusioned with Cartesianism and began to reject it "as a sect."[144] But perhaps the loudest anti-Cartesian voice mentioned by Borch in his journal was that of Gilles Personne de Roberval. Since his arrival in Paris in the fall of 1663, Borch had attended Roberval's classes of mechanics a number of times, at either the Collège Royal, the meetings of Pierre Bourdelot, or the Collège de Clermont.[145] In these lectures Roberval inveighed strongly against Cartesianism, often calling Descartes the "king of chimeras" [*Rex Chimaerarum*].[146] Roberval had been attending the intellectual salons of Paris since the early days of Mersenne's meetings. He was known for important discoveries in geometry, trigonometry, and mechanics, and Steno probably admired him because of his collaborations with Pecquet, as detailed in the latter's *Experimenta nova anatomica*.[147] There is no direct evidence that Steno and Roberval met, although that would not be improbable. Regardless, Roberval became one of the founding members of the Académie des Sciences, which speaks to his cultural influence in the Parisian circles that Steno frequented.

The history of Cartesianism in seventeenth-century Paris, however, is much more complex. There were also influential disciples of Descartes in Paris at that time, such as Claude Clerselier (1614–1684), who edited several of Descartes's works, Étienne de Villebressieu (c. 1607–1674), a

---

141. Chapelain to Huet, 6 April 1665, in *Lettres de Jean Chapelain*, 2:393.

142. Thomas M. Lennon, "Pierre-Daniel Huet, Skeptic Critic of Cartesianism and Defender of Religion," in Nadler, Schmaltz, and Antoine-Mahut, *The Oxford Handbook of Descartes and Cartesianism*, 780–790, esp. 780.

143. Pierre Daniel Huet, *Censura philosophiae Cartesianae* (Paris: Daniel Horthemels, 1689); and Huet, *Nouveaux mémoires pour servir à l'histoire du Cartesianisme* (Paris: Henry Desbordes, 1692).

144. Sophie Roux, "The Condemnations of Cartesian Natural Philosophy Under Louis XIV (1661–91)," in Nadler, Schmaltz, and Antoine-Mahut, *The Oxford Handbook of Descartes and Cartesianism*, 755–779, esp. 772–776.

145. OBI, 3:134, 7 December 1663, and 3:215, 28 January 1664.

146. OBI, 3:290, 10 March 1664. For other Roberval criticisms of Descartes, see OBI, 3:377, 28 April 1664, and 4:32, 14 July 1664. Sylvius also spoke of chimeras when criticizing Descartes; see Ragland, "Mechanism, the Senses, and Reason," 198.

147. On Roberval, see Kokithi Hara, "Roberval, Gilles Personne de," in *Complete Dictionary of Scientific Biography*, vol. 11 (Charles Scribner's Sons, 2008), 486–491.

friend and collaborator of Descartes, and Jacques Rohault (1618–1672), who held many doctrines of Cartesian natural philosophy, such as the nonexistence of the vacuum.[148] In July 1665, when Steno and Thévenot were away from Paris, Graindorge also reported that "nothing is done at Bourdelot's, where a Cartesian named Phillipeaux discoursed according to the principles of Descartes."[149] Graindorge's disdain for Cartesianism either undergirded or was fostered by his admiration for Steno. Borch had also mentioned the "Cartesian" Philippeaux's leading a dissection of the brain in one of the meetings of the surgeon Morel, as already shown. Finally, Christiaan Huygens, a Cartesian natural philosopher, despite his criticisms of Descartes, was chosen as another founding member of the Académie.[150]

What united these Cartesian scholars with the so-called anti-Cartesians of Paris was an interest in experimentation. As shown in previous chapters, "Cartesianism" was often a label associated with those who dismissed observations in the sciences in the 1650s and 1660s. Steno's criticism of Descartes pointed in that direction, such as when he said, also in the discourse on the brain, that "the description of Monsieur Descartes is not in agreement with the dissection of the bodies."[151] Yet, Cartesian scholars such as Borel, Rohault, and Huygens also favored a strong experimental program of research.[152] In many ways they agreed with Sorbière's plea for the king to support an experimental program. Not surprisingly, also in this case, Steno checked the box. In his lecture on the anatomy of the brain,

---

148. Clerselier edited Descartes's *L'homme* (Paris: Charles Angot, 1664). For Clerselier in Thévenot's meetings, see OBI, 4: 297, 13 April 1665. On Villebressieu, see Thomas M. Lennon, "Villebressieu, Étienne de," in *The Cambridge Descartes Lexicon*, ed. Lawrence Nolan (Cambridge: Cambridge University Press, 2016), 747; and OBI, 4:275, 285. On Rohault as Cartesian, see Mihnea Dobre, "Jacques Rohault and Cartesian Experimentalism," in Nadler, Schmaltz, and Antoine-Mahut, *The Oxford Handbook of Descartes and Cartesianism*, 388–401. On Rohault knowing Steno, see Jacques Rohault, *Traité de physique, II. III. & IV. Partie* (Amsterdam, 1672), 414.

149. Graindorge to Huet, 1 July 1665, in Tolmer, "Vingt-deux lettres," 288–289.

150. On Huygens as Cartesian, see Struik, *The Land of Stevin and Huygens*, 96–98.

151. Steno, *Discours sur l'anatomie du cerveau*, 14.

152. On Borel's sympathies toward Cartesianism, see Jean-François Maillard, "Descartes et l'alchimie: Une tentation conjurée?," in *Aspects de la tradition alchimique au XVIIe siècle: Actes du colloque international de l'Université Reims-Champagne-Ardenne, Reims, 28 et 29 novembre 1996*, ed. Frank Greiner (Paris: Arché, 1998), 95–109, esp. 105. Borel's interest in observations is confirmed by his book of *observationes*, the *Historiarum centuriæ IV*. On Rohault's experimental program, see Dobre, "Jacques Rohault and Cartesian Experimentalism." For Huygens, see Struik, *The Land of Stevin and Huygens*, 92–96.

Steno reiterated the need for royal patronage of experiments in anatomy, saying that "in every century, chymistry has had individuals and princes who built laboratories for it, but few people have applied themselves with similar fervor for anatomy. It is not as if it has depended on princes."[153] In the months after Steno delivered this discourse, what followed was a battle of princes, or rather between a king and a grand duke, for Steno.

In September 1665, at the time of his departure, Steno left French scholars longing for more. On 16 September Graindorge wrote that Steno "left this morning for Italy" but that "he promised me that he would pass by Caen."[154] In a letter from December to Steno, who was already in Montpellier, Chapelain suggested that Steno postpone the scheduled trip to Italy and return to Paris in the fall of 1666—precisely the time the Académie would officially begin:[155] "To speak frankly of the interest that we would have in that," Chapelain said, "you would delight us with your presence."[156] Colbert's formation of the Académie Royale des Sciences included members from outside of France, most notably the two highest-paid members, Christiaan Huygens (who was a Protestant) and, in 1669, Gian Domenico Cassini (1625–1712).[157] Were there plans to invite another non-French, non-Catholic scholar to join the Académie?

Chapelain's letters suggest that Colbert and Louis XIV were indeed ready to make Steno an offer to become a founding member. In a hint at his real purpose, in March 1666 Chapelain told Steno that the invitation to have him back in Paris "only looked at our advantage, [and] was more [self-]interested than discreet."[158] But, as the next chapter shows, by April Steno was already courting the Medici grand duke in Florence. Indeed, Steno had turned down Chapelain's invitation to return to Paris before arriving in Florence, saying that he was already "too far along the coast of Italy."[159] When news of Steno's arrival in Florence reached Paris in the summer, it complicated any possible plans to hire Steno for the Académie

---

153. Steno, *Discours sur l'anatomie du cerveau*, 33.
154. Graindorge to Huet, 16 September 1665, in Tolmer, "Vingt-deux lettres," 321.
155. Chapelain to Steno, 8 December 1665, in Larroque, *Lettres de Jean Chapelain*, 2:424.
156. Chapelain to Steno, 8 December 1665, in Larroque, *Lettres de Jean Chapelain*, 2:424.
157. Alice Stroup, *A Company of Scientists: Botany, Patronage, and Community at the Seventeenth-Century Parisian Royal Academy of Sciences* (Berkeley and Los Angeles: University of California Press, 1990), 35–36. Cassini was hired only in 1669; see ibid., *A Company of Scientists*, 8, 26.
158. Chapelain to Steno, 15 March 1666, in Larroque, *Lettres de Jean Chapelain*, 2:447.
159. Chapelain to Steno, 15 March 1666, in Larroque, *Lettres de Jean Chapelain*, 2:447.

des Sciences.¹⁶⁰ Prince Leopoldo wrote to Thévenot mentioning Steno's arrival in Florence and asking him to publish the manuscript of Steno's dissection of the brain. But Thévenot replied that he was reticent to use "the freedom that Mr. Steno gave me to handle his most exact discourse on the anatomy of the brain." Instead, he wanted to show it to Steno one more time before publishing it, since he "may have acquired new lights [*nuovi lumi*] in the Accademia del Cimento."¹⁶¹ Thévenot wanted to bring Steno back to Paris. In December 1666 or early 1667 Thévenot admitted that he had been "charged with a special invitation to Mr. Steno."¹⁶² But it was already too late: Steno was not returning to Paris. Chapelain finally confessed that "I would not have despaired of succeeding if you had been in this court when the resolution was taken to form an assembly of real physicians and excellent anatomists."¹⁶³ Indeed, "we preferred the present to the absentees, and Monsieur Pecquet . . . was admitted in your absence."¹⁶⁴

If we take Chapelain at his word, we see that Steno's prestige surpassed that of Pecquet, who had been the leading anatomist of France and a major influence to Steno himself. Pecquet's social position, however, had suffered in previous years: he had been imprisoned for political reasons and had also become an alcoholic.¹⁶⁵ Pecquet's earlier political choices put him at odds with Huygens's suggestion that at the Académie's meetings "there will never be a discussion of . . . religion or the affairs of state."¹⁶⁶ And his alcohol problem shows that manners, at which Steno excelled, mattered not just for formalities, but also to keep vices away. On top of this, Paris seemed devoid of French anatomists at the level of Steno and Pecquet in the 1660s.¹⁶⁷ Rather than diminishing Steno's merits, these circumstances emphasize how much Chapelain and Colbert needed Steno in Paris. Chapelain could have recruited Borch, Swammerdam, or even de

---

160. Thévenot to Prince Leopoldo, 2 August 1666, in BNCF, Gal. 315, fols. 1038r–1039r.
161. Thévenot to Prince Leopoldo, 16 October 1666, in BNCF, Gal. 277, fol. 348r.
162. Thévenot to Prince Leopoldo, December 1666(?), in BNCF, Gal. 280, fol. 117r. The letter is undated, but Thévenot mentions that "s'è dato finalmente principio alle sessioni della nuova Academia."
163. Chapelain to Steno, 31 March 1667, in Larroque, *Lettres de Jean Chapelain*, 2:507.
164. Chapelain to Steno, 31 March 1667, in Larroque,, *Lettres de Jean Chapelain*, 2:507.
165. On naming Pecquet to the Académie, see Guerrini, *The Courtiers' Anatomists*, 90–91. Pecquet had been working for Nicolas Fouquet (1615–1680), one of Colbert's political rivals, and was in prison with him from 1661 to 1665; see Guerrini, *The Courtiers' Anatomists*, 83–84.
166. Hahn, *The Anatomy of an Institution*, 10–12, esp. 12.
167. Guerrini, *The Courtiers' Anatomists*, 84.

Graaf. De Graaf had just arrived in Paris and, unlike Steno, was Catholic, making him more at home in the city's religious environment.[168] But as far as we know, none but Steno was considered.

※

The desire to appoint Steno to a permanent paid position in the new Académie Royale des Sciences shows the extent to which Steno had gained the hearts and minds of the French elite and, more specifically, of Jean Chapelain.[169] More important, it shows the perfect match between Steno and the cultural and scientific environment of Paris at the time of the founding of the Académie des Sciences. Not only was Steno one of the stars of anatomy in those years, but his cross-disciplinary interests and methods meant that he held many of the same ideas as the Académie's founding members. He knew how to apply physico-mathematical knowledge to anatomy and was very open to collaborative research, as his friendship with Swammerdam demonstrates. Above all, he was a vocal critic of Cartesianism. Furthermore, Steno's public dissections displayed the value of a more hands-on study of nature, one based on observations. He even spoke of the need for princely support of empirical research in anatomy. He did all this with gentle manners and with a good command of the French language; evidence of this is that he corresponded in French and his *Discours sur l'anatomie du cerveau* was published in French.[170] Although the book was not published until 1669, Steno had left the manuscript with Thévenot before leaving Paris. Thévenot immediately wrote of sending it to Christiaan Huygens in the Netherlands, a hint that the book was close to being ready and may have circulated as a manuscript before being published.[171]

---

168. On de Graaf's arrival in Paris, see Chapelain to Steno, 15 March 1666, in Larroque, *Lettres de Jean Chapelain*, 2:447.

169. See also the dedicatory letter from the printer Robert de Ninville to the Royal Physician Mons. de la Chambre in Steno, *Discours sur l'anatomie du cerveau*.

170. Thévenot edited the text, but the original manuscript was likely ready when Steno left, because in September 1665 Thévenot sent a copy to Huygens. Steno's first known letter in French is from less than a year later; see Steno to William Croone, 23 May 1666, in *Epistolae*, 1:185. Although it is an English translation, the original was likely in French because another letter to Croone is in French; see Steno to Croone, November (?) 1666, in London, Royal Society of London, EL/S1/92.

171. Thévenot to Christiaan Huygens, 18 September 1665, in Huygens, *Oeuvres complètes*, 5:488.

Steno's success in Louis XIV's Paris proves that he had mastered the dynamics of the new scientific institutions of seventeenth-century Europe. His manners, dissection skills, and scientific insights attracted much prestige not only in Protestant Copenhagen and Leiden, but also in Catholic Paris. In many ways, he was ready to do the same anywhere. And that was precisely what happened when he traveled south to Montpellier. Montpellier had one of the oldest medical schools of Europe and was an attractive place for Protestants in seventeenth-century France. Pierre Borel, a Protestant himself, had studied in Montpellier before going to Paris.[172] For this reason, Montpellier attracted many English scholars. Martin Lister (1639–1712), who was to become an important physician in England, wrote in his journal that he had "the honour, at a lecture of anatomy and at some particular dissections, to assist Mr. Steno the Dane" in the fall of 1665.[173] Philipp Skippon (1641–1691), who also witnessed Steno's dissections, noted that Steno "is very happy in some anatomical discoveries."[174] In Montpellier Steno also met several founding members of the Royal Society of London, such as John Ray (1627–1705), William Croone (1633–1684), and Robert Bruce (c. 1626–1685), the earl of Aylesbury, at whose lodgings Steno dissected.[175] Croone had also been working on muscles and animal motion, and possibly because of this joint interest, Steno continued to correspond with him. In subsequent years Martin Lister and John Ray would publish books on the origin of fossils, but there is no evidence that they spoke about this matter with Steno. As in Paris and Leiden, Steno's anatomical skills and his "friendly and agreeable" conversation shaped the positive image that English intellectuals made of him.[176] His next stop in Florence, however, would be the one that would shape Steno's career in more ways than he could have foreseen.

---

172. Marie-Rose Carré, "A Man Between Two Worlds: Pierre Borel and His *Discours nouveau prouvant la pluralité des mondes* of 1657," *Isis* 65, no. 3 (1974): 322–335.

173. Oxford, Bodleian Library, MS Lister 5, fols. 224v–226v. See also Anne Marie Roos, *Web of Nature: Martin Lister, the First Arachnologist*, History of Science and Medicine Library 22/Medieval and Early Modern Science 16 (Leiden: Brill, 2011), 67–68.

174. Skippon's journal as quoted in F. N. L. Poynter, "Nicolaus Steno and the Royal Society of London," in Scherz, *Steno and Brain Research*, 273–280, here 274.

175. Roos, *Web of Nature*, 67.

176. Oxford, Bodleian Library, MS Lister 5, fols. 224v–226v.

CHAPTER FIVE

Anatomy and Mathematics at the Medici Court

Italy brought Steno lasting fame in the history of science. His research on fossils made him known as the "founder of modern geology," and his conversion to Catholicism turned him into a leading actor in debates on science and religion.[1] These achievements in geology and religion, however, seem radically detached from his early career as an anatomist. How did these changes come about? The following three chapters answer this question by looking closely at Steno's arrival in Italy. Their main argument is that it is only by understanding Steno in his own historical context, as an anatomist with interests in various disciplines, that his rising interests in geology and religion can be fully captured. In these days, Steno remained committed to searching for certainty in anatomy through observations and ideas from physico-mathematics. This search culminated with the publication of *Elementorum myologiæ specimen* (Florence, 1667), in which Steno proposed a geometrical model for muscle motion. This book, published more than ten years before the more famous *De motu animalium* (Rome, 1680–1681), by Giovanni Borelli, was a pioneering work in the mathematization of anatomy. In addition to contributing to studies of the body, Steno's book also included his first ideas about fossils, launching a series of important contributions to Earth studies. In this chapter I explain how this apparently disjointed book came about by looking at Steno's social and intellectual life in his first Italian months.[2] I show that his scientific claims were key to his successful navigation of the social networks and court culture of Florence. Above all, this and the following chapters show

---

1. Gary D. Rosenberg, "Preface," in Rosenberg, *The Revolution in Geology from the Renaissance to the Enlightenment*, vii; and Kermit, *Niels Stensen*.
2. On Steno's system of patronage in Italy, see Jakob Bek-Thomsen, "Nicolaus Steno and the Making of an Early Modern Career: Nature, Knowledge and Networks at the Court of the Medici c. 1650–1675" (Ph.D. diss., Aarhus University, 2012).

that Steno's activities in Florence were enabled only by the friendships that he made there. His friends, scientists or not, helped him not only to dissect, think, and write, but also to eat, laugh, and even pray with persons of other confessions and cultures.

Steno arrived in Italy in the final years of the Accademia del Cimento, the academy founded in 1657 by Prince Leopoldo de' Medici (1617–1675) and his brother, Grand Duke Ferdinand II of Tuscany—as mentioned earlier, Galileo's last patron.[3] The Accademia's goal was to foster the study of natural philosophy primarily through experiments, new instruments, and scientific collaboration. Its only publication, the *Saggi di naturali esperienze* (Florence, 1667), stressed this focus on experiments by famously refusing any interpretation of experimental observations, instead only describing experiments and their results. This detachment from philosophical discussion has broadly been interpreted by most historians in one of two ways. The first is that it was a way to avoid theological conflicts with the Catholic Church in the aftermath of its condemnation of Galileo.[4] The second is that it was a way for Prince Leopoldo to maintain "the politeness of philosophical etiquette," in which princely patrons did not, as a rule, commit to their clients' ideas.[5] However, these reasons are not what the sources, including the members' correspondence and the book's preface, indicate. In their own words, the *Saggi* authors wanted to avoid the "slight hint of speculation" because "man, improperly fitting causes to effects . . . forms a false science within his own mind."[6] According to Marco Beretta, the academicians saw

---

3. On the academy's foundation, see W. E. Knowles Middleton, *The Experimenters: A Study of the Accademia del Cimento* (Baltimore, MD: Johns Hopkins University Press, 1971), 17–26. This is still the standard study on the Accademia del Cimento. See also Paolo Galluzzi, "L'Accademia del Cimento: 'Gusti' del principe, filosofia e ideologia dell'esperimento," *Quaderni storici* 16, no. 48(3) (1981): 788–844; Marco Beretta, "At the Source of Western Science: The Organization of Experimentalism at the Accademia del Cimento (1657–1667)," *Notes and Records: The Royal Society Journal of the History of Science* 54, no. 2 (2000): 131–151; Luciano Boschiero, *Experiment and Natural Philosophy in Seventeenth-Century Tuscany: The History of the Accademia del Cimento* (Dordrecht: Springer, 2007); and Marco Beretta, Antonio Clericuzio, and Lawrence Principe, eds., *The Accademia del Cimento and Its European Context* (Sagamore Beach, MA: Science History Publications, 2009). For a useful historiography, see Paula Findlen, "Academies, Networks, and Projects: The Accademia del Cimento and Its Legacy," *Galilaena: Journal of Galielean Studies* 7, no. 7 (2010): 277–298.

4. Galluzzi, "L'Accademia del Cimento," 800–801, 823–832; and Boschiero, *Experiment and Natural Philosophy*, 192, 229–231.

5. Biagioli, *Galileo, Courtier*, 359.

6. [Lorenzo Magalotti,] *Saggi di naturali esperienze fatte nell'Accademia del Cimento* (Florence, 1667), "Preface to the Readers," trans. in Middleton, *The Experimenters*, 92, 89.

that "nature in its infinitude manifested an unattainable complexity," so any commitment to a specific interpretation was futile and was therefore left out of the Accademia's published research program.[7] Moreover, the experimental discovery of phenomena that were hard to explain within the traditional Aristotelian worldview, such as the vacuum, highlighted how much of the world and its phenomena were unknown and how existing systems of natural philosophy were inadequate to explain its complexity. In other words, the Accademia's focus on observations was a response to the same epistemic problems that Steno encountered in his own studies.

Steno's swift integration into Florentine intellectual circles reflects the resonance between his approach to the study of nature and the interests of the academicians. As seen in previous chapters, his approach included the social skills and intellectual methods he developed in Leiden and Paris, such as courteous manners, an interest in mathematics and mechanics, an explicit rejection of Cartesianism, a constant recourse to observations, and an openness to scientific collaboration. The role of manners in early modern Italian science has already been studied extensively.[8] Steno had such good etiquette, in fact, that Lorenzo Magalotti (1637–1712), the Accademia del Cimento's secretary and author of the *Saggi*, commented that his manners followed "the style of France."[9] Good manners, however, were not enough. In the scientific circles of the Medici princes, an interest and proficiency in mathematics mattered a great deal. Steno's intentional recourse to mathematics as a way to obtain the patronage of the Medici family is a good illustration of the interplay between intellectual ideas and social factors in the history of science. As for Cartesianism, there is still much to be written on its spread in seventeenth-century Italy. But recent research suggests that Italian scholars at the time were not very interested in it, with some indeed rejecting it altogether.[10] In Italy, Steno digested his rejection of Cartesianism into a rejection of a dogmatic and nonexperimental approach to natural philosophy. He did so by dissecting often at court and highlighting unresolved problems in anatomy. The first three sections of this

---

7. Beretta, "At the Source of Western Science," 140–141; and Domenico Bertoloni Meli, "Shadows and Deception: From Borelli's *Theoricæ* to the *Saggi* of the Cimento," *The British Journal for the History of Science* 31, no. 4 (1998): 383–402, esp. 399.

8. Galluzzi, "L'Accademia del Cimento;" Findlen, "Controlling the Experiment"; and Biagioli, "Etiquette, Interdependence, and Sociability."

9. Lorenzo Magalotti to Francesco Redi, 25 February 1667, in BML, Redi 206, fol. 28r: "A Mr. Stenone vi prego a baciar le mani affettuosamente. Anzi la fronte ancora all'usanza di Francia, essendo egli veramente una sposa."

10. Bertoloni Meli, *Mechanism, Experiment, Disease*, 84–88; and Conforti, "'Se fusse meno cartesiano lo stimarei molto.'"

chapter show exactly how Steno put these factors into play in the few months in which he moved to Florence, Rome, and back again to Florence.

Once he had secured Medici patronage, Steno's interest in mathematics was given free rein, where he published his magnum opus on the geometry of muscles. This chapter's fourth section shows that the *Elementorum myologiæ specimen* falls squarely into the genre of *observationes anatomicæ* that Steno had mastered during his early days in Copenhagen. My argument is based on a novel reading of the book's original manuscript, still preserved at the Royal Library of Copenhagen. Finally, I end with Steno's relationship with the Accademia del Cimento and Vincenzo Viviani (1622–1703), the grand duke's mathematician, who referred to himself as Galileo's "last disciple."[11]

Travels Across the Networks of Florence

Why did Nicolaus Steno travel to Italy in the first place? From the events that followed, several reasons can be deduced: to learn mathematics in the group of Galileo's disciples, to enjoy the scientific patronage of the Medici family, to visit geological formations in the Italian peninsula, and even to learn more about the Catholic Church. But all these reasons are the fruit of judgments made from the perspective of Steno's later life and thus run the risk of being anachronistic. To put it simply, Italy was only meant to be the last stretch of Steno's intellectual travels through Europe.

As this book stresses, the *peregrinatio medica* that was an important component of the education of young Danish scholars is the main reason Steno traveled across Europe, including Italy. Just like Steno, most Danes spent several months in Italy in their travels, including Borch and Thomas Bartholin.[12] Borch had left Paris for Italy several months before Steno. As he says in an autobiographical text, he passed through Milan, Bologna, Rome, Padua, and also Florence, the most "flourishing and elegant" city in Italy.[13] A French nobleman recommended that Borch become acquainted with Francesco Redi (1626–1697), the chief physician of Grand Duke Ferdinand and a well-established presence of the Republic of Letters.[14] Steno and Borch

---

11. Florence, BNCF, Gal. 243, fol. 119r, translated in Boschiero, *Experiment and Natural Philosophy*, 50.
12. Bartholin, *De peregrinatione medica*, 11.
13. Borch's autobiography as translated in Henrik Schepelern, "Introduction," in OBI, 1:vii–xliii, at xviii.
14. L'hospital de St. Mesme [Alexandre de L'Hôpital] to Francesco Redi, 20 June 1665, in BML, Redi 203, fol. 129r. On Redi, see Walter Bernardi, "Uno scienziato Aretino

were fully aware of Redi's recent work on the venom of vipers because it had been discussed in the meetings of Bourdelot and Thévenot in Paris.[15] Redi's conclusion that viper venom could be drunk and was not lethal unless it came into contact with the blood in its circulatory system made him known as an apologist for the new anatomy. A recently discovered letter from Borch to Redi praised him for fighting "with Herculean audacity ... against the ancients and modern writers alike, even against Galen himself."[16] Borch stayed in Florence during the fall of 1665 but forced himself to leave because, as he later wrote, "his [university] chair awaited him at home."[17] This was the same reason Thomas Bartholin had departed from Italy decades earlier.[18]

Unlike Bartholin and Borch, Steno did not have a deadline to return home, since he did not yet have a university appointment. Steno's lack of a professorship was probably due to the fact that the University of Copenhagen prioritized the relatives of current faculty, such as members of the Bartholin family.[19] Regardless, when Steno left Paris he was clearly thinking of returning to Copenhagen, or so he said to friends there. According to the *Journal des sçavans*, Steno was planning "to treat [some anatomical] matters more extensively, when his return to Denmark will give him more leisure than he is able to meet during his travels."[20] Steno's decision to stay in Florence, therefore, is more complex than his biographies tend to portray and sheds light on the centrality of itinerancy in Steno's career.

As it turns out, the Florentines already knew a great deal about Steno and his anatomical publications before he arrived. Steno had barely left Paris when Prince Leopoldo asked Alessandro Segni (1633–1697), a traveling diplomat of the Medici court, about Steno's passage in Lyon.[21] Most

---

protagonista della nascita della modernità," in *Francesco Redi Aretino: Atti del convegno*, ed. Lorella Mangani and Giuseppi Martini (Arezzo: Accademia Petrarca di lettere arti e scienze, 1990), 17–36.

15. OBI, 4:173, 18 November 1664; and Graindorge to Huet, 19 August 1665, in Tolmer, "Vingt-deux lettres," 308.

16. Ole Borch to Francesco Redi, 26 December 1665, in BML, Redi 203, fol. 231r–v: "Herculea audacia opus aggressus es, contra tot Antiquos, tot recentes scriptores imo ipsum quoque Galenum oculos attolendo, demonstrandoque non eundum."

17. Borch's autobiography in OBI, 1:xviii.

18. Skavlem, "The Scientific Life of Thomas Bartholin," 68.

19. Grell, "Caspar Bartholin and the Education of the Pious Physician," 91.

20. Anon., "Review of Steno's *De musculis et glandulis*," 139.

21. See Alessandro Segni to Leopoldo de' Medici, 11 December 1665, in Alfonso Mirto, *Alessandro Segni e gli Accademici della Crusca: Carteggio* (Florence: Accademia della Crusca, 2016), 104–106. On Steno in Lyon, see Graindorge to Huet, 16 September 1665, in Tolmer, "Vingt-deux lettres," 319–322. Steno had already left Lyon by the time Segni arrived.

likely Borch informed the Medici court of Steno's travels, since he often used Steno's discoveries to attract audiences. He did so in Paris, before Steno's arrival in France, and in Turin, where "he had a friendly dispute about the true structure of the heart and the recently discovered salivary ducts."[22] There is also evidence that Steno corresponded with Borch when they were not together, because Borch once wrote notes on a dissection carried out by Steno in Copenhagen that Borch could not have known otherwise.[23]

Steno's anatomical work was known in the Medici court even before Borch's arrival. In February 1663, a year after Steno published *Observationes anatomicæ*, the anatomist Lorenzo Bellini (1643–1704) tried to demonstrate the "salivary ducts of Steno" to the grand duke.[24] Borelli, who worked as a mathematician for the Medici and was interested in anatomy, witnessed these dissections and mentioned them to his friend Malpighi, one of Italy's leading anatomists, who would soon meet and befriend Steno in Rome.[25] Both Borelli and Malpighi tried to follow Steno's publication record closely, though they were not always successful. In the summer of 1665, while traveling through southern Italy, Malpighi wrote to a friend in Bologna asking about a supposedly new book published by Steno.[26] As for Borelli, a 1663 list of book requests that he sent to Antonio Magliabechi (1633–1714), chief librarian for the Medici, included Steno's books among those that he wanted to read.[27] Interestingly, Borelli also read Anton Deusing's (1612–1666) attacks on Steno.[28] For this reason, he thought that Steno had plagiarized the discovery of the salivary duct from Blasius, which greatly affected his attitude towards Steno in their first meetings in Florence in 1666.

---

22. See OBI, 3:340, 4 April 1664; and Borch's autobiography in OBI, 1:xvii.

23. OBI, 3:392, 7 May 1664. Steno's dissection on the ray was only published after June 1664; see Steno, *De musculis et glandulis*, 84.

24. Giovanni Borelli to Marcello Malpighi, 15 February 1663, in Marcello Malpighi, *The Correspondence of Marcello Malpighi*, ed. Howard Adelmann, 5 vols. (London: Cornell University Press, 1975), 1:147–149.

25. See Borelli to Malpighi, 15 February 1663, and Borelli to Malpighi, 22 March 1663, in Malpighi, *The Correspondence of Marcello Malpighi*, 1:147–149, 152–154.

26. Malpighi to Silvestro Buonfigliuoli, 5 August 1665, in Malpighi, *The Correspondence of Marcello Malpighi*, 1:274–275.

27. Borelli to Antonio Magliabechi, 22 November 1663, in BNCF, Magl.VIII.518, fol. 128r: "che V.S. . . . mandi con la prima comoditá . . . tutte le osservazioni anatomiche dello Steno." On Magliabechi, see Massimiliano Albanese, "Magliabechi, Antonio," in *Dizionario biografico degli italiani*, 100 vols. (Rome: Treccani, 1960–2020) (hereafter cited as DBI), vol. 67, 422–427.

28. Borelli to Malpighi, 10 July 1665, in Malpighi, *The Correspondence of Marcello Malpighi*, 1:266–267.

Malpighi's and Borelli's diligence in following the scientific developments taking place elsewhere in Europe shows that they could do it rather easily, especially Borelli from Florence. Borelli's access to new knowledge was possible because the Medici court had books readily available to scholars.[29] Moreover, Prince Leopoldo was actively interested in extending his intellectual networks beyond the Alps, which meant that scientific knowledge also arrived by letter. According to Knowles Middleton, the Galileo collection at the National Library of Florence alone has about three hundred letters between Prince Leopoldo and scholars in Paris and other countries such as England and Germany.[30]

On the opposite side of the Alps, scholars also had a positive image of Medici-sponsored science, in part because of Leopoldo's efforts to inform others across Europe of what happened in Florence, often by sending books published there. This was the case in all three places where Steno spent most of his life: Copenhagen, Leiden, and Paris. In Copenhagen, the brothers Thomas and Erasmus Bartholin, Steno's professors of anatomy and mathematics at the university, respectively, corresponded with scholars of the Medici court, such as Viviani and Carlo Dati (1619–1676).[31] Thomas Bartholin even dedicated a large history of Danish medicine to Grand Duke Ferdinand II a few years before Steno's arrival in Florence.[32] In his preface to Ferdinand, Bartholin highlighted one of the coveted resources of Florence, namely "the inestimable treasure of the Medici Library," famous for its rare manuscripts.[33] In Leiden, Steno probably heard about this library through Golius, his professor of mathematics, who had been closely following the studies of an Arabic manuscript held at the Medici Library that contained the lost books of Apollonius's *Conics*, a famous work of mathematics from the second century BC.[34] The final three books of the *Conics*, which dealt with the theory of ellipses, had been

---

29. Alfonso Mirto, *La biblioteca del Cardinal Leopoldo de' Medici: Catalogo* (Florence: Leo S. Olschki Editore, 1990), 42.

30. Middleton, *The Experimenters*, 281–308; and Robert A. Hatch, "The Republic of Letters: Boulliau, Leopoldo, and the Accademia del Cimento," in Beretta, Clericuzio, and Principe, *The Accademia del Cimento*, 165–181.

31. Thomas Bartholin to Carlo Dati, 30 August 1662 and 10 July 1663, in EMC IV, 101–102, 474–475. On the letters between Erasmus Bartholin and Viviani, see BNCF, Gal. 254.

32. Thomas Bartholin, *Cista medica Hafniensia* (Copenhagen, 1662).

33. Bartholin, *Cista medica Hafniensia*, vi.

34. Alessandra Fiocca and Andrea del Centina, "Borelli's Edition of Books V–VII of Apollonius's *Conics*, and Lemma 12 in Newton's *Principia*," *Archive for History of Exact Sciences* 74, no. 3 (2020): 255–279, esp. 261.

lost since antiquity, but two Arabic translations resurfaced in the early modern period, one held by the Medici and the other by Golius himself.[35] Borelli was in charge of translating the Medici copy with the help of an Arabist in Rome. Besides Leiden, news of the ongoing translation also reached Paris, where Thévenot made himself available to publish Borelli's work there.[36] Steno was working with Golius and later with Thévenot right after this communication triangle between Florence, Paris, and Leiden emerged. Steno probably did not know all the details of these texts, but he was certainly aware of the widespread interest of scholars in Arabic manuscripts and of the riches of the Medici library, of which he would make extensive use.[37]

It is also relevant for Steno's travels south that the letters written between Paris and Florence were the most significance scientific correspondence that Prince Leopoldo had at the time. When Steno arrived in Paris, most scholars from Thévenot's circle had corresponded with Leopoldo for years, such as Thévenot himself, Chapelain, Christiaan Huygens, Ismael Boulliau (1605–1694), and the mathematician Pierre Petit.[38] In 1659 Huygens even dedicated his book on the rings of Saturn to Leopoldo.[39] Chapelain praised this decision, noting that Leopoldo "is very knowledgeable in these matters and is regarded in Italy as the only support of learning."[40] When things started to go wrong at the Montmor academy,

---

35. The German Christianus Ravius (1613–1677) also owned and translated an Arabic paraphrase of the lost books; see Fiocca and Centina, "Borelli's Edition," 256.

36. Luigi Guerrini, "Matematica ed erudizione: Giovanni Alfonso Borelli e l'edizione fiorentina dei libri V, VI e VII delle Coniche di Apollonio di Perga," *Nuncius* 14, no. 2 (1999): 505–568, esp. 511. On Thévenot's interest in Arabic manuscripts, see Nicholas Dew, *Orientalism in Louis XIV's France* (Oxford and New York: Oxford University Press, 2009), 81–130.

37. In the spring of 1665 Steno helped Chapelain and Thévenot read an Arabic manuscript, albeit unsuccessfully; see Jean Chapelain to Isaac Vossius, 29 April 1665, in Larroque, *Lettres de Jean Chapelain*, 2:395. Borch also mentioned the "exquisite Medicean Library" in his autobiography; see OBI, 1:xviii. On Steno's use of the Medici library, see Scherz's biography in Kardel and Maquet, *Nicolaus Steno: Biography and Original Papers*, 310.

38. Giulia Giannini, "An Indirect Convergence Between the Accademia del Cimento and the Montmor Academy: The 'Saturn Dispute,'" in *The Institutionalization of Science in Early Modern Europe*, ed. Mordechai Feingold and Giulia Giannini (Leiden: Brill, 2020), 83–108; on letters between Leopoldo and Chapelain, Huygens, Boulliau and Petit in 1664 and 1665, see BNCF, Gal. 277.

39. Christiaan Huygens, *Systema saturnium* (The Hague: Adriaan Vlacq, 1659).

40. Jean Chapelain to Christiaan Huygens, 15 October 1659, translated in Middleton, *The Experimenters*, 302.

mentioned in the previous chapter, scholars often turned to Florence as an example to follow. As Christiaan Huygens learned from his brother in 1660, "it seems to me that those gentlemen in Florence are worth much more than these Parisians and treat things with forethought and modesty."[41] It is thus not unexpected that, according to David Sturdy, the new Académie des Sciences in Paris, founded in the fall of 1666, "replicated or expanded upon practices of the Accademia del Cimento."[42] In short, when Steno was in Paris, several aspects of Florentine science were coveted by French scholars, especially the royal patronage of a research program centered on experiments. However, it should be noted that the networks between Leopoldo and Paris were mostly centered around mathematical and astronomical topics—Leopoldo's main interests—and not on anatomy.[43] Regardless, when the time came for Steno to travel south, there were already sturdy bridges between Paris and Florence, and Steno was about to make good use of them.

The First Italian Month: Dissecting and Searching for Certainty

Steno arrived in Pisa in the spring of 1666 just in time to find the Medici court spending the end of the winter season there.[44] As soon as he arrived, Steno interacted with a variety of scholars from the University of Pisa and the court, and he quickly became the talk of the town. On 4 April the Flemish professor of rhetoric at the University of Pisa Petrus van den Broecke (1619–1675) reported the arrival of Steno, "a learned anatomist who excellently put forth here some clear experiments of anatomical things before Our Princes."[45] Just as in Paris, Steno spent most of his time dissecting, in this case before the Medici court in Pisa. Prince Cosimo de' Medici (1642–1723), the son of Grand Duke Ferdinand II

---

41. Constantyn to Huygens, 18 November 1660, translated in Middleton, *The Experimenters*, 298–299.

42. David J. Sturdy, "The Accademia del Cimento and the Académie Royale des Sciences," in Beretta, Clericuzio, and Principe, *The Accademia del Cimento*, 181–194, esp. 190.

43. Giannini, "An Indirect Convergence," 101.

44. On the Medici winter sojourn in Pisa, see Eric Cochrane, *Florence in the Forgotten Centuries, 1527–1800: A History of Florence and the Florentines in the Age of the Grand Dukes* (Chicago: University of Chicago Press, 1973), 250, 269.

45. Petrus van den Broecke to Stefano Gradi, 4 April 1666, in Petrus van den Broecke, *Epistolarum libri tres* (Lucca: Hyacinthus Pacius, 1684), 171–172. Biographical details about van den Broecke can be found in a scribal note in a copy of this book at Ghent University Library.

and future Grand Duke Cosimo III, wrote about these dissections in a letter to Magliabechi. The young prince mentioned that "we have here Mr. Nicolaus Steno, famous anatomist who printed, as you know, the book of glands."[46] Steno dissected a dog there and showed the place "where Pecquet's duct [i.e., the thoracic duct] is inserted into the [subclavian] vein and where a lymphatic vessel is inserted into the same duct."[47] With this dissection, Steno showed his command of the new anatomy, since the thoracic duct and the wider lymphatic vessels had been unknown to the ancients, having been discovered only in the 1650s. The fact that Steno had studied directly under Thomas Bartholin, the discoverer of the lymphatics, lent even more authority to Steno's hand and words. Moreover, even though Prince Cosimo did not say it, this link between a lymphatic vessel, the thoracic duct, and a vein exhibited the relationship between Steno's work and that of Harvey and the circulation of the blood. The icing on the cake, however, was Steno's demonstration of the "structure of the heart" in "spiraling muscles as if they were fleshy fibers."[48] As seen in previous chapters, Steno had published a book where he argued that the heart was made of muscular flesh, denying the traditional claim that it was the seat of the soul or the source of heat for the body, as originally held by Harvey.[49] Thus, in his very first days in Tuscany, Steno placed himself fully on the side of the moderns by appearing to be even more modern than Harvey.

By the end of the month Bellini, who had recently been hired as a professor of anatomy at Pisa, was also praising Steno.[50] Lest Magliabechi think

---

46. Prince Cosimo to Antonio Magliabechi, before 10 April 1666, in BNCF, Magl. VIII.643, fol. 120r: "Noi qui habiamo il Sig. Niccolo Stenone famoso anatomico che a stampato come voi sapete il libro delle Glandole." This letter is undated, but since Cosimo is introducing Steno, it was probably written when Steno arrived in Pisa. Steno left Pisa on 10 April (see below). This is also the dating given by Johan Nordström, "Antonio Magliabechi och Nicolaus Steno: Ur Magliabechis Brev till Jacob Gronovius; with a Summary in English, 'Antonio Magliabecchi and Nicolaus Steno: From Magliabecchi's Letters to Jacob Gronovius,'" *Lychnos* 20 (1962): 1–42, esp. 34.

47. Prince Cosimo to Magliabechi, before 10 April 1666, in BNCF, Magl.VIII,643, fol. 120r–v: "dove il dotto del Paqueto si inserisce nella vena e dove pure si inserisce vel nel medesimo dotto un vaso linfatico."

48. Prince Cosimo to Magliabechi, before 10 April 1666, in BNCF, Magl.VIII,643, fol. 126: "di piu ha mostrato la strutura del quore . . . a' muscoli spirali si come se ne fibre sforera carnivoze."

49. Steno, *De musculis et glandulis*, 22, 27; and Harvey, *De motu cordis*, 59.

50. Giulio Coari and Claudio Mutini, "Bellini, Lorenzo," in DBI, vol. 7, 713–716.

Bellini did not know who Steno was, Bellini wrote him that he already "revered and loved [Steno] for his good and rare qualities, which I clearly knew to be found in him from his little works."[51] Now, however, he was able to "enjoy Steno's most gentle conversation, [and] create a friendship."[52] Bellini also mentioned that Borelli and the physician Carlo Fracassati (1630–1672) were there in Pisa at the time, an indication of the range of people whom Steno met in those first few days.[53] On 11 April Steno sailed up the Arno to Florence to be received by Magliabechi, and by the courtiers Dati and Lorenzo Magalotti, at the direct request of Prince Leopoldo.[54]

Steno stayed in Florence for about a month. The little we know of his first time there comes from a letter from Magalotti to a friend in Rome about a dissection that took place before 18 May 1666.[55] This dissection, although very simple, is worth following in detail, because it reveals the various approaches that Steno displayed to his Florentine audiences. As was usual, several animals that had been fished from the Mediterranean Sea arrived in Florence from its closest port city, Livorno. Steno was given a lobster, known by "the Latins [as] *Astacus marinus*," to dissect. Unlike most crustaceans, which when alive are usually "very dark and similar to mud," this specific lobster was "more purple in color, and in some places tinged with such a lively blue, that every ultramarine blue loses [against it]." And yet, despite this color difference, after being cooked, the lobster became "red like all the others," a phenomenon well-known to every seafood enthusiast. But why, in fact, do crustaceans become red after being cooked? The answer, Steno told Magalotti, had to do with a "thick and viscous matter like mustard" spread "along the back of these animals" whose color is of "a purple so deep that in the body it looks dark, but when stretched and thinned out it lightens up." This viscous substance was the dye that, "spread through invisible paths to the external regions, groans

---

51. Lorenzo Bellini to Magliabechi, a few days before 26 April 1666 (Easter), in BNCF, Magl.VIII.432, fol. 11r: "riverito ed amato per le sue buone e rare qualitá, che ritrovarsi in lui dalle sue operette chiaramente conoscevo."

52. Bellini to Magliabechi, before 26 April 1666, in BNCF, Magl.VIII.432, fol. 11r: "ho potuto godere della dolcissima sua conversazzione, e far l'amicizzia."

53. Bellini to Magliabechi, before 26 April 1666, in BNCF, Magl.VIII.432, fol. 11r.

54. Prince Leopoldo to Magliabechi, 10 April 1666, in BNCF, Magl.VIII.718, fol. 5r. And Magliabechi to Prince Leopoldo, 11 April 1666, in BNCF, Fondo Nazionale, II.IV.539, fol. 37r. For Dati and Magalotti, see Carlo Dati to Prince Leopoldo, 13 April 1666, in BNCF, Gal. 313, fol. 323r–v (original in BNCF, Aut. Pal. III, 25).

55. Magalotti to Ottavio Falconieri, 18 May 1666, in Angelo Fabroni, *Delle lettere familiari del Conte Lorenzo Magalotti e di altri insigni uomini a lui scritte* (Florence: S.A.R., 1769–1775), 1:172–173.

and drips into the substance of the scales, and ... pierces the blue ... that shines." Magalotti said Steno supported this claim with simple experiments. First, Steno spread this purple substance "subtly with a pencil on a paper sheet." Then, when drawn near a flame, "still only feeling the heat," the substance dried up and became red, so that "it looked like a piece of boiled lobster scales." The same phenomenon happened when this substance was "bathed with *acquarzente*" and "spirit of vitriol"—that is, alcohol and sulfuric acid. They then compared the three final colors and "could not see the slightest difference."[56]

These experiments, of course, only showed the various means by which a purple substance became red. They did not explain why exactly seafood turns red when cooked. In fact, the redness of seafood began to be better understood only in the early twentieth century and is still a matter of research today.[57] But Steno had something else in mind with these observations other than simply explaining the redness. In a later letter to the physician William Croone, Steno mentioned that he had "conducted an experiment regarding colors, the various ways of changing which always trouble me."[58] In short, these observations on the color of lobsters displayed some of the epistemic problems of natural research in the second half of the seventeenth century. First, it showed that visible phenomena are illusive. The fact that most people see a red lobster on their plates has nothing to do with the lobster's real blue or gray color; therefore, visible colors do not necessarily teach something about nature. Steno had previously exposed this problematic nature of colors when he argued that the uniform color of blood hid its complex composition.[59] Second, just as he used chymical reactions in Leiden as an analogy to show that blood carried different substances regardless of color, so in Florence, by exposing the lobster flesh to heat, alcohol, and sulfuric acid he used different chymical reactions to produce the same phenomenon. One could ask which reaction really happened when cooking a lobster. But the point was not to propose one process over the other, since no one eats a lobster after exposing it to acid. On the contrary, these three mechanisms showed the

---

56. Magalotti to Falconieri, 18 May 1666, in Fabroni, *Delle lettere familiari*, 1:172–173.

57. Shamima Begum, Michele Cianci, Bo Durbeei, et al., "On the Origin and Variation of Colors in Lobster Carapace," *Physical Chemistry, Chemical Physics* 17, no. 26 (2015): 16723–16732.

58. Steno to William Croone, after 18 October 1666, in London, Royal Society, EL/S1/92: "J'ai fait ces jours passes une experience touchant les couleurs dont les divers maniers de se changer me donne toujours de la pene." Steno mentions the October dissection of the great white shark, thus the dating.

59. Steno, *Observationes anatomicae*, 30.

complexity of the inner workings of nature. Finally, these chymical reactions also exposed the unreliability of traditional research methods. In Magalotti's provoking terms, "let a Peripatetic [i.e., an Aristotelian] come to offer me his qualities and let him explain with reason how from the heat of a flame, the *acquarzente*, and the acid of vitriol one can equally produce in the same substance the same quality." Not only were these processes difficult to explain with Aristotelian qualities, but they also eluded the senses as arbiters of what is actually taking place. Steno circumvented this frailty of the senses by explaining that the original blue dye of the lobster spread to the scales "through invisible paths [*per invisibili vie*]."[60]

When Steno finished his dissection, the process by which lobsters turn red remained unknown. Perhaps giving voice to Steno's rhetoric on certainty, Magalotti finished his letter with an ironic question: "Now you would like to understand how purple can turn into vermilion, right? Me too, but unfortunately this is one of those many things that I do not know."[61] The lack of a definitive answer, to which neither Steno nor Magalotti suggested a hypothetical solution, signaled that much of the natural world was still unknown. Yet, if Magalotti seemed to enjoy the inescapable uncertainty of nature, Steno's attitude, on the other hand, was not to give up.[62] Magalotti does not mention it, but Steno also spent a couple of hours looking at the muscles in the lobster's claws. An isolated sheet among Steno's papers in Florence contains an unfinished sketch of the muscles of a lobster claw, possibly drawn during this dissection (fig. 5.1).[63] This claw muscle later became one of the examples that Steno used in his upcoming book to illustrate muscle fibers (fig. 5.2).[64] Moreover, the only mechanical representation of muscles as levers in the book is also from a lobster claw (fig. 5.3). To conclude, Steno's first month in Tuscany gave him opportunities to demonstrate the epistemic power of his dissections to the Medici court and, perhaps, to think further about his geometrical model for muscle motion with lobster muscles. Yet, Steno's time in Florence was short, and a few days after dissecting the lobster he hit the road to Rome.[65]

---

60. Magalotti to Falconieri, 18 May 1666, in Fabroni, *Delle lettere familiari*, 173. See also Magalotti to Archbishop Ascanio Piccolomini, 9 November 1666, in Siena, Biblioteca Communale di Siena, D. V. 7, fols. 18r–19v, esp. 19v. Given the letter to Piccolomini and Croone, it is possible that Fabroni dated the Falconieri letter wrongly.

61. Magalotti to Falconieri, 18 May 1666, in Fabroni, *Delle lettere familiari*, 173.

62. On Magalotti's attitude toward natural philosophy, see Cochrane, *Florence in the Forgotten Centuries*, 253–254.

63. BNCF, Gal. 291, fol. 98r.

64. Steno, *Elementorum myologiæ specimen*, 40.

65. For Steno in Rome in late May or early June, see Malpighi, *Opera posthuma*, 43.

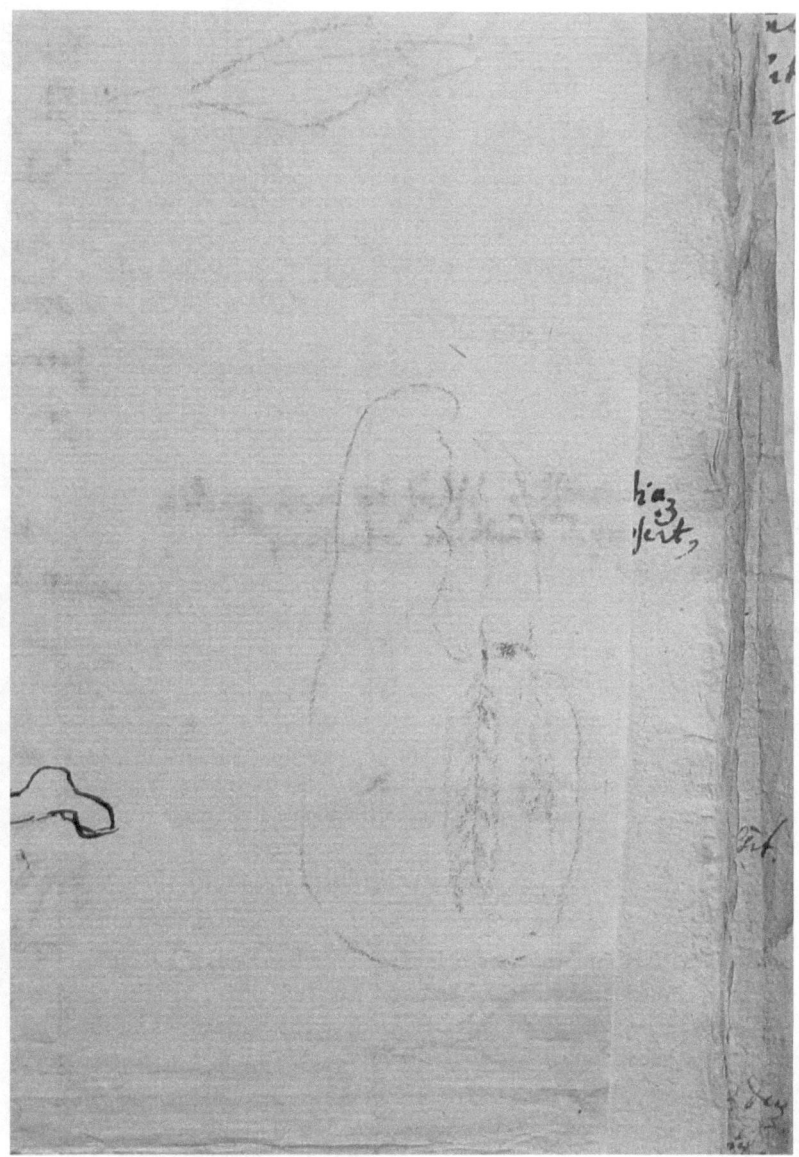

FIGURE 5.1. Steno's lightly drawn sketch with two muscles inside a lobster claw in Florence, BNCF, Gal. 291, fol. 98r. Reproduced with the permission of the Ministero della Cultura/Biblioteca Nazionale Centrale di Firenze.

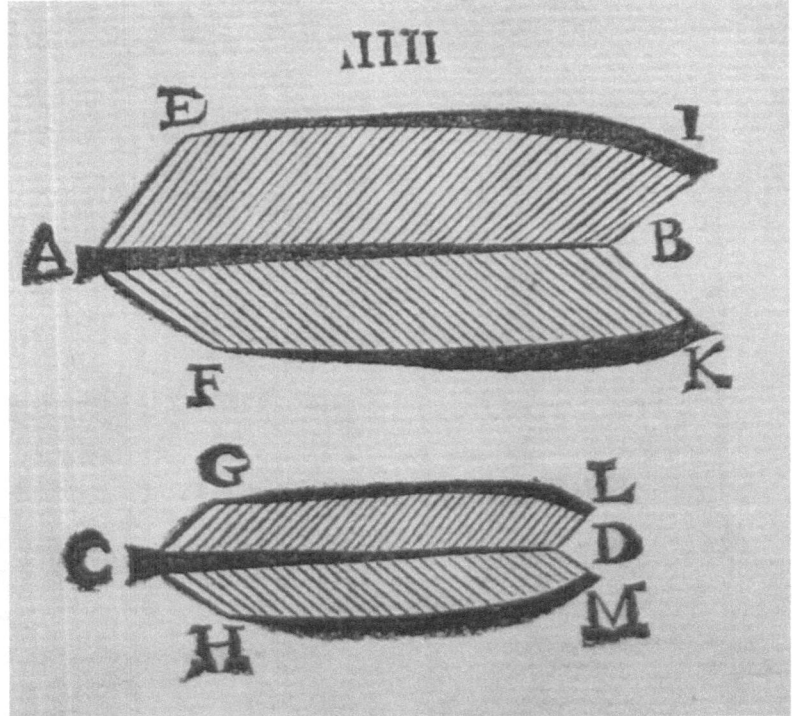

FIGURE 5.2. Lobster claw muscles, in Steno, *Elementorum myologiæ specimen* (Florence, 1667), tabula III. Reproduced with the permission of the Ministero della Cultura/Biblioteca Nazionale Centrale di Firenze.

Mathematics Mattered in Rome

Steno's trip to Rome was already in his plans when he arrived in Pisa, since van den Broecke asked Steno to carry books for him to Stefano Gradi (1613–1683), one of the priestly librarians at the Vatican.[66] Just like Florence, Rome was a stage on the Danish *peregrinatio medica*. Borch was in Rome from October 1665 to March 1666 and then "rather unwillingly" returned to Denmark.[67] Even though Borch and Steno already knew the

---

66. Van den Broecke to Stefano Gradi, 4 April 1666, in van den Broecke, *Epistolarum libri tres*, 171–172. On Gradi, see Tomaso Montanari, "Gradi, Stefano," in DBI, vol. 58 361–363; and Francesco Bustaffa, "Michelangelo Ricci (1619–1682): Biografia di un cardinale innocenziano" (Ph.D. diss., Università di San Marino, 2010), 133 n. 61.

67. Borch's autobiography in OBI, 1:xviii.

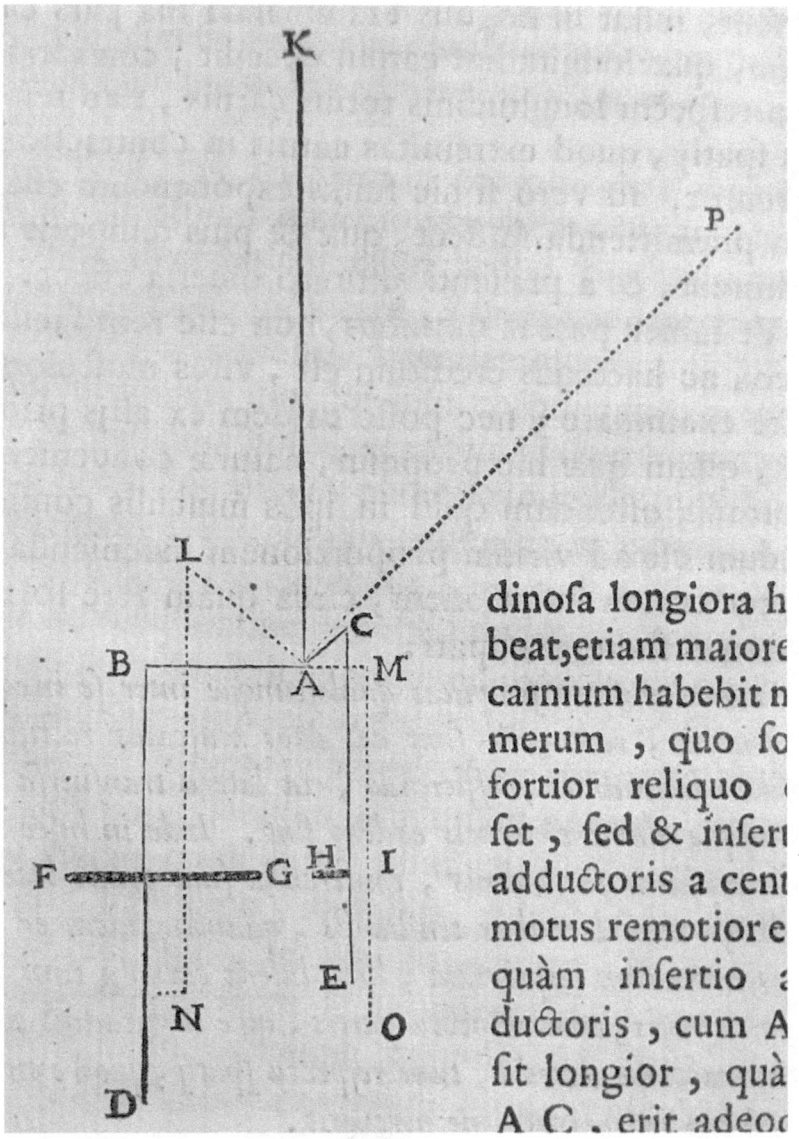

dinosa longiora h
beat, etiam maiore
carnium habebit n
merum , quo fo
fortior reliquo
set , sed & inser
adductoris a cent
motus remotior e
quàm insertio
ductoris , cum A
sit longior , quà
A C . erit adeo

FIGURE 5.3. Geometrical representation of the lobster claw muscles in Steno, *Elementorum myologiæ specimen* (Florence, 1667), 41. Reproduced with the permission of the Ministero della Cultura/Biblioteca Nazionale Centrale di Firenze.

splendor of Catholic art from their travels throughout Europe, Rome must have made a special impression on them. In the early modern period, the papacy continually sponsored works of art around the city, including the almost-finished colonnade around St. Peter's Square by Gian Lorenzo Bernini (1598–1680).[68] They also found a thriving intellectual community in Rome. Borch wrote of being on "friendly terms" with intellectuals such as Kircher and the mathematician Michelangelo Ricci (1619–1682).[69] He also attended the informal academy hosted by his fellow Scandinavian (and convert to Catholicism) Queen Christina of Sweden (1626–1689) in order to "communicate to her the secrets of chymistry."[70] Steno also met all these scholars and attended Queen Christina's academy.[71]

Most significantly, in Rome, Steno crossed paths with Malpighi, who was also visiting the city on his way back from Naples to Bologna.[72] Malpighi immediately mentioned the encounter with Steno to Tommaso Cornelio (1614–1684) in Naples, who told him that he also had "a great desire to meet Mr. Steno, especially for what Your Lordship speaks of him."[73] By then both Malpighi and Steno knew that they shared very similar approaches to the study of anatomy, since both of them had published books based on mechanical analogies and numerous dissections. Malpighi later commented that "Steno has always been a dear friend and guest for me."[74] The two anatomists met during a dinner at the Villa Ludovisi, where they also shared the table with the Roman physician Guglielmo Riva (1627–1677).[75] Although not a well-known figure, Riva became physician to Pope Clement IX (r. 1667–1669) and personally reproduced Thomas Bartholin's dissection of the lymphatics, making him a good candidate to quickly become friends with Steno.[76] In the *Elementorum myologiæ*

---

68. The elliptical colonnade was in construction from 1657 to 1667, see Carla Keyvanian, "Papal Urban Planning and Renewal: Real and Ideal, c. 1471–1667," in *A Companion to Early Modern Rome*, ed. Pamela M. Jones, Barbara Wisch, and Simon Ditchfield (Leiden: Brill, 2019), 305–323, esp. 322.

69. Borch's autobiography in OBI, 1:xviii.

70. Borch's autobiography in OBI, 1:xviii.

71. For Steno at Queen Christina's academy, see Benedetto Milino to Prince Leopoldo, 4 June 1667, in BNCF, Gal. 314, fols. 952r–953r: "Io conobbi lo Stenone l'altra volta che fu in Roma nell'anticamera della Regina."

72. Malpighi, *Opera posthuma*, 43.

73. Tomaso Cornelio to Malpighi, 29 June 1666, in Malpighi, *The Correspondence of Marcello Malpighi*, 1:317.

74. Malpighi, *Opera posthuma*, 43.

75. Malpighi, *Opera posthuma*, 43.

76. Michele Augusto Riva, Marta Benedetti, Francesca Vaglienti, Chiara Torre, Gaspare Baggieri, and Giancarlo Cesana, "Guglielmo Riva and the End of

specimen, the book that he would publish a year later in Florence, Steno mentioned Riva when reporting an observation of the eyes, possibly because the two anatomists dissected them together in Rome.[77] These encounters with Malpighi and Riva confirm Steno's ongoing commitment to anatomy in Rome. It also marks the beginning of a close and deep relationship that led Steno to tell Malpighi that, "after God, it seems to me that there is nothing holier in the world than true friendship."[78]

Interestingly, Steno's anatomical interests also drew him closer to scholars who were not necessarily interested in anatomy, such as the Jesuit priest and natural philosopher Honoré Fabri (1608–1688), one of the most brilliant minds in Rome at the time.[79] Fabri had already published several novel ideas on natural philosophy, mathematics, and theology and had an anatomy book coming out.[80] In this new book, Fabri defended the circulation of the blood and, more important, argued that the heart was "of the likeness of a muscle," just as Steno had argued.[81] Not surprisingly, Steno and Fabri got along very well. A year later Fabri praised Steno's work on the muscles to Prince Leopoldo, saying that "those new things of muscle fibers ... are not new to me, because the author was with me in the previous summer [i.e., June 1666] and told me."[82] Steno's visit to Fabri is not surprising, given his and Borch's custom of visiting Jesuit colleges throughout Europe. As this book's last chapter will show, this meeting with Fabri would also have an important role in Steno's conversion to Catholicism.

Ultimately, Steno found many people interested in his ideas about the body during this short but intense visit to Rome. Indeed, in addition to

---

Hepatocentrism: A 17th-Century Painting," *Vesalius: Acta Internationalia Historiæ Medicinæ* 20, no. 2 (2014): 69–72.

77. Steno, *Elementorum myologiæ specimen*, 82.

78. Steno to Malpighi, 30 September 1670, in *Epistolae*, 1:248–250.

79. Marin Mersenne called Fabri "a veritable giant in science"; see Mersenne to René Descartes, spring 1646, as translated in John Heilbron, *Electricity in the 17th and 18th Centuries: A Study of Early Modern Physics* (Berkeley and Los Angeles: University of California Press, 1979), 195. On Fabri, see Stefania Tutino, *Uncertainty in Post-Reformation Catholicism: A History of Probabilism* (Oxford and New York: Oxford University Press, 2018), 223–264.

80. Honoré Fabri, *Tractatus duo: Quorum prior est de plantis et de generatione animalium et posterior de homine* (Paris: François Muguet, 1666). This book's first copy was printed in Paris on 12 May 1666; see the *imprimatur* page.

81. Fabri, *Tractatus duo*, 371. Blood circulation is mentioned in several parts of the book.

82. Fabri to Prince Leopoldo, 7 July 1667, in BNCF, Gal. 278, fol. 31r–v: "nova illa fibrarum musculi ... non tamen nova mihi accidit, nam æstate præterita, mecum illam auctor communicaverat."

reports about attending meetings at Queen Christina's academy and the Villa Ludovisi, two other accounts mention Steno's visits to the libraries of the Vatican and of Cardinal Flavio Chigi (1631–1693).[83] Cardinal Chigi was a nephew of the current pontiff, Pope Alexander VII (r. 1655–1667), who also hosted scholars regularly at his palazzo.[84] At the Vatican Steno shared his research with two officials from the papal curia: Michelangelo Ricci and his friend the Vatican librarian and priest Stefano Gradi.[85] Although not a priest, Ricci worked in the Vatican's Congregation of the Index and would be made a cardinal a few years later.[86] In the year that Steno visited Rome, Ricci published a book on mathematical maximums and minimums. Ricci's book contributed to a trendy mathematical topic that culminated with the development of infinitesimal calculus by Newton and Leibniz only a few years later.[87] Gradi, on the other hand, became prefect of the Vatican Library and then published a collection of four "physico-mathematical dissertations," one of which also treated infinitesimals.[88] More important, both Ricci and Gradi corresponded with scholars in Florence, many of whom Steno already knew.[89] No source from those days suggests that Steno had a particular religious interest in these library tours and meetings, since the resonances between him and other scholars were mostly scientific. However, as I will show later, although Steno was not yet considering conversion, this early trip to Rome contributed significantly to his positive view of Catholicism.

Like other Danish travelers, Steno left Rome after about a month. However, instead of continuing his travels, Steno returned to Florence.[90] Three

---

83. Testimonies of Stefano Gradi and Thomas de Juliis for *Processus super vita, moribus et qualitatibus*, 27 August 1677, in *Epistolae*, 2:935–937.

84. See William E. Knowles Middleton, "Science in Rome, 1675–1700, and the Accademia Fisicomatematica of Giovanni Giustino Ciampini," *British Journal for the History of Science* 8, no. 2 (1975): 138–154, esp. 139.

85. Steno's meetings are mentioned in Michelangelo Ricci to Prince Leopoldo, 2 July 1666, BNCF, Gal. 277, fol. 322r; and in Stefano Gradi to Prince Leopoldo, 18 June 1667, in BNCF, Gal. 278, fol. 28r–v.

86. Bustaffa, "Michelangelo Ricci," 131, 408.

87. On Ricci as a mathematician, see Andersen and Bos, "Pure Mathematics," 718–721.

88. Stefano Gradi, *Dissertationes physico-mathematicæ quatuor* (Amsterdam: Daniel Elsevier, 1680). On Gradi as a mathematician, see Ivica Martinović, "Stjepan Gradić on Galileo's Paradox of the Bowl," *Dubrovnik Annals* 1 (1997): 31–69.

89. On Ricci and Gradi as members of the Holy Office, see Bustaffa, "Michelangelo Ricci," 131–133. On the Florentine correspondence, see Alfonso Mirto, "Lettere di Stefano Gradi ai Fiorentini: Viviani, Dati, Redi, Leopoldo e Cosimo III de' Medici," *Studi secenteschi* 49 (2008): 371–404.

90. Magalotti to Segni, 24 August 1666, in Mirto, *Alessandro Segni e gli Accademici della Crusca*, 372–375.

letters of recommendation carried by Steno to Prince Leopoldo reveal the reasons for his return. The letters were written in Rome by some of the scholars that Steno met there: Ricci, Fabri, and Kircher.[91] All of them spoke of or alluded to Steno's intellectual skills and friendliness. Fabri told Leopoldo that Steno, "although young, is gifted with the wisdom of the elders and [with] a singular modesty; and I have no doubt that he would be warmly received by your most Serene Highness not just due to your humanity toward all, but especially due to that love you have for the learned."[92] And Ricci wrote that Steno was to arrive in Florence, "willing to display his talents where they may be better known and appreciated, and to gain that prize proper of virtue, which is the esteem and praises of those who are competent judges, like Your Most Serene Highness, whether by universal skill or by the protection that [you] give to virtuous men..., [and thus] to have place under the shadow of your protection."[93] As for Kircher, although his letter is now lost, we know that he asked Steno to deliver one of his books to Leopoldo. The Medici prince acknowledged the letter, telling Kircher that Steno was "meritorious for his value and of those praises that are justly attributed to him."[94]

These three letters confirm that Steno was undoubtedly seeking the patronage of the Medici family, and especially of Prince Leopoldo. However, the trio of recommenders whom Steno chose seems unusual at first glance because none of them was an anatomist. Anatomical circles had

---

91. Michelangelo Ricci to Prince Leopoldo, 2 July 1666, BNCF, Gal. 277, fol. 322r; Fabri to Prince Leopoldo, 2(?) July 1666, in BNCF, Gal. 281, fol. 84r-v; and Athanasius Kircher to Prince Leopoldo, 3 July 1666, in BNCF, Aut. Palat. IV, 49, as quoted in Alfonso Mirto, "Le lettere di Athanasius Kircher della Biblioteca Nazionale Centrale di Firenze," *Atti e memorie dell'Accademia toscana di scienze e lettere "La Colombaria"* 54 (1989): 127–165, esp. 164–165.

92. Fabri to Leopoldo, 2 (?) July 1666, BNCF, Gal. 281, fol. 84rv: "Junior quidem est, senili tamen sapientia et singulari modestia præditus et nullus dubito, quin Celsit.do Tua Seren.ma pro sua in omnes humanitate, sed præsertim pro eo quo est in Literatos amore, benigne illum expectum sit."

93. Ricci to Leopoldo, 2 July 1666, BNCF, Ms. Gal. 277, fol. 322r: "desideroso di far' apparire i suoi talenti, dove possono essere più conosciuti e graditi; e de riportarne quell premio, qui è proprio della virtù, cioè la stima e la lode di coloro, que ne sono giudizi competenti, qual' è V.A.S. si per la universal perizia, come per la protezione che tiene de' Virtuosi, onde à ciascuno è Gloria nel cospetto del mondo, aver luogo sotto l'ombra della protezione di lei."

94. Prince Leopoldo to Kircher, 10 July 1666(?), BNCF, Gal. 282, fol. 137r: "meritevole per il suo valor, di quelle lode che giustamente le vi deve." The letter is undated, but its content suggests it has the same date of Leopoldo's letter to Ricci, on the previous folio.

been Steno's main environment in his first weeks in Italy, as shown by his friendships with Bellini, Malpighi, and Riva, and by his dissections in Pisa, Florence, and Rome. Yet, even though neither Ricci, Fabri, nor Kircher was an anatomist, they were all regular correspondents of Leopoldo, sometimes even referred to as correspondent members of the Accademia del Cimento.[95] More important, Ricci was a close friend of Thévenot and had been Thévenot's original point of contact with the Cimento years before.[96] Given Ricci and Thévenot's familiarity, at least to judge by their letters, it is likely that Ricci was Steno's first point of contact in Rome and possibly his host during his stay.[97]

The links with Fabri and Kircher are not as easy to identify as those with Ricci, but there are some leads about them from Steno's French period. On his way out of France Steno passed through Lyon, where there was a Jesuit priest who was a "mathematician . . . busy in anatomical things" and interested in the circulation of the blood, according to the diplomat Alessandro Segni, who passed through Lyon a few weeks after Steno.[98] It is not clear who this Jesuit was, but Segni described him as "friends with Father Fabri, continuously corresponds with him, [and] loves the good way of doing philosophy."[99] Given the closeness of interests between Steno and this Jesuit from Lyon, as well as the latter's correspondence with Fabri, this French Jesuit could have been the one who introduced Steno to Fabri. As to Kircher, Steno probably met him when visiting his museum at the Collegio Romano, which was not only a typical stop on intellectual tours of Rome but was also designed as a way to respond to the seventeenth-century search for certainty in science, of which Steno was fully aware.[100] As mentioned in chapter 1, in Copenhagen Steno had studied carefully Kircher's work on magnets, especially the section on hydraulic machines. Therefore, finding that Steno was interested in the subject and learning of his plans to return to Florence, Kircher may simply have asked him to bring a book to Leopoldo, whose accompanying

---

95. Middleton, *The Experimenters*, 298. Kircher is not usually listed as a "correspondent member," but he sent as many letters as others; see Mirto, "Le lettere di Athanasius Kircher della Biblioteca Nazionale," 130–136.

96. Giannini, "An Indirect Convergence," 85–87.

97. Steno explicitly acknowledges Thévenot's intercession in *Elementorum myologiæ specimen*, 48.

98. Segni to Leopoldo, 11 December 1665, in BNCF, Gal. 277, fols. 221r–222r: "un Padre gesuita, il matematico . . . si affatica in cose anatomiche"

99. Segni to Leopoldo, 11 December 1665, in BNCF, Gal. 277, fols. 221r–222r: "Questo buon Padre è amico del padre Fabbri, con esso carteggia di continuo, ama il buon modo di filosofare. . . ."

100. Waddell, *Jesuit Science and the End of Nature's Secrets*, 89–91.

letter would serve as a letter of recommendation. However, it is also possible that the German physician Martin Fogel (1634–1675) recommended Steno to Kircher as part of their ongoing correspondence.[101] Steno probably met Fogel in his last month in Paris, since the two were there at the same time and started corresponding soon after.[102]

The dynamics of early modern letters of recommendation or introduction is a well-mapped topic.[103] But these letters were more than just a mechanism of social integration. Just like Leopoldo de' Medici, these three recommenders were all interested in mathematics. Ricci had just finished writing his *Geometrica exercitatio de maximis et de minimis* (Rome, 1666), and both Fabri and Kircher were respected authors in the field of physico-mathematics. It was probably Ricci who guided Steno to the Vatican Library and to Gradi, to whom Ricci dedicated his short *Geometrica exercitatio*.[104] It was also Ricci who spoke of Steno's interests in mathematics more explicitly in his letter, praising Steno as "an excellent anatomist and well-versed in all the parts of mathematics."[105] Leopoldo's interests in the mathematical sciences shaped much of his European network of correspondence, as the names of Steno's recommenders confirm. In turn, Steno's successful use of this network reveals his willingness and ability to adapt himself to this network, in large part because he too was interested in applying mathematics to anatomy.

Intersecting Muscles with Mathematics in Florence

Leopoldo de' Medici was happy to have Steno in Florence again and wrote back to Ricci, Fabri, and Kircher about "how good it was to receive a virtuoso of his condition and truly worthy of the praises that you attribute

---

101. Martin Fogel to Athanasius Kircher, 24 July 1664 and 22 September 1664, in Rome, Archivio della Pontificia Università Gregoriana, Ms. 563, fols. 127r–v, 196r–197v.

102. For Fogel in Paris, see Harcourt Brown, "Martin Fogel e l'idea accademica Lincea," *Rendiconti della R. Accademia nazionale dei Lincei* 11, nos. 11–12 (1936): 814–833, esp. 821. On Fogel and Steno, see Fogel to Carlo Dati, 17 August 1667, in BNCF, Baldov. 258, as quoted in Brown, "Martin Fogel," 833.

103. Paul D. McLean, *The Art of the Network: Strategic Interaction and Patronage in Renaissance Florence* (Durham, NC: Duke University Press, 2007), 90–120; and Paula Findlen, *Possessing Nature: Museums, Collecting, and Scientific Culture in Early Modern Italy* (Berkeley and Los Angeles: University of California Press, 1994), 365–392.

104. The dedication to Gradi is dated 8 July 1666; see Michelangelo Ricci, *Geometrica exercitatio* (Rome: Nicolaus Angelus Tinassius, 1666), ii.

105. Ricci to Leopoldo, 2 July 1666; BNCF, Gal. 277, fol. 322r: "Niccolò Stenoni, eccelente anatomista, e versato in tutte le parti della matematica."

to him."¹⁰⁶ In Florence Steno was able to develop his anatomical research with princely patronage for the first time, something that he, like other scholars in Paris, had coveted.¹⁰⁷ The Medici princes gave Steno access to animal corpses, dissection materials, and a courtly audience, all of which he put to good use.¹⁰⁸ At the end of August Magalotti wrote that Steno, "whom we have here in Florence for two months[,] . . . has shown various beautiful experiments to the grand duke."¹⁰⁹ As before, Steno's experiments highlighted ideas and methods of the new anatomy. Magalotti described Steno's "insertion of the lymphatic vessels in the subclavian vein," the opening of an eel with a special vessel "whose function has not yet been understood," and several "jets of fluids [injected] into the veins of a dog that died due to the sudden coagulation of blood."¹¹⁰ By the end of November, Steno, who moved "forward every day in the esteem of the princes and the court," had fully established himself in Florence.¹¹¹

But material resources were not the only thing that Steno was seeking in Florence. While in Paris, Steno heard that the Accademia del Cimento mathematician Borelli was also working on muscles.¹¹² Indeed, he was already working on what would become one of the most famous works in the history of science, the *De motu animalium* (Rome, 1680–1681), where he applied physico-mathematics to animal motion.¹¹³ An August 1665 letter to Viviani from Paolo dell'Ara (fl. 1660), secretary of the Medici embassy in Paris, says that Thévenot wanted "to know whether Mr. Borelli had written an anatomical treatise on the muscles," or, if not, what was "the opinion of the aforementioned Mr. Borelli about the above matter."¹¹⁴ The

---

106. Leopoldo to Ricci, 10 July 1666, in BNCF, Gal. 282, fol. 138r–v: "quanto mi sia stato caro il rivedere un virtuoso della di lui condizione e veramente degno di quelle lodi che V.S. le attribuisce." On the letters to Fabri and Kircher on the same date, see BNCF, Gal. 282, fols. 136v–137r. These are drafts of letters.

107. Steno, *Discours sur l'anatomie du cerveau*, 33.

108. Findlen, "Controlling the Experiment."

109. Magalotti to Segni, 24 August 1666, in Mirto, *Alessandro Segni e gli Accademici della Crusca*, 372–375.

110. Magalotti to Segni, 24 August 1666, in Mirto, *Alessandro Segni e gli Accademici della Crusca*, 372–375. Steno included this observation of the eel in *Elementorum myologiæ specimen*, 73.

111. Magalotti to Segni, 27 November 1666, in Mirto, *Alessandro Segni e gli Accademici della Crusca*, 438–439.

112. On Borelli, see Federica Favino and Giulia Giannini, "Borelli Reloaded: Contexts and Networks in 17th-Century Italy," *Physis* 57, no. 2 (2022): 289–300.

113. See Boschiero, *Experiment and Natural Philosophy*, 84–85.

114. Paolo dell'Ara to Vincenzo Viviani, 22 August 1665, in BNCF, Gal. 254, fol. 275r: "sapere se il Sig. Borelli abbia fatto un tratado d'Anatomia intorno a muscoli. . . .

reason for this request, which dell'Ara also wrote, was to "satisfy the curiosity or increase the knowledge on similar science of a certain Dane, Mr. Steno, ... [, who,] according to the most knowledgeable in this profession, is the most learned and skillful of France in such things."[115] As soon as Steno arrived in Florence in July 1666, he looked for and met Borelli at the Medici court. Borelli mentioned this meeting to Malpighi, stating that Steno "wanted to come up here to me and that he wants me to teach him something about geometry."[116] Steno perhaps hoped to do with Borelli what he had done with Swammerdam, namely to visit him at his home and share his research on the muscles with him. Borelli, however, was not "so credulous that he [Steno] was the ideal of modesty and good manners, as they proclaim of him at the palace."[117] Probably influenced by his readings of Anton Deusing, one of Steno's adversaries in the Netherlands, Borelli said that "those little letters [*epistolette*] that he [i.e., Steno] printed give a clear indication of his avidity to absorb all things."[118] More seriously, he said, "I know that these ultramontanes [i.e., non-Italians] come here to us prepared and most disposed to be watchful and cautious in such a way that by means of tricks they beat us at the game [*ci ponno dar quaranta, e la mano*], so that in the long run we end up below [them]."[119] Because Borelli was already working on his book on animal motion, which also dealt with muscles and mechanism, he probably feared that Steno would steal his ideas.[120] So Borelli was not as helpful as Steno would have liked.

Nonetheless, in less than four months Steno completed the text that became his most innovative work in anatomy, in which he described muscular contraction with a mathematical model. The book's title was

---

O quando il Sig. Borelli non abbia fatto il predetto trattato desidera ... l'opinione del prefato Sig. Borelli sopra la sudetta materia."

115. dell'Ara to Viviani, 22 August 1665, in BNCF, Gal. 254, fol. 275rv: "sia per appagare la curiosita o accrescere cognizione di simile scienza, d'un tal Monsú Stenon Danese. ... Uomo al parere degli intellegentissimi in questa professione, il piú dotto e il piu abile della Francia in simil genere."

116. Giovanni Borelli to Marcello Malpighi, 17 July 1666, in Malpighi, *The Correspondence of Marcello Malpighi*, 1: 318.

117. Borelli to Malpighi, 17 July 1666, in Malpighi, *The Correspondence of Marcello Malpighi*, 1:318.

118. Borelli to Malpighi, 17 July 1666, in Malpighi, *The Correspondence of Marcello Malpighi*, 1:318.

119. Borelli to Malpighi, 17 July 1666, in Malpighi, *The Correspondence of Marcello Malpighi*, 1:319. Borelli is likely referring to the card game "scala quaranta."

120. Giovanni Borelli, *De motu animalium*, 2 vols. (Rome: Angeli Bernabo, 1680–1681). For Borelli being already at work on this book, see Boschiero, *Experiment and Natural Philosophy*, 84–85.

Elementorum myologiæ specimen seu musculi descriptio geometrica (*A Specimen of the Elements of Myology or a Geometrical Description of the Muscle*; Florence, 1667).[121] Steno said he wanted to display this specimen of elements—an obvious reference to Euclid's *Elements*—because he thought there was no other way to do it unless "myology became part of mathematics [*nisi Matheseos pars Myologia fieret*]."[122] By "mathematics" Steno meant physico-mathematics because, he asked rhetorically, "why can we not give to the muscles that which astronomers give to the skies, [and] geographers to the Earth?"[123] Steno was able to write the *Elementorum myologiæ specimen* quickly because he had already introduced mathematical elements in his previous work, *De musculis et glandulis*. In this earlier book, Steno suggested that muscle fibers form an "oblique parallelogram or the figure of a rhomboid."[124] Steno arrived at this conclusion after developing a new method of dissecting muscles, in which he cut "along the course of the fibers ... not in a plane cutting everything transversely through the middle," but in a plane in which "the tendons remain intact with the flesh."[125]

Steno did not accompany this first text on muscle motion with images, making it hard to understand what he meant.[126] Part of his time in Paris was thus dedicated to drawing and thinking about images, and it is possible that some were already done when he arrived in Florence. Chapelain mentioned trying "to force him to give us a treatise of his new discoveries with images for greater clarity," and a few months later Steno taught Graindorge how to draw a muscle fiber (fig. 5.4).[127] Indeed, Steno's new book, published in Florence, was filled with images, mostly geometrical diagrams of parallelograms in two and three dimensions. In the first new illustration, on the third page, he represented "a muscle through a collection of motor fibers arranged in such a way that the middle flesh forms an oblique parallelepiped, but the tendons form two opposite tetragonal prisms" (fig. 5.5).[128]

---

121. The book was published in April 1667, but Steno had most of it ready by November 1666, as seen from the 27 October 1666 imprimatur.
122. Steno, *Elementorum myologiæ specimen*, iii.
123. Steno, *Elementorum myologiæ specimen*, iii–iv.
124. Steno, *De musculis et glandulis*, 15.
125. Steno, *De musculis et glandulis*, 15. For more on this longitudinal cut, see Castel-Branco and Kardel, "Drawing Muscles with Diagrams."
126. There are no images, possibly because "an unexpected event drew me away from my papers and dissections," in Steno, *De musculis et glandulis*, 3.
127. Chapelain to Huet, 6 April 1665, in Larroque, *Lettres de Jean Chapelain*, 2:393 n. 3; and Graindorge to Huet, 29 July 1665, in Copenhagen, Royal Library of Denmark, NKS 4660 kvart, 47.
128. Steno, *Elementorum myologiæ specimen*, 3. Italics in the original.

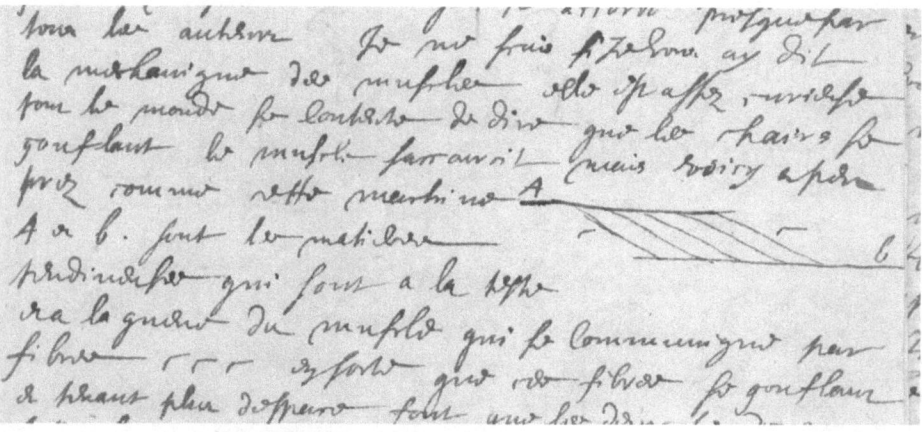

FIGURE 5.4. An ordo of muscle fibers in Graindorge's letter to Huet. Copenhagen, Royal Library, NKS 4660 kvart, 47.

Tellingly, Steno had not yet decided how to represent the tendons in 1664, but now he opted for representing them with prisms. Moreover, after extensive geometrical explanations, Steno said that he had "to demonstrate their certainty with examples taken from Nature herself [*ex ipsa Natura*]." Thus, he included "figures of different muscles ... displayed at the magnitude at which I have measured them in cadavers" (fig. 5.6).[129]

These and other differences between Steno's work in Copenhagen and that in Florence are significant and speak to the influence of the intellectual environment he found in the latter city.[130] Domenico Bertoloni Meli has argued, correctly, that Steno used life-sized images not as mere illustrations but rather as demonstrations, in the same way that Galileo used experiments to prove his arguments.[131] Another major difference was the new definitions that Steno provided to describe muscle fibers. In *De musculis et glandulis*, Steno named a "rectilinear series of fibers" as an *ordo* and a "transverse" group of this series as *versus*.[132] In *Elementorum*

129. Steno, *Elementorum myologiæ specimen*, 34.

130. See Troels Kardel, *Steno on Muscles: Introduction, Texts, Translations*, Transactions, American Philosophical Society 84, pt. 1 (Philadelphia: American Philosophical Society, 1994), 15–18; and Bertoloni Meli, "The Collaboration Between Anatomists and Mathematicians," 696–706.

131. Bertoloni Melli, "The Collaboration Between Anatomists and Mathematicians," 705–706.

132. Steno, *De musculis et glandulis*, 15. For a brief account of the translation of *ordo* and *versus*, see Kardel, *Steno on Muscles*, 229 n. 12. For the sake of faithfulness to the original and because of the lack of an obvious English equivalent, I use the original terms.

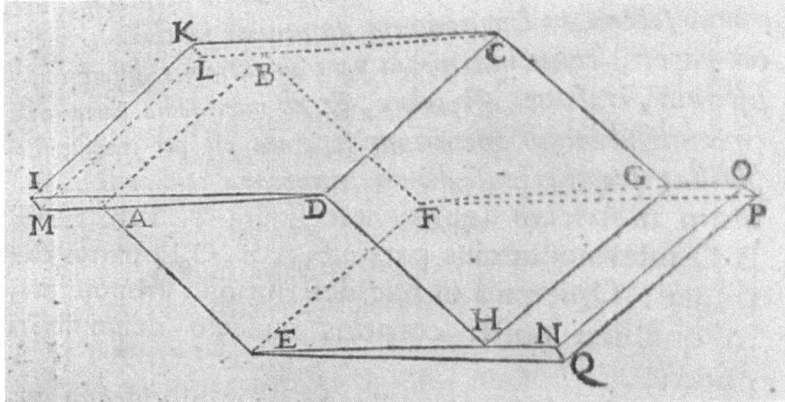

FIGURE 5.5. A muscle as an oblique parallelepiped, in Steno, *Elementorum myologiæ specimen*, 3. Courtesy of the Institute of the History of Medicine, Johns Hopkins University.

myologiæ specimen, he defined *ordo* anew as "a series of *unequally equal* rectilinear motor fibers inflected at the same angle" (fig. 5.7).[133] Steno used terms such as "unequally equal [*inæqualiter æqualium*]" and "equally equal [*æqualiter æqualium*]" many times throughout the book. It has been rightly suggested, therefore, that he borrowed the mathematical language Galileo and Viviani used to speak about uniformly accelerated motion, based on the medieval tradition of the Oxford Calculators.[134] But the most striking novelty in Steno's work, which distinguished it from all previous anatomical works, including his own, was the book's geometrical formalism.[135] Steno explained that he wanted to introduce "all the terms, as is usually done by Geometers, in a synthetic order and with a table of definitions."[136] Like Euclid's *Elements*, the *Elementorum myologiæ specimen* was structured with a series of definitions, lemmas, and corollaries. In the single proposition of this book, written after many lemmas, Steno argued that geometry helped him to exclude the role of animal spirits in muscle contraction, a problem that will be further explained in the

---

133. Steno, *Elementorum myologiæ specimen*, 10. My italics.
134. Bertoloni Meli, "The Collaboration Between Anatomists and Mathematicians," 705–706.
135. See esp. Raphaële Andrault, "Mathématiser l'anatomie: La myologie de Stensen (1667)," *Early Science and Medicine* 15, nos. 4–5 (2010): 505–536.
136. Steno, *Elementorum myologiæ specimen*, 4.

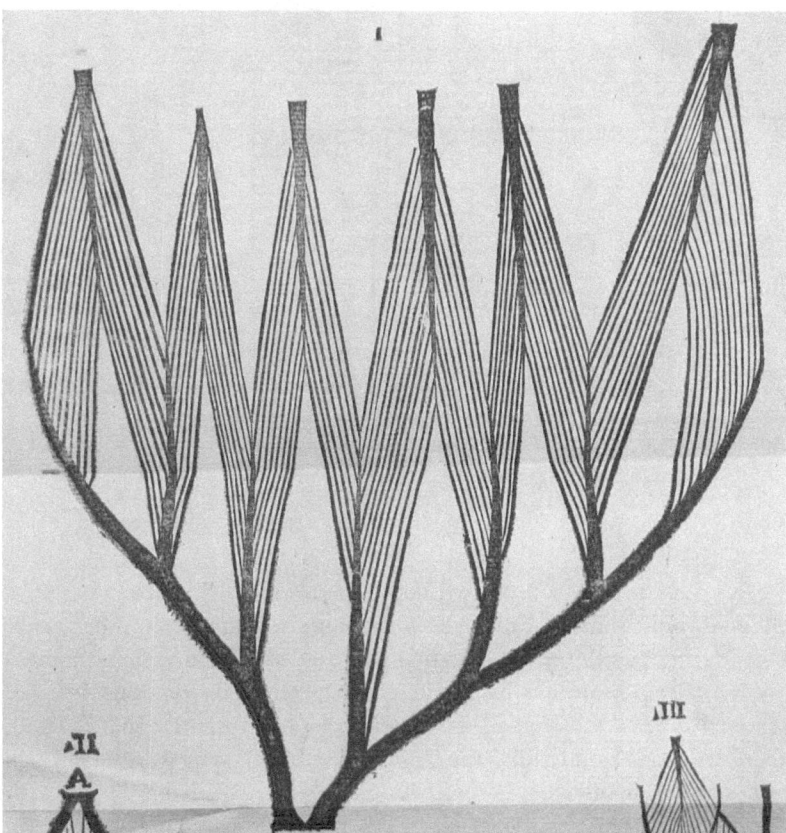

FIGURE 5.6. "The delicate structure of the deltoid [shoulder] muscle in which twelve single muscles are counted," in Steno, *Elementorum myologiæ specimen*, tabula III, fig. I. Quote from ibid., 37. Courtesy of the Institute of the History of Medicine, Johns Hopkins University.

following chapter.[137] In short, Steno's new book differed greatly from his previous anatomical works thanks to new kinds of images, explanations, and structure. Although rarely acknowledged as such, Steno's work was an important step toward a renewed interest in the study of fibers and mathematics among eighteenth-century anatomists.[138]

137. Steno, *Elementorum myologiæ specimen*, 25.
138. Tobias Cheung, "Omnis Fibra Ex Fibra: Fibre Œconomies in Bonnet's and Diderot's Models of Organic Order," in "Transitions and Borders Between Animals, Humans and Machines 1600–1800," [special issue,] *Early Science and Medicine* 15 (2010): 66–104.

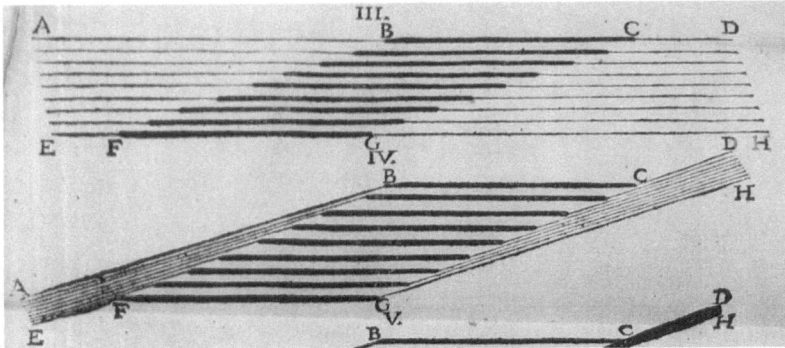

FIGURE 5.7. "Figure 3 shows the rectilinear, unequally equal motor fibers..., figure 4 shows the same motor fibers equally inflected," in Steno, *Elementorum myologiæ specimen*, tabula I, figs. III and IV. Quote from ibid., 35. Courtesy of the Institute of the History of Medicine, Johns Hopkins University.

But did the *Elementorum myologiæ specimen* really create a break in Steno's publication record?[139] This problem becomes even more acute considering that the book's two final treatises have little to do with muscles and seem more like his works of anatomical observations. Indeed, it was in one of those treatises, on Steno's dissection of a shark's head, that he wrote about fossils. Thus, one may also ask, why was the book in which Steno made a name for writing on fossils entitled *A Specimen of the Elements of Myology, or A Geometrical Description of the Muscle*? My research on the book's original manuscript, preserved at the Royal Danish Library, solves this problem and integrates this apparently disruptive publication with Steno's other books.[140] In the book's very last sentence, Steno mentioned his "closest friend, Vincenzo Viviani, Mathematician of the Most Serene Grand Duke, who, more than a spectator, was present in these and other things contained in it."[141] This statement confirms Viviani's contribution as a witness to Steno's observations, a common practice in anatomical treatises.[142] However, the original manuscript is more revealing of Viviani's role: there, Steno wrote that Viviani had been "*very frequently*

---

139. Andrault, "Mathématiser l'anatomie," 507–508.
140. MsElem. The only study of the manuscript is in Danish; see Gustav Scherz, "Danmarks Stensen-Manuskript," *Fund og Forskning* (1959): 19–33.
141. Steno, *Elementorum myologiæ specimen*, 119.
142. See Bertoloni Meli, "The Collaboration Between Anatomists and Mathematicians," 699–700.

present in ... the things contained in this book *both in advice as in help*."¹⁴³ These two final compliments are struck out in the manuscript, possibly by Viviani himself. Viviani had direct access to the original manuscript because he oversaw its publication in Florence while Steno was in Pisa; he even copied one of the sections himself.¹⁴⁴ Furthermore, in a letter dated January 1667 and sent from Florence to the Medici court in Pisa, Viviani mentioned that as soon "as the printing of the book is under way," Steno may go to Pisa to dissect.¹⁴⁵ Thus, there is no doubt that Viviani was directly involved in the editing and perhaps the composition of the manuscript. But Viviani also left other marks on the book.

Perhaps the most revealing change lies in the book's title. When Steno departed for Pisa, the original title of the book, written in Steno's hand, was "Another specimen of anatomical observations [*observationes anatomicæ*] on muscles and other animal parts in which a specimen of the elements of myology and the dissected head of a shark are proposed, illustrated by various observations and figures" (fig. 5.8).¹⁴⁶ This original title reveals that Steno did not see this book as a break with his previous works. On the contrary, in his mind, this book fitted perfectly with his other written works in the genre of *observationes anatomicæ*. As explained in previous chapters, Steno kept on crafting his own style of writing anatomy, always framing it within this genre of *observationes*. A list of Steno's titles in Latin makes this continuity clearer:

Observationes anatomicæ *quibus varia oris, oculorum, et narium vasa describuntur* (1662)
De musculis et glandulis **observationum [anatomicarum] specimen** (1664)
De musculis aliisque animalium partibus **observationum anatomicarum specimen alterum** (1667)¹⁴⁷

---

143. MsElem, fol. 99v: "qui hisce aliisque præsenti libro contentis et consilio et auxilio persæpe adfuit." My italics highlight the differences with the printed version.

144. MsElem., fol. 59r–v. Most Viviani letters were in Italian and in a loose hand. Nonetheless, there are many similarities, especially in the capital letters. Scherz also suggests it is Viviani's hand in "Danmarks Stensen-Manuskript," 30.

145. Viviani to Bruto della Molara, 26 January 1667, in BNCF, Gal. 252, fols. 109r–110v, esp. 110r: "poiche incamminata che sarà la stampa del libro potra il Sig. Niccolo ò rimanere quà o venire."

146. MsElem, fol. 1r: "De musculis aliisque animalium partibus observationum Anatomicarum specimen alterum quo Elementorum myologiæ specimen et Canis carachariæ dissectum caput variis observationibus et figuris illustrata proponuntur."

147. The frontispiece of the Copenhagen edition of *De musculis et glandulis* (1664) is entitled "Observationum anatomicarum specimen."

Steno's original title also explains better how the account of the dissection of the shark's head fits with the rest of the book. Indeed, Steno's book was just "another sample of anatomical observations" on muscles and other animal parts, such as those of sharks. It is also interesting that, when Steno returned to Florence in late April or early May 1667, he still decided to add an observation about fossils that he had heard of in his previous trip to Lucca. But since the book was already complete, possibly already in print, he could only include this detail in the book's index, which he wrote with his own hand in the final manuscript.[148] This continuous collage of observations—first the dissection of an entire shark in Pisa and then the fossil news from Lucca—confirms that, regardless of the title change, Steno kept intervening in the book as if it were part of the genre of *observationes*, just as he did when taking notes as a young student in Copenhagen.

The book's final title, probably chosen by Viviani, highlights the mathematical section on muscles over the others, which, to be fair, corresponds to the book's longest part. But it also corresponds to the interests of Viviani and the Medici family, whose publications at the time were mostly on mathematical topics. Two other scientific works published under Medici sponsorship in this year also underwent intense editorial intervention at the hands of Prince Leopoldo, namely Borelli's *Theoricæ mediceorum planetarum* (Florence, 1666) and the Cimento's *Saggi di naturali esperienze*.[149] Therefore, it is possible that both Viviani and Leopoldo edited Steno's entire book while Steno was with the court in Pisa and Livorno in the spring of 1667. In the end, the book was published as ***A Specimen of the Elements of Myology, or A Geometrical Description of the Muscles*** to Which Are Added the Dissected Head of a Shark and a Dissected Fish of the Species of Sharks.[150] This was not the only time Viviani edited a manuscript by Steno: he would also edit the *De solido intra solidum naturaliter contentum* (Florence, 1669), Steno's second book sponsored by the Medici.[151]

---

148. MsElem, fols. 100r–101v; and Steno, *Elementorum myologiæ specimen*, 122. For the travel details, see, e.g., Bruto della Molara to Viviani, 11 April 1667, in BNCF, Gal. 163, fol. 110v.

149. See Bertoloni Meli, "Shadows and Deception," 399–400.

150. *Elementorum myologiæ specimen seu musculi descriptio geometrica cui accedunt canis carchariæ dissectum caput et dissectus piscis ex canum genere.*

151. Gal. 291, fols. 1r–27r, esp. fol. 1r: "Questo fu stampato sotto la mia cura." Viviani's hand is much clearer in this case.

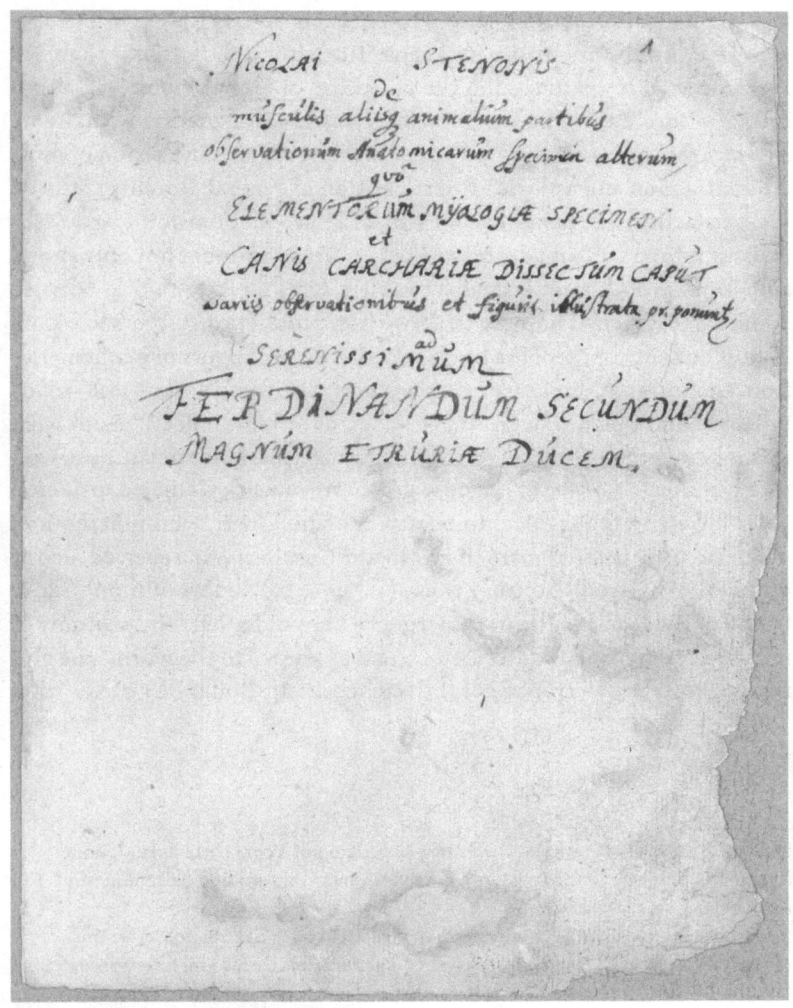

FIGURE 5.8. Title page of the manuscript of Steno's *Elementorum myologiæ specimen* in Copenhagen, Royal Library of Denmark, NKS 4019 kvart, fol. 1r.

Regardless of how it was made, Prince Leopoldo appreciated Steno's mathematical approach to the muscles from very early on. In late October 1666, when the bulk of the manuscript was almost complete, Leopoldo acknowledged Steno's mathematical skill.[152] As soon as the book

152. Prince Leopoldo to Alessandro Segni, 27 November 1666, in Copenhagen, Royal Library of Denmark, Acc. 2019/54, fol. 1v: "il S. Stenone Danese, Anatomico

was published, Leopoldo sent it to contacts across Italy and Europe, many of whom were mathematicians. The earliest letter that I found in Florentine archives related to the reception of Steno's book was, ironically, by Borelli, who had previously rejected collaborating with Steno. After learning that Steno's book dealt with a very specific action of muscle contraction and not with animal motion in general, Borelli expressed his satisfaction and acknowledged that "I have no occasion to grieve as that sumptuous title made me suspect."[153] This comment not only shows that Borelli too acknowledged a problem with the book's title but also confirms that he had nothing to do with its production. Others to whom Leopoldo sent copies of the book include two professors of mathematics from the order of the Jesuati (not Jesuits), Urbano d'Aviso (1618–1686) in Rome and Stefano degli Angeli (1623–1697) in Venice.[154] Both were former students of the more famous Jesuate mathematician Bonaventura Cavalieri (1598–1647), whose geometrical work Galileo considered second only to that of Archimedes.[155] Stefano Gradi, whom Steno met in Rome, also studied mathematics with Cavalieri and received a copy of the book.[156] And Giovanni Bona (1609–1674), a devout monk, close friend of Ricci and of the pope, wrote to Leopoldo that "the anatomy of muscles geometrically described ... makes known to the world" the glories of Florence's patronage of the sciences.[157] It should be noted that at

---

gioviane d'età ma insigne nel suo mestiere corredato poi d'ogni sorte di laudazione, e geometra bravo il che molto li giova al suo mestiere et il vero tipo della modestia." The letter begins "D. Aless[andro]."

153. Borelli to Prince Leopoldo, 16 May 1666, in BNCF, Gal. 281, fol. 100v: "mi rallegro di non aver occasione di dolermi come quel titolo fastoso mi faceva sospettare." On Borelli's further engagement with Steno, see Nuno Castel-Branco, "Who Was Borelli Responding To? Nicolaus Steno in *De motu animalium* (Rome, 1680–1681)," *Physis* 57, no. 2 (2022): 431–448.

154. Urbano d'Aviso to Prince Leopoldo, 28 May 1667, in BNCF, Gal. 278, fol. 23r; and Stefano degli Angeli to Prince Leopoldo, 4 June 1667, in BNCF, Gal. 278, fol. 27r–v.

155. See Franco Aurelio Meschini, "Davisi, Urbano Giovan Francesco," in DBI, vol. 33, 171–173; and Mario Gliozzi, "Angeli, Stefano degli," in DBI, vol. 3, 204–206. On Cavalieri, see Ettore Carruccio, "Cavalieri, Bonaventura," in *Dictionary of Scientific Biography*, vol. 3 (Detroit, MI: Charles Scribner's Sons, 2008), 149–153, esp. 149.

156. Stefano Gradi to Prince Leopoldo, 18 June 1667, in BNCF, Gal. 278, fol. 28r–v.

157. Giovanni Bona to Prince Leopoldo, 4 July 1667, in BNCF, Gal. 312, fols. 98r–99r: "l'anatomia de muscoli descritta geometricamente dallo Stenone ... fà conoscere al mondo. ..."

this time Steno was still a Protestant author, suggesting that these praises were centered on scientific merit alone. By the end of May, Chapelain was also aware of the book in Paris.[158] In England, Oldenburg and Robert Boyle (1627–1691) would have to wait until February 1668 for their copy to arrive, which Prince Leopoldo sent through the hands of the French mathematician Auzout.[159] But, perhaps not surprisingly, the greatest acclaim came from those Roman friends who had originally recommended Steno to Leopoldo. Fabri, for instance, was ecstatic, confessing that "I will not say [that] I read, but devoured [the book]."[160] And Ricci was also in awe at Steno's mathematical myology, comparing it to "the geometrical method of Galileo," who had also described the motion of projectiles with geometry.[161]

Friendships, Patronage, and the Accademia del Cimento

This positive reception of the *Elementorum myologiæ specimen* confirms that Steno met the expectations of his Medici patrons. The book's success was due in large part to the princely patronage of Grand Duke Ferdinand II. As Steno wrote in the book's preface: "You [i.e., the grand duke] have given me hospitable reception in the city and allotted me everything which might possibly be of service for my studies and experiments." Indeed, the grand duke took a personal interest in Steno's research by giving him "some share in those hours in which you relax your mind, . . . [when] you take delight in the mysteries of art and nature which others would seek in games and pleasure." But one final key aspect to the book's success, as well as his positive relationship with the Medici, was, I argue, the friends that he made in Florence and with whom he shared his interests. These friends that Steno made at "the court blooming with arts" were all members of the Accademia del Cimento.[162] Viviani, whose role in Steno's book was mentioned earlier, was one of his closest friends and may have played a role in writing this preface to the grand duke. Indeed, one January night, when Steno and

158. Jean Chapelain to Steno, 27 May 1667, in Larroque, *Lettres de Jean Chapelain*, 2:514.
159. Henry Oldenburg to Robert Boyle, 4 February 1668, in Michael Hunter, Antonio Clericuzio, and Lawrence Principe, *The Correspondence of Robert Boyle*, 6 vols. (London: Pickering & Chatto, 2001), 4:21–23.
160. Fabri to Leopoldo, 7 July 1667, in BNCF, Gal. 278, fol. 31r-v: "non dicam evolvi, sed devoravi."
161. Ricci to Leopoldo, 30 May 1667, in BNCF, Gal. 278, fol. 24r: "che ridotta a metodo geometrico dal Galileo, com'è ridotta a la materia de' muscoli dallo Stenone."
162. Steno, *Elementorum myologiæ specimen*, ii.

Viviani were talking, they found that "it was very easy and quick to reach the topic of praising the singularity of the Most Serene Grand Duke over any other Prince in [his] delight and affection for the sciences and most noble arts."[163] Other scholars that Steno met at the Medici court were the "most respected Lorenzo Magalotti," who was an eyewitness to many of Steno's dissections and observations, as well as his "most illustrious friend Francesco Redi" and the "most learned friend Carlo Dati," both of whom shared important manuscripts from their personal libraries with Steno as he was writing his book.[164] Redi and Dati, who were prolific writers of Italian literature, must have passed on their literary interests to Steno, who was nominated for membership in the Accademia della Crusca on 16 July 1668.[165] Even Magalotti extended his help beyond scientific matters, helping Steno settle a problem with his first hostess in Florence.[166]

In the long run, it was Viviani who looked after Steno by introducing him to Florentine daily life and fighting to keep him in Florence for longer. On 26 January 1667 Viviani explained to Bruto della Molara, the grand duke's secretary, that Steno's hesitations about staying in Florence were "entirely out of respect for his [Danish] king, taking care to not give there the slightest shadow" of not wanting to go back.[167] After all, like those of all other Danes, Steno's European travels would come to an end, and he would have to return home. Yet, despite these expectations, Steno's heart wavered toward Florence. As he "sighed" to Viviani, in Florence there was "the great commodity that, with a mere wink of His Highness, it is possible

---

163. Viviani to Molara, 26 Janury 1667, in BNCF, Gal. 252, fol. 109r: "fù molto facile, e pronto l'incontrar materia di commendare la singolarità del Serenissimo Gran Duca sopra ogni altro gran Principe nel diletto et affezzione alle scienze e alle nobili arti."

164. Steno, *Elementorum myologiæ specimen*, 108, 109, 70, respectively. On Dati, see Magda Vigilante, "Dati, Carlo Roberto," in DBI, vol. 33, 24–28. On Redi, see Findlen, "Controlling the Experiment."

165. Biblioteca dell'Accademia della Crusca, Ms. 33 (BCF–Cat.1a), as quoted in www.accademicidellacrusca.org/scheda?IDN=2148. On this academy, see Mirto, *Alessandro Segni e gli Accademici della Crusca*.

166. Magalotti to Redi, 21 January 1667, in BML, Redi 206, fol. 18r. This problem continued well into March, as Magalotti's letters to Redi show; see BML, Redi 206, fols. 27r–28v, 48r. Steno was a guest at an inn known as "le mome"; see also Folco Rinuccini to Alessandro Segni, 10 December 1666, in Mirto, *Alessandro Segni e gli Accademici della Crusca*, 447–450.

167. Viviani to Bruto della Molara, 26 January 1667, in BNCF, Gal. 252, fol. 109v: "io ritrassi che il suo [parlare] era tutto rispetto verso il suo Rè, a premura di non dar colà minima ombra."

to carry out observations and experiments of all kinds, especially in anatomy, with the dissection of various animals and especially of the corpses from our hospital."[168] Viviani also said that Steno "added various reasons why similar opportunities are not found [elsewhere]," nor can they be enjoyed as they are in Florence.[169] Fortunately for Steno, as Viviani said, "it was up to him to give up or keep such a beautiful opportunity, because His Highness would certainly" offer it to him.[170]

In the end, the deal was that Steno would travel south and, upon his return, would stay for another period in Florence.[171] In addition, Viviani convinced Steno "to postpone his travel to Naples until the upcoming winter and to stay here" for the time being.[172] As mentioned earlier, Chapelain had made a similar proposal to Steno in Paris, albeit unsuccessfully. But Viviani, unlike Chapelain, was a natural researcher and better able to understand Steno's intellectual anxieties. In his letter to della Molara, Viviani stated that Steno's main concern was in "making use of it [the grand duke's support] to obtain new knowledge."[173] Therefore, in April 1667, Viviani started looking for a better place for Steno to stay, "since the grand duke besides the twenty-five scudi per month wants to continue further the commodity of a lodging."[174] As he wrote to della Molara: "For now, I think of the Palazzo Vecchio or elsewhere where it might please more His Highness, between the Palazzo and [the Hospital of] Santa

---

168. Viviani to Bruto della Molara, 26 January 1667, in BNCF, Gal. 252, fol. 109r: "e tra l'altre sospirò il Sig. Stenone le gran comodità che a un solo cenno di V.A. l'ottengono di poter far osservazioni et esperienze in ogni genere, et in particolare nella Anatomia, con la sezzione di vari animali e principal[mente] sopra i cadaveri di questo nostro ospedale."

169. Viviani to Bruto della Molara, 26 January 1667, in BNCF, Gal. 252, fol. 109r: "adducendomi diverse ragioni per le qual simili occasioni non se trovano, nè si possono goder altrove."

170. Viviani to Bruto della Molara, 26 January 1667, in BNCF, Gal. 252, fol. 109v: "Io risposi che il lasciare o ritenere questa cosi bella occasione stava in lui perche per la parte di V.A. egli era certo d'esser gradito."

171. Viviani to Bruto della Molara, 26 January 1667, in BNCF, Gal. 252, fol. 109v: "compresi che volentieri doppo il ritorno dal viaggio di Napoli, si sarebbe fermato qua di nuovo a servire a V.A. per qualche tempo con intraprendere l'osservazioni del sangue."

172. Viviani to Bruto della Molara, 26 January 1667, in BNCF, Gal. 252, fol. 110r: "ma col differire il viaggio de Napoli al futuro Inverno, a rimanere adesso."

173. Viviani to Bruto della Molara, 26 January 1667, in BNCF, Gal. 252, fol. 109v: "che per approfittarssene con l'acquiste di nuove cognizioni."

174. Viviani to Bruto della Molara, 1 April 1667, in BNCF, Gal. 158, fol. 179r: "che il Sermo. Gran Duca oltra a 25 scudi il mese vuol continuargli ancora la comodità del quartiere."

Maria Nuova, where he will often have to be in the summer. And if there were some separate and free rooms where he could make some chymical or other experiments on his own, I know it would be very welcome to him."[175] One of the reasons for this request was Steno's previous lodging problems in Florence. But the other reason was the ability to fulfill his intellectual ambitions, the details of which Viviani was well aware of. Steno had told him that "to open the way for any useful knowledge, besides the investigation of the true structure of [bodily] parts, a diligent examination of the fetus and the nature of fluids and blood was required, calling in the help of some curious operations of chymistry."[176] This access to a private room and a chymical laboratory in the center of Florence, to which the grand duke agreed, enabled Steno to explore such problems as the chymical constitution of blood and the precipitation of solids from liquids, which he would use in his work on fossils, as the next chapter will show.[177] Moreover, this private room also gave Steno the solitude he needed to pray and study more about his religious anxieties, which would eventually lead to his conversion in November 1667.

Yet, despite this relative intellectual freedom, the following months were a time of intensive work in Florence. As a new member of the scientific circles of the Medici, Steno had extra responsibilities. Around November 1667 the *Saggi di naturali esperienze* was published. This book was an account of the most important experiments done at the Accademia del Cimento, and scholars in Europe had been waiting years for it.[178] For instance, in December 1665 Segni told Redi that the *Saggi* were "infinitely desired" in Paris.[179] Steno had almost certainly heard about it from his time in Paris. Indeed, Steno acknowledged to Leopoldo that the *Saggi* came

---

175. Viviani to Bruto della Molara, 1 April 1667, in BNCF, Gal. 158, fol. 179r–v: "e per ora mi sovviene in Palazzo Vecchio, o altrove dove più piacesse a V.A. in sito medio tra Palazzo e S.Maria Nuova dove nell'estate avvenire averà egli bisogno d'andare spesso. E se vi fusse qualche stanza appartata e libera da potervi far qualche esperienze chimica o altro da se solo, so che gli sarebbe gratissimo."

176. Viviani to Bruto della Molara, 26 January 1667, in BNCF, Gal. 252, fol. 109r–v: "per aprire la strada a qualche cognizione d'utilità, oltre alla perquisizione della vera struttura delle parti, si richiedeva al parer suo una diligente esamine del feto e della natura de' fluidi e del sangue, chiamandovi in aiuto qualche curiosa operazione della chimica."

177. Bruto della Molara's response to Viviani confirms this. See Molara to Viviani, 11 April 1667, in BNCF, Gal. 163, fol. 110r–v.

178. See Boschiero, *Experiment and Natural Philosophy*, 181–193; and Middleton, *The Experimenters*, 65–80.

179. Segni to Redi, 11 December 1665, in Mirto, *Alessandro Segni e gli Accademici della Crusta*, 102–104.

just in time, when "the greatest minds of all experimental academies show a particular ambition to celebrate the noble enterprise of the Accademia del Cimento."[180] In addition, Steno commented, "among all the books I have seen on this kind of experiments, I have not seen a counterpart" to this one.[181]

As a member of the Medici court, Steno mailed copies of the *Saggi* to his friends in northern Europe, including the king of Denmark and colleagues in Leiden.[182] Moreover, as objections to the book arrived in Florence, Steno also stepped up to respond to some of them, even though they had little to do with anatomy. Indeed, they were not sent without Viviani's reviewing them beforehand.[183] These objections involved experiments on the vacuum, such as using syringes to inject air into Torricellian tubes.[184] Thanks to his studies on the circulation of the blood, especially through Pecquet's *Experimenta nova anatomica* and his own research on respiration, Steno knew some of these experiments with Torricellian tubes. In addition, he was familiar with the injection of fluids into the blood circulation of animals, and so he also knew how to carefully maneuver similar instruments.[185] Ultimately, Steno's ability to write on topics of physicomathematics such as the vacuum, his proficiency in the new anatomy, and his more recently acquired mastery of the Italian language made him much similar to the figure of Borelli, who had left Florence and the Accademia del Cimento a few months before.

But did Steno replace Borelli as a member of the Accademia del Cimento? The answer to this question is as complex as the history of the Accademia. Even though this academy was fully funded by the Medici princes, its daily operations were more similar to those of the informal

---

180. Steno to Cardinal Leopoldo, 14 January 1668, in BNCF, Gal. 278, fols. 4r–5r (also in *Epistolae*, 1:196): "dove li più grandi ingenii di tutte le accademie esperimentali mostrano una particolar abizione nel celebrare la nobile impresa dell'Accademia del Cimento."

181. Steno to Cardinal Leopoldo, 14 January 1668, in BNCF, Gal. 278, fols. 4r–5r: "fra tutti i libri sin ora da me visti nel genere delle esperienze non ho trovato pari a questo."

182. Steno to Cardinal Leopoldo, 14 January 1668, in BNCF, Gal. 278, fols. 4r–5r; and Lodovico Magalotti to Steno, 21 January 1668, in BNCF, Gal. 291, fol. 225r (also in *Epistolae*, 1:197).

183. Steno to Viviani, 4 February 1668, in BNCF, Gal. 269, fols. 216r–v, 223r–224v.

184. Fabrizio Guastaferri to Steno, January 1668, in BNCF, Gal. 269, fols. 219r–220v.

185. Redi uses the word *schizattoio* for Steno's injection instruments, which is also in the *Saggi*; see Nuno Castel-Branco, "Friendship Fostered by Poisons: The Collaboration of Nicolaus Steno and Francesco Redi," in *Poison: Knowledge, Uses, Practices*, ed. Caterina Mordeglia and Agostino Paravicini (Florence: SISMEL–Edizioni del Galluzzo, 2022), 325–346.

academies of Thévenot and Bourdelot than to those of modern scientific academies. For example, it is not always clear exactly who the academy's members were, where they worked, and how regularly they met—besides the meetings mentioned in the extant diaries of the Cimento.[186] Dati, for example, is absent from these diaries but is known to have made important contributions to the experiments carried out at the Cimento meetings.[187] Moreover, Redi, in two letters from around 1660, mentions a few experiments performed "in the Accademia del Cimento," which are also not mentioned in the diaries.[188] Therefore, the fact that Steno is not mentioned in the Cimento diaries does not necessarily mean that he was not a member. In fact, in light of Steno's research in Florence and his work in publicizing the *Saggi*, several historians have concluded that Steno did become an academician.[189] The problem with this conclusion is that the timing of Steno's arrival in Florence coincided with the official dissolution of the academy. As Knowles Middleton argues, the Cimento ended in 1667 with the departure of three of its leading members from Tuscany, especially Borelli.[190] Early in January 1668 Lorenzo Magalotti tried to salvage it with Steno, writing to Leopoldo de' Medici that before "the present scattering of our Academy by the departure of Borelli, Oliva, and Rinaldini it is my belief that nothing more desirable [than hiring Steno] could happen."[191] But the reality about the academy's end comes from Steno himself, who in a letter to Leopoldo praising the *Saggi* mentioned his "great pain" at seeing the end of the meetings of the Cimento.[192] That is, in Steno and Magalotti's eyes, the Accademia had ceased all activities by 1668, and there

---

186. Middleton, *The Experimenters*, 27. On the extant diaries and notes, see ibid., 42–45.

187. Middleton, *The Experimenters*, 27.

188. Middleton, *The Experimenters*, 50–52. These letters also show that the name Cimento was used before the publication of the *Saggi*.

189. Maria Luisa Righini Bonelli, "The Accademia del Cimento and Niels Stensen," in Scherz, *Steno and Brain Research*, 253–260, esp. 260; Scherz's biography in Kardel and Maquet, *Nicolaus Steno: Biography and Original Papers*, 171–172; and Miniati, *Nicholas Steno's Challenge for Truth*, 194.

190. Middleton, *The Experimenters*, 316–319.

191. Magalotti to Leopoldo, in BNCF, Gal. 278, fols. 119r–120r, as quoted in Middleton, *The Experimenters*, 317.

192. Steno to Leopoldo, 14 January 1668, in BNCF, Gal. 278, fols. 4r–5r: "celebrare la nobile impresa dell'Accademia del Cimento, non senza grandissimo dolore riguardando da non sperati acidenti impedito uno studio, dal quale i grandissimi avanzi fatti in poco tempo nelli ricercamenti naturali danno ad ognuno chiari indizi di quello che dalla sua continuazione sarebbe stato da sperare."

is no evidence that Leopoldo continued using its name.[193] This does not mean that scientific research in Medici Florence stopped. Far from it, as the work of Steno, Viviani, and Redi, often in collaborative ways, shows. The truth is that Steno was like an academician, but the Accademia del Cimento was no more.

※

The *Elementorum myologiæ specimen* was Steno's most innovative and last anatomy book. But his anatomical practice did not vanish. On the contrary, he continued to make anatomical observations in the following months by dissecting at the Florentine court and with friends. Redi mentioned some of these observations in his books on the generation of insects, in which he argued against the spontaneous generation of larvae in rotten meat.[194] In early 1669 Steno finally started his return trip to Denmark. Yet, even on his way north, he continued dissecting. In Innsbruck, at the court of Anna de' Medici (1616–1676), archduchess of Austria and sister of Grand Duke Ferdinand II, Steno dissected a "monstrous calf," as described by the archduchess in a letter to Ferdinand.[195] When Steno finally settled in Denmark in 1672, it was as Royal Anatomist.[196] Thomas Bartholin, who wanted to start a publication in Copenhagen similar to the *Journal de sçavans* and the *Philosophical Transactions*, asked Steno to contribute to it.[197] Steno accepted and shared with Bartholin notes from observations that he had made in Florence and subsequent travels. Steno's nine series of anatomical notes appeared in the first two volumes of

---

193. Interestingly, those who argue that Steno was a member of the Cimento also argue that the Cimento stopped only with Leopoldo's death in 1675.

194. Redi, *Esperienze intorno alla generazione degl'insetti*, 75, 129; and Francesco Redi, *Osservazioni intorno agli Animali viventi che si trovano negli Animali viventi* (Florence: Piero Matini, 1684), 159. See also Castel-Branco, "Friendship Fostered by Poisons," 329–335.

195. Anna de' Medici to Ferdinand II, 16 June 1669, in ASF, MdP, 1020, fols. 664r–665r. See also Troels Kardel, "Steno on Hydrocephalus: Introduction to Niels Stensen's Letter 'On a Calf with Hydrocephalus,' with a Short Biography," *Journal of the History of Neurosciences* 2 (1993): 171–178.

196. For Steno as "Anatomicus Regius," see Thomas Bartholin, ed., *Acta medica et philosophica Hafniensia*, 5 vols. (Copenhagen, 1673–1680), 1:200. On Steno's return to Denmark, see Scherz's biography in Kardel and Maquet, *Nicolaus Steno: Biography and Original Papers*, 343–349.

197. Skavlem, "The Scientific Life of Thomas Bartholin," 80.

the *Acta medica et philosophica Hafniensia* (Copenhagen, 1674–1679).[198] These notes were of all kinds and included accounts of observations in bullet points, dissection narratives, a letter that he wrote to Ferdinand II on the dissection of a calf's head (and which Bartholin translated into Latin), and even notes for a speech at a public anatomical demonstration in Copenhagen. This material could perfectly have been compiled into a book of *observationes*. The fact that it ended up in Bartholin's journal confirms that the epistemic genres of *observationes* and note-taking practices were the root of the new scientific journals of the early modern period.[199]

Steno published his final scientific book in 1669. Rather than being about the body, it was about rocks and the history of the Earth, and it required a lot of traveling. Indeed, after the publication of *Elementorum myologiæ specimen*, Steno's new research on fossils started to take more and more of his time. In 1667 and 1668, under the auspices of the Medici family, Steno traveled to Livorno, the island of Elba, Volterra, and Arezzo (Redi's hometown) for purposes of geological research.[200] Archduchess Anna also told her brother Ferdinand that after dissecting the calf's head in Innsbruck, Steno gave a "sound lecture ... after having seen the salt mines of Halle and the mines of Schwaz."[201] Finally, in Copenhagen, Steno taught his anatomy students the history of the Earth's strata, as confirmed by the lecture notes of one of them.[202] In short, in 1667 anatomy ceased to be the focus of his interdisciplinary research, but it did not disappear altogether. The truth is that Steno's interest in the Earth was not entirely separate from his interest in the body, as the following chapter argues.

---

198. Bartholin, *Acta medica et philosophica Hafniensia*, 1:200–203, 203–207, 249–262; 2:81–92, 141, 210–218, 240–241, 320–345, 359–366.

199. Pomata, "Sharing Cases," 225–226. See also David Banks, *The Birth of the Academic Article: "Le Journal des Sçavans" and the "Philosophical Transactions," 1665–1700* (Sheffield: Equinox, 2017).

200. On Steno's Italian travels, see Scherz's biography in Kardel and Maquet, *Nicolaus Steno: Biography and Original Papers*, 284–291. See also Francesco Redi to Giovanni Battista Redi, 26 August 1668, in BML, Acq. e doni 859, fol. 15r–v: "Questa mattina sabato è passato di qui il Sr. Stenone per andare a Vallobrosa. ..."

201. Anna de' Medici to Ferdinand II, 16 June 1669, in *Epistolae*, 1:17.

202. Axel Garboe, *Niels Stensens (Stenos) geologiske Arbejdes Skaebne: Et Fragtment af Dansk Geologis Historie* (Copenhagen: Danmarks Geologiske Undersøgelse, 1948), 16–25.

CHAPTER SIX

Thinking the Earth with the Body

In October 1666 news arrived in Florence of an extraordinarily large white shark caught in Livorno, the port city of Tuscany. Having heard this, Grand Duke Ferdinand II de' Medici asked that the shark's head be shipped up the Arno to Florence and to have Nicolaus Steno, the new anatomist at court, dissect it. Steno had settled in the Florentine court that summer. Since dissection was something that Steno did regularly, he accepted the invitation. Moreover, why not use the shark to explore those very same questions about anatomy that he was already researching about muscles, glands, and the reproductive organs?[1] The shark's dissection took place around 18 October and became known for igniting Steno's geological research, which has made him known today as the founder of modern geology.[2] In short, so the story goes, Steno realized during the dissection that shark's teeth were identical to a kind of fossil often found far from the sea.[3] This led him to argue that the Earth has a history, which can be known through a series of rules still taught today as Steno's principles of stratigraphy.[4] In the words of the historian of

---

1. Only the head of the large white shark arrived in Florence, but Steno dissected another shark in which he observed the sexual organs a few months later; see Steno, *Elementorum myologiæ specimen*, 111–119.

2. See, e.g., Stephen Jay Gould, *Hen's Teeth and Horse's Toes: Further Reflections in Natural History* (New York: W. W. Norton, 1983), 72; and Cutler, *The Seashell on the Mountaintop*, 187.

3. See Scherz's biography in Kardel and Maquet, *Nicolaus Steno: Biography and Original Papers*, 196; Angeli, *Niels Stensen*, 151–152; Martin Rudwick, *The Meaning of Fossils: Episodes in the History of Palaeontology* (London: Macdonald, 1972), 50; and Cutler, *Seashell on the Mountaintop*, 53–56.

4. See, e.g., Michael E. Brookfield, *Principles of Stratigraphy* (Malden, MA: Blackwell, 2004), 115–117. The history of how these principles came to be known as Steno principles has yet to be written.

geology Martin Rudwick, Steno's examination of this shark's teeth "had a striking catalytic effect" on the history of science and the study of fossils.⁵

The problem with associating this shark's dissection with Steno's research on fossils, however, is that none of the eyewitnesses to the dissection mentioned the shark's teeth nor anything related to fossils. One of the eyewitnesses was Lorenzo Magalotti, secretary of the Accademia del Cimento. About a week after the dissection took place, Magalotti wrote to the intellectually curious archbishop of Siena that the shark was examined by "the finest knife of Mr. Nicolaus Steno."⁶ He also highlighted three different aspects of Steno's anatomical dissection, none of which was related to shark's teeth.⁷ Carlo Dati, another intellectual at the Medici court, also mentioned to a friend that the shark was explicitly brought to Florence to allow Steno to make anatomical observations of it.⁸ Steno himself wrote to William Croone that he had seen "various curiosities in the head of a fish called Canis Carstrarius [sic], about which I will soon give a report and will make sure to send it to you when it is finished."⁹ In that letter, Steno wrote about "water to remove stains [*l'eau pour lever les taches*]" and his color experiments with the lobster, both topics related to his long-standing chymical interests. Finally, five weeks later, in a recently discovered letter, Prince Leopoldo de' Medici, the main sponsor of the Accademia del Cimento, praised Steno's broad knowledge but never mentioned fossils or the Earth.¹⁰ Therefore, at least in the five weeks after this dissection took place, it seems that no one associated the dissection of this shark with fossils. More important, in the months that followed there is no reference in any of Steno's writings to fossils.

---

5. Rudwick, *The Meaning of Fossils*, 49–100, esp. 49–50.

6. Lorenzo Magalotti to Archbishop Ascanio Piccolomini, 26 October 1666, in Siena, Biblioteca Communale di Siena, D. V. 7, fol. 16v: "La testa è venuta a Firenze sotto l'esame del finissimo coltello del Sig. Niccolò Stenon."

7. Magalotti mentioned vessels of skin glands, the connection between arteries, veins, and cartilage, and an optical nerve.

8. Carlo Dati to Ottavio Falconieri, 25 October 1666, in Michele Mercati, *Metallotheca vaticana* (Rome: Johannes Maria Salvioni, 1719), xxxiv–xxxv, esp. xxxiv.

9. Steno to William Croone, after 18 October 1666, in London, Royal Society, EL/S1/92: "Nous avons veu icy diverses curiosites dans la teste d'un poison, qui s'appelle Canis Carstrarius, desquelles ja donneray bien tost une relation et je ne manqueray pas de vous l'envoys quand elle sera achevee."

10. Leopoldo to Segni, 27 November 1666, in Copenhagen, Royal Library of Denmark, Acc. 2019/54, fol. 1v: "il S. Stenone Danese, Anatomico gioviane d'età ma insigne nel suo mestiere corredato poi d'ogni sorte di laudazione, e geometra bravo il che molto li giova al suo mestiere et il vero tipo della modestia."

How, then, did Steno go from performing an anatomical dissection to making claims about rocks?[11] Two recent attempts to answer this question show how difficult it is, yet how significant, to contextualize Steno's shift from anatomy to Earth history. One attempt is the explanation of continuity.[12] This explanation argues that Steno was already working on fossils long before performing this dissection. It is a thesis that also resonates with Steno's familiarity with tongue stones and fossils since his formative years in Copenhagen.[13] Steno's encounter with English naturalists in Montpellier in 1665 also suggests that they may have discussed fossils before the shark's dissection, since all of them would go on to write upon the topic. The problem with this explanation, however, is that no primary sources show that Steno wanted to write about fossils before publishing on them for the first time, six months after the dissection of the shark.[14] Stefano Dominici convincingly demonstrates that Steno visited geologically rich sights in Tuscany before dissecting the shark. And Steno was certainly familiar with the debate and observations on fossils, but familiarity does not translate to expertise.

The other attempt to explain Steno's turn to fossils is the theory of discontinuity.[15] It argues that Steno completely shifted his work from the body to the Earth because of the needs of the Medici court. It claims that Steno adopted a method named *historia*, based on observations and detached from philosophical interpretation, in order to make his research more similar to works sponsored by the Medici court, such as the *Saggi di naturali esperienze*.[16] The *Saggi* famously described experiments carried out at the Medici court while explicitly avoiding any interpretation of their

11. Steno once wrote that it was "a shark of prodigious size" that led him away from his anatomical projects; see Steno, *De solido*, 4. This was a broad, retrospective comment because he never associated observing this specific shark's teeth with fossils, as he associated, e.g., this shark's skin glands with sudoriferous glands; see Steno, *Elementorum myologiæ specimen*, 72–75.

12. Stefano Dominici, "A Man with a Master Plan: Steno's Observations on Earth's History," *Substantia* 5, no. 1, suppl. 1 (2021): 59–75.

13. Troels Kardel, "Prompters of Steno's Geological Principles: Generation of Stones in Living Beings, Glossopetrae and Molding," in Rosenberg, *The Revolution in Geology from the Renaissance to the Enlightenment*, 127–134, esp. 129–130.

14. On the primary sources related to the Montpellier meetings, see Poynter, "Nicolaus Steno and the Royal Society of London," 274; and Roos, *Web of Nature*, 67–68.

15. Jakob Bek-Thomsen, "From Flesh to Fossils: Nicolaus Steno's Anatomy of the Earth," in Duffin, Moody, and Gardner-Thorpe, *A History of Geology and Medicine*, 289–305; and Bek-Thomsen, "Steno's *Historia*: Methods and Practices at the Court of Ferdinando II," in Andrault and Lærke, *Steno and the Philosophers*, 233–258.

16. Bek-Thomsen, "Steno's *Historia*," 242–248.

results. In this light, the argument continues, Steno's writings on fossils were of the same genre of the *Saggi* and thus served as a suitable "patronage gift to the Medici court."[17] This argument creates a definite break between Steno's research in anatomy and the history of the Earth. In the words of Jakob Bek-Thomsen, Steno "never used *historia* as deliberately and methodologically" as he did at the Medici court.[18] Steno's books of anatomical *observationes*, however, belie this claim.

This chapter challenges both theories of continuity and discontinuity to argue that the answer lies somewhere in the middle. The trigger that turned Steno's attention to the Earth rather than dissecting a shark was reading a manuscript that challenged his research methods. This trigger, I demonstrate, was intrinsically linked not only to his anatomical research but also to his focused interdisciplinary method. Steno's search for reliable knowledge amid the uncertainty of the natural sciences led him to rely heavily on observations (mostly dissections) and on conclusions from physico-mathematics and chymistry. Therefore, I also argue that Steno's turn to fossils is a result of his search for certainty and of the specific methods that he developed before arriving in Italy. Taking Steno's anatomy seriously, as this book has been doing, thus leads to a more sophisticated understanding of his history of the Earth as well as his integration into the Medici court. With this in mind, this chapter starts in Steno's first year in Florence, 1666. This is the year in which Steno expanded his research from the body to include fossils. I then follow Steno's steps from his dissection of the shark in Florence to the publication of the book *De solido*, considered his masterpiece in geology, three years later. Throughout I add flashbacks to important aspects of his anatomical career that, as I demonstrate, are key to grasping his shift from the body to the Earth.

The Delayed Manuscript

When Steno dissected the famous shark in October 1666, he had just completed the manuscript on his new mathematical model of muscle contraction.[19] This is known because the book's first imprimatur, the typical

---

17. Bek-Thomsen, "From Flesh to Fossils," 303; and Bek-Thomsen, "Steno's *Historia*," 256. On the *Saggi* and its avoidance of philosophical discussions, see Biagioli, *Galileo, Courtier*, 359; Galluzzi, "L'Accademia del Cimento," 800; and Beretta, "At the Source of Western Science," 141. On gifts of patronage in early modern science, see Biagioli, *Galileo, Courtier*; and Findlen, *Possessing Nature*, 346–392.

18. Bek-Thomsen, "Steno's *Historia*," 248.

19. Steno, *Elementorum myologiæ specimen*, 123.

ecclesiastical approval of a book, is dated 27 October 1666.[20] However, in light of his dissection of the shark, Steno decided to place the printing of the book on hold so that he could add to it his anatomical remarks on the shark. The book was therefore published in April 1667 under the already mentioned title of *Elementorum myologiæ specimen* (*A Sample of the Elements of Myology*, myology being the study of muscles), with an additional description of the shark's head. As mentioned earlier, it was also in this description that Steno included his first treatise on fossils.[21] Historians have assumed that the text on fossils was complete by the end of 1666, two months after the dissection took place.[22] This is because a second, final imprimatur, dated December 1666, is preceded by a censor's report mentioning the shark's head.[23] Yet, if the text on fossils was complete by the end of 1666, then Steno would have written his first treatise on fossils in the month and a half that followed his dissection. This would indicate either that Steno was a genius who wrote a major treatise on fossils in only a few weeks, or that he had already written it before the shark's dissection.

My research on the book's original manuscript indicates that this imprimatur is misleading and that the time between the approval of the book and its publication is longer than previously thought. The book's manuscript shows that the original December approval refers only to "a work on the elements of myology."[24] Indeed, the censors only reviewed the text on the shark's head at the end of February 1667, two months later.[25] Strikingly, Steno's new ideas on fossils were not included in the original manuscript reviewed by the censors in February. This can be known because the February approval appears in a page before the folios on fossils begin.[26] More important, the original manuscript that was reviewed by the censors mentions only one copperplate: that of the shark's head.[27] The final printed version, on the other hand, refers to *two* copperplates. A second plate, on fossils that looked like shark's teeth, known as tongue stones or

---

20. The book's first *imprimatur*, from 27 October 1666, refers only to "Anatomica geometrice demonstrata." See Steno, *Elementorum myologiæ specimen*, 123.
21. Steno, *Elementorum myologiæ specimen*, 91.
22. Dominici, "A Man with a Master Plan," 64.
23. Steno, *Elementorum myologiæ specimen*, 123.
24. MsElem, fol. 103v: "Opus elementorum myologiæ . . . vidi." This passage is crossed out and points to an approval from 3 March 1667 on fol. 102r.
25. MsElem, fol. 75v: "Tractatus hic de capite canis carchariæ nil contra fidem . . . exhibet . . . die 23. Feb.1667."
26. The text on fossils begins in MsElem, fol. 78r.
27. MsElem, fol. 60v: "meis usibus **eam concessit**, qua Lamiæ caput et dentes expressos vides."

glossopetrae, was added later.[28] In short, this protracted process of the manuscript's submission and reviews shows that Steno's treatise on fossils was not yet ready four months after he dissected the shark. That is, Steno probably started writing on fossils after he dissected the shark, but he had a longer period to develop his ideas than historians have assumed until now.

Comparisons in Anatomy

Why, then, did Steno decide to write on fossils? To answer this question, it is useful to take a step back and revisit Steno's research in anatomy up to the publication of *Elementorum myologiæ specimen*. As mentioned in previous chapters, anatomy in the seventeenth century was Aristotelian in the sense that it relied on the comparative method of Aristotle's works on animals.[29] Anatomists developed new knowledge about an organ's *historia* by explaining similarities between the same organ in different animals. But what happened when two supposedly different organs shared too many similarities? This problem appeared before Steno multiple times, and his solution was to take anatomy's comparative method a step further by declaring that different things sometimes were the same. This was the step that would also lead him to start writing on fossils.

Since his young years in Copenhagen, Steno had played around with mechanical devices that looked like the heart. In the "Chaos" manuscript he wrote not only about a hydraulic device designed by Kircher that simulated the motion of the heart, but also about building a mechanical heart with a real bladder and blood.[30] Later, in 1663, when dissecting muscles for the first time, he noticed a great similarity between the tissues of the heart and muscle fibers.[31] Steno explored this similarity in his second book, *De musculis et glandulis*. There he wrote that in muscles, as "in all the substance of the heart, nothing else occurs than *arteries, veins, nerves, fibers, membranes*."[32] Therefore, he continued, "if it is certain" that the heart and muscles are made of the same substance, then "truly the

---

28. Steno, *Elementorum myologiæ specimen*, 70: "meis usibus **eas concessit**, quibus Lamiæ caput et dentes atque glossopetras maiores expressas vides."

29. Cunningham, *The Anatomical Renaissance*, 27, 175–177. See also Gian Battista Vai, "The Scientific Revolution and Nicholas Steno's Twofold Conversion," in Rosenberg, *The Revolution in Geology from the Renaissance to the Enlightenment*, 187–208, esp. 190 for a brief mention of comparative anatomy in Steno's shift to geology.

30. Gal. 291, fols. 39r (22 March), 59r (18 May), in Steno, *Chaos*, 122, 285.

31. Steno to Bartholin, 30 April 1663, in EMC VI, 414–421, esp. 417. A rabbit as the first observation is confirmed in Steno, *De musculis et glandulis*, 11.

32. Steno, *De musculis et glandulis*, 22. Italics in the original.

heart must be greeted with the name of muscle."[33] This claim, also made by a few others at the time, challenged the traditional understanding of the heart—originally shared by Harvey—that it was a unique source of energy for the entire body.[34] But for Steno, this equivalence of substance was enough to convince him that the heart was just a muscle. After this book came out, Steno presented his case before several audiences in Paris in 1664. He then responded to criticisms in a public letter to Thévenot published in the *Elementorum myologiæ specimen*.[35] In it Steno mentioned similarities not just between appearances (*historia*) but also between the action (*actio*) of muscles and heart. He concluded that since "a movement much evident to the senses is observed in the same way in the heart as in what are commonly called muscles, I sufficiently hope to have demonstrated clearly . . . that the heart is a muscle."[36]

In Florence, when writing his dissection account of the shark's head, Steno continued to rely on similarities as a method. First, he confirmed that what he dissected "was the head of a great white shark," because it matched "everything that is narrated by zoographers [*a zoographis narrantur*] about this fish."[37] Then, Steno turned to the "excretory vessels" in the skin of the fish in order to improve the knowledge on perspiration, which had resurfaced in the early modern period with Santorio.[38] This was a theme that had long interested Steno owing to his youthful obsession with measuring temperature and body weight, as shown in chapter 1.[39] As he traveled across Europe and changed research projects, Steno was often able to bring this interest in the anatomy of sweat into them. In *Observationes anatomicæ* Steno had promised to show not only that perspiration existed, as Santorio did, but also that sweat was produced by glands spread everywhere in the skin.[40] Yet, when he summarized most of his discoveries on glands in *De musculis et glandulis*, he did not mention anything about skin glands. The reason was, as he later wrote, that he was not yet sure how exactly glands were distributed in the skin. However, when describing the shark's skin, Steno argued that his observations "seem to establish to no small extent my opinion about the human skin"—that is, that perspiration

---

33. Steno, *De musculis et glandulis*, 27, 26.
34. Fabri, *Tractatus duo*, 371; Malpighi, *Opera posthuma*, 3; and Harvey, *De motu cordis*, 59.
35. Steno, *Elementorum myologiæ specimen*, 45, 55–56.
36. Steno, *Elementorum myologiæ specimen*, 59.
37. Steno, *Elementorum myologiæ specimen*, 69–70.
38. Steno, *Elementorum myologiæ specimen*, 74.
39. Gal. 291, fols. 56v (3 May), 57r (11 May), in Steno, *Chaos*, 270, 274–275.
40. Steno, *Observationes anatomicæ*, 99.

vessels had their origin in skin glands.⁴¹ In other words, the similarity between his observations of the shark and ideas about human skin confirmed his theory about human skin glands.

Finally, when he was already writing on fossils, Steno again used similarities between different organs to demonstrate something new about the female reproductive system. Two years earlier he had become intrigued by similarities between the uterus of mammals and the eggs of a ray he dissected in Copenhagen. Harvey had indeed suggested that mammalian uteri performed the same functions as certain eggs, thus the dictum "everything comes from an egg [*ex ovo omnia*]."⁴² However, because Steno knew the mammalian uterus had more functions than an egg, he concluded that no analogy could be used "unless as completely misapplied."⁴³ Now, in Tuscany, after dissecting bodies of various mammalian species, Steno observed that "female testicles" contained internal eggs. Many early modern anatomists, including Harvey, thought that the ovaries, then called female testicles, contributed little to animal reproduction. They existed only as vestiges in women, just like nipples in men. Challenging this claim, Steno concluded that the "testicles of females are analogous [*analogi sint*] to an ovary," as in fishes and other animals.⁴⁴ Steno published this observation as a side comment to a larger description of another shark that he dissected. Yet, for Steno, this small note and its accompanying illustration (see fig. 6.1)—the *historia* of the ovaries—were enough to "show the analogy of the genital parts [*exponere partium genitalium analogia*] and remove this error by which it is believed that the genitals of women are analogous to the genitals of men."⁴⁵ Steno promised to do more work on this topic, but his emerging interest on fossils led him away from it. In the end, Steno's Dutch friends Regnier de Graaf, who described the development of ovarian or Graafian follicles, and the famous microscopist Swammerdam took the lead in exploring these matters further.⁴⁶ Indeed, when Swammerdam

---

41. Steno, *Elementorum myologiæ specimen*, 74.
42. Bertoloni Meli, *Mechanism, Experiment, Disease*, 210–213.
43. Steno, *De musculis et glandulis*, 61, 63.
44. Steno, *Elementorum myologiæ specimen*, 117.
45. Steno, *Elementorum myologiæ specimen*, 117–118.
46. Regnier de Graaf, *De mulierum organis generationi inservientibus* (Leiden: Hack, 1672), 183; and Swammerdam, *Miraculum naturae*, 20. Steno published two later articles on ovaries in which he mentioned "observationes amicorum"; see Nicolaus Steno, "Observationes anatomicæ spectantes ova viviparorum," in Bartholin, *Acta medica et philosophica Hafniensia*, 2:210–218 (1675); and Steno, "Ova viviparorum spectantes observationes aliæ," in Bartholin, *Acta medica et philosophica Hafniensia* 2: 219–232 (1675). See also Jorink, "Cartesian Sex."

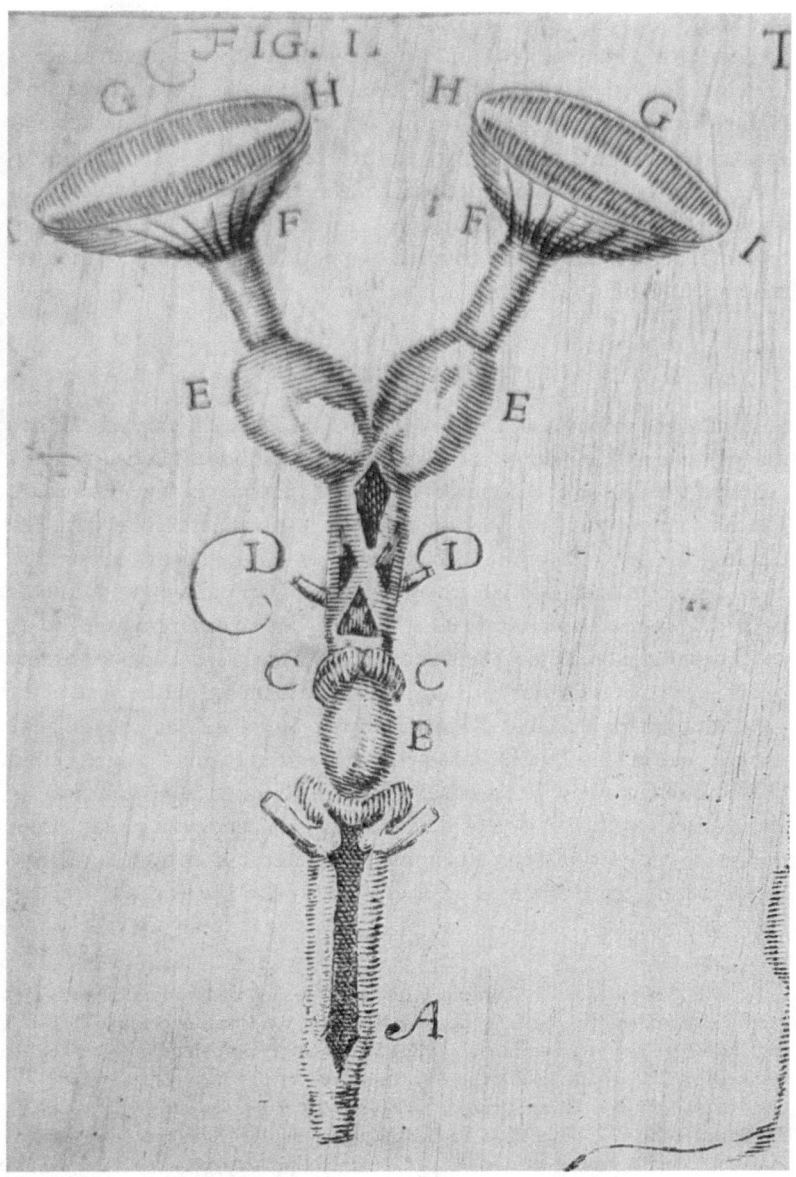

FIGURE 6.1. Drawing of a fish's ovaries and uterus. Steno, *Elementorum myologiæ specimen* (Florence, 1667), tabula VII. Courtesy of the Institute of the History of Medicine, Johns Hopkins University.

told Steno that van Horne had observed the ovaries a few weeks before he had, Steno replied that he would have been happy to accept van Horne's priority and even quote him, had he known.[47] The difference between this kind reply and his reaction to Blasius's claim to priority is remarkable. It speaks not only to the accuracy of van Horne's observations and conclusions, but also to the importance of friendship in the scientific interactions between these anatomists. One thing is certain: Steno's firm commitment to comparative anatomy contributed to a new understanding of the ovaries and, as I will now show, to a new history of the Earth.

Comparisons in Earth History

Soon after the shark dissection in October 1666, Dati, a literary scholar and one of the eyewitnesses to the dissection, had an idea. He had recently acquired the almost-one-hundred-year-old manuscript of the *Metallotheca Vaticana*, an extensive description of the Vatican's mineral collection. The text had been written by the collection's curator in the sixteenth century, the papal physician Michele Mercati (1541–1593).[48] Dati wanted to publish the book in Florence and dedicate it to the pope, but he never managed to do so.[49] For this reason, he saw Steno's new work on the shark as an opportunity to display one of the beautiful images from the manuscript: the head of the great white shark. Dati lent the manuscript to Steno and suggested that he use the plate in his publication, which Steno agreed to do (see fig. 6.2). But Steno also focused on the text itself, especially on what Mercati had to say about shark's teeth. What he read dramatically changed his research career.

In his manuscript, Mercati claimed that, because of their similarity, fossils and the teeth of sharks "are confused even by the learned," and that

---

47. Whether van Horne had priority is not clear. He supposedly observed ovaries a few weeks before Steno, on 21 January 1667. However, van Horne wrote that female testicles looked like ovaries *only after* reading Steno's work (without quoting it): Johannes van Horne, *Suarum circa partes generationis . . . observationum prodromus* (Leiden, 1668), 8. For the exact date of van Horne's dissection, see Jan Swammerdam, *The Book of Nature* (London: C. G. Seyffert, 1758), vii. For the whole chronology, see Swammerdam, *Miraculum naturae*, 46–53. On Steno being happy to accept van Horne's priority, see Steno to Swammerdam, 18 March 1668, ibid., 51.

48. For Dati's account, see Dati to Falconieri, 25 October 1666, in Mercati, *Metallotheca vaticana*, xxxiv–xxxv.

49. The book was printed in 1717 by Giovanni Maria Lancisi (1654–1720) in Rome; see Alix Cooper, "The Museum and the Book: The *Metallotheca* and the History of an Encyclopaedic Natural History in Early Modern Italy," *Journal of the History of Collections* 7, no. 1 (1991): 1–23.

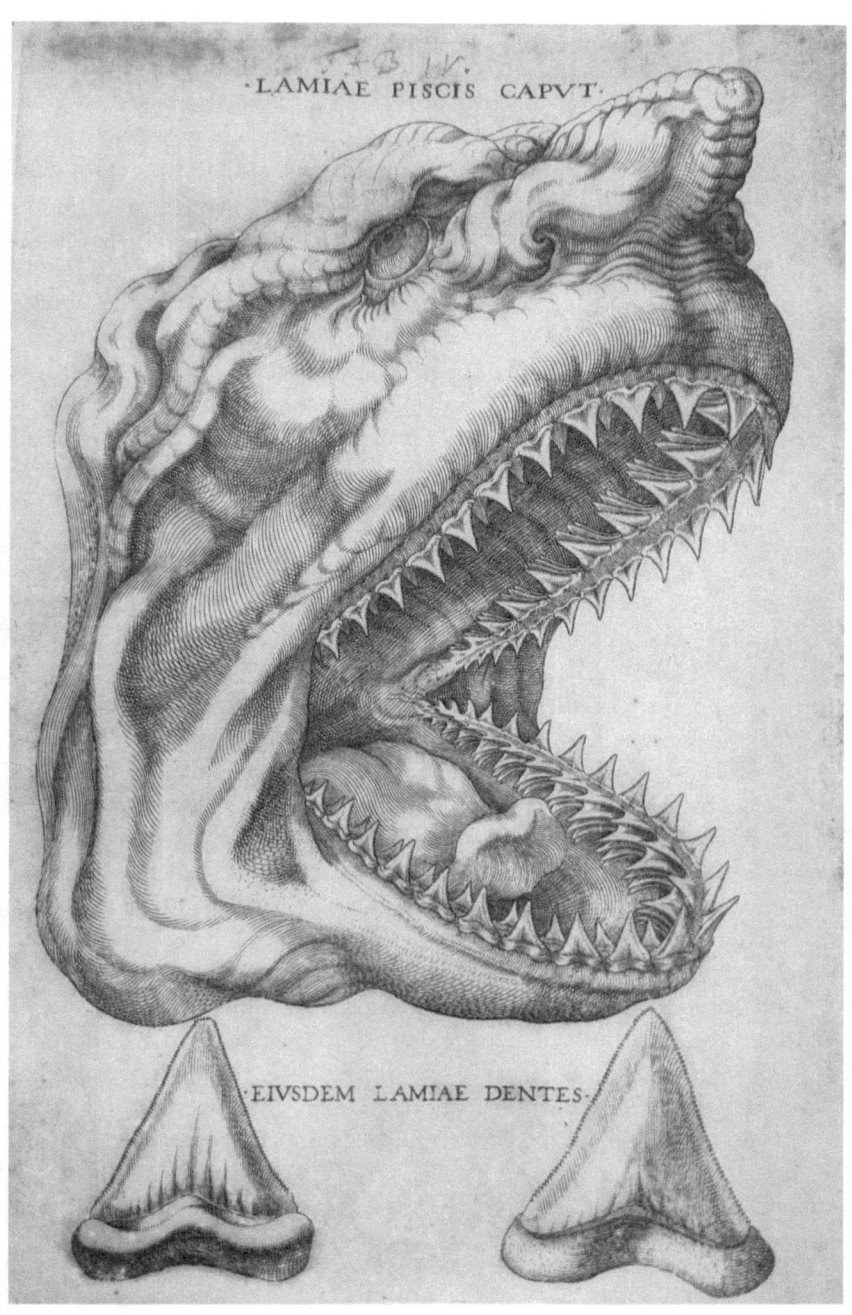

FIGURE 6.2. Steno, *Elementorum myologiæ specimen*, tabula IV. Courtesy of the Institute of the History of Medicine, Johns Hopkins University.

it was an error to say that they were the same.⁵⁰ That is, for Mercati, despite their strong resemblance to one another, fossils and shark's teeth were not the same thing. On its own, this claim directly contradicted Steno's radical exploration of analogies of equivalence in anatomy. Whereas Steno claimed that two distinct but similar organs were of the same substance, like muscles and the heart, Mercati concluded the exact opposite in a different field. Moreover, from Steno's perspective, there was another problem with Mercati's text. Mercati mentioned a few "very little" differences between tongue stones and shark's teeth. One of them was that shark's teeth had "a single and unbroken white [color], or yellowish with age," whereas tongue stones often appeared in more colors.⁵¹ But, as explained in previous chapters, Steno knew from his research that colors were often epistemically unreliable. Thus, Mercati's attempt to differentiate fossils from teeth based on color was, to Steno's way of thinking, unconvincing.

By this time Steno was fully engaged in using comparative methods in anatomy. In addition, he often associated this method with an intense desire to fight uncertainty in anatomy. Yet, Mercati also claimed to fight uncertainty in natural history. He wrote that he wanted to "be as clear as possible" and "not only to teach but also to remove" false and supposed things.⁵² How could Mercati aim for scientific rigor in the same terms as Steno and yet conclude that shark's teeth and fossils were different despite their obvious similarities? Faced with this stark methodological difference, Steno had to respond. Therefore, he added to his account of the shark's head what he called a "digression on the origin of fossils and on the soils themselves where these fossils are found."⁵³

In this digression, which occupied half of the text on the white shark, Steno scrupulously applied the methods he had developed as an anatomist to study fossils.⁵⁴ He described the various soils where fossils resembling parts of marine animals are found. And, as a good anatomist, he entitled this observational description a *historia*. It was in this *historia* that Steno first wrote that these soils are "composed of layers imposed on each other and inclined to the horizon"—ideas that resemble the modern principles of superposition of strata and of original horizontality.⁵⁵ In the end, this *historia* of soils allowed Steno to apply his comparative method to fossils.

---

50. Steno, *Elementorum myologiæ specimen*, 70–71.
51. Steno, *Elementorum myologiæ specimen*, 71.
52. Steno, *Elementorum myologiæ specimen*, 70.
53. Steno, *Elementorum myologiæ specimen*, 91.
54. The two parts are Steno, *Elementorum myologiæ specimen*, 69–90, 90–110.
55. Steno, *Elementorum myologiæ specimen*, 91. On these principles see Cutler, *The Seashell on the Mountaintop*, 110–111.

He argued that fossils "are teeth of the great white shark" because of their great similarity, "since planes are most similar to planes, sides to sides, [and] base to base."[56] Ironically, he relied on Mercati's copperplate to support his point visually (see fig. 6.3). But if these fossils were really the teeth of sharks, how did they end up on mountaintops?[57] To answer this, Steno channeled his quest for certainty in anatomy to explain the formation of soils and of fossilization. That is, he wrote a history of the Earth with the same methods that he used in his *historia* of the body.

The Search for Certainty in Anatomy

As mentioned in previous chapters, Steno's deep interest in finding certain or reliable knowledge about nature had been present since his early days studying the body. This interest was a response to the rise of new philosophies and the search for certainty in his time. Steno was well aware that new systems that explained natural phenomena in a sound way often differed to their core. He faced this same problem in anatomy and became obsessed with solving it, in part by trying to understand efficient causes better. Indeed, Steno began his major research project on salivary glands because he wanted to understand the action and use of the parotid salivary duct, which he had just discovered. When studying lacrimal glands, he also noticed that their action was affected by muscles around arteries and their blood flow.[58] This observation thus made him want to study the muscles and heart more seriously. It was during this research that Steno also linked the motion of the intercostal muscles to the role of respiration in blood circulation.[59] And, since the muscles are connected to the brain through the nervous system, Steno also undertook a serious study of the brain while in Paris in 1665. In short, one research question led him to another: Steno always wanted to go deeper.

For this search for certainty to succeed, Steno also thought that good research should rule out that which is not certain. For instance, in his dissection of the brain, Steno unveiled the errors of Descartes's and Thomas Willis's recent works on the brain. Interestingly, he did that by comparing each of Descartes's claims with his own dissections of

---

56. Steno, *Elementorum myologiæ specimen*, 109. This argument resonates with Steno's comment in *Observationes anatomicæ*, viii: "Observation must lead to reasoning and the thing itself must be examined in all its parts as far as possible."

57. I am borrowing from Cutler, *The Seashell on the Mountaintop*, which draws from Ovid, *Metamorphoses*, Book XV:264–265.

58. Steno, *Observationes anatomicæ*, 92–97.

59. Steno, *De musculis et glandulis*, 9.

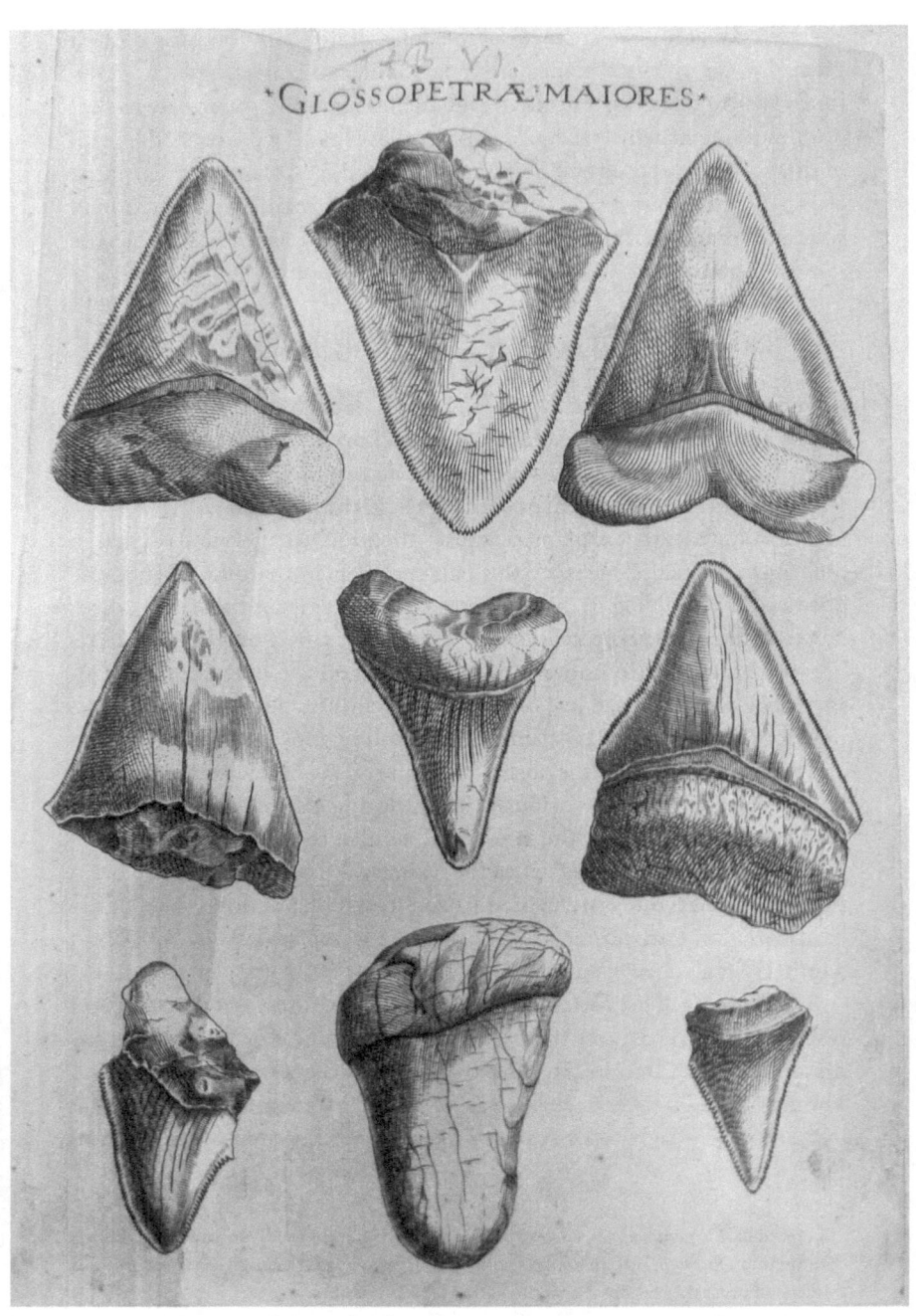

FIGURE 6.3. Steno, *Elementorum myologiae specimen*, tabula VI. Courtesy of the Institute of the History of Medicine, Johns Hopkins University.

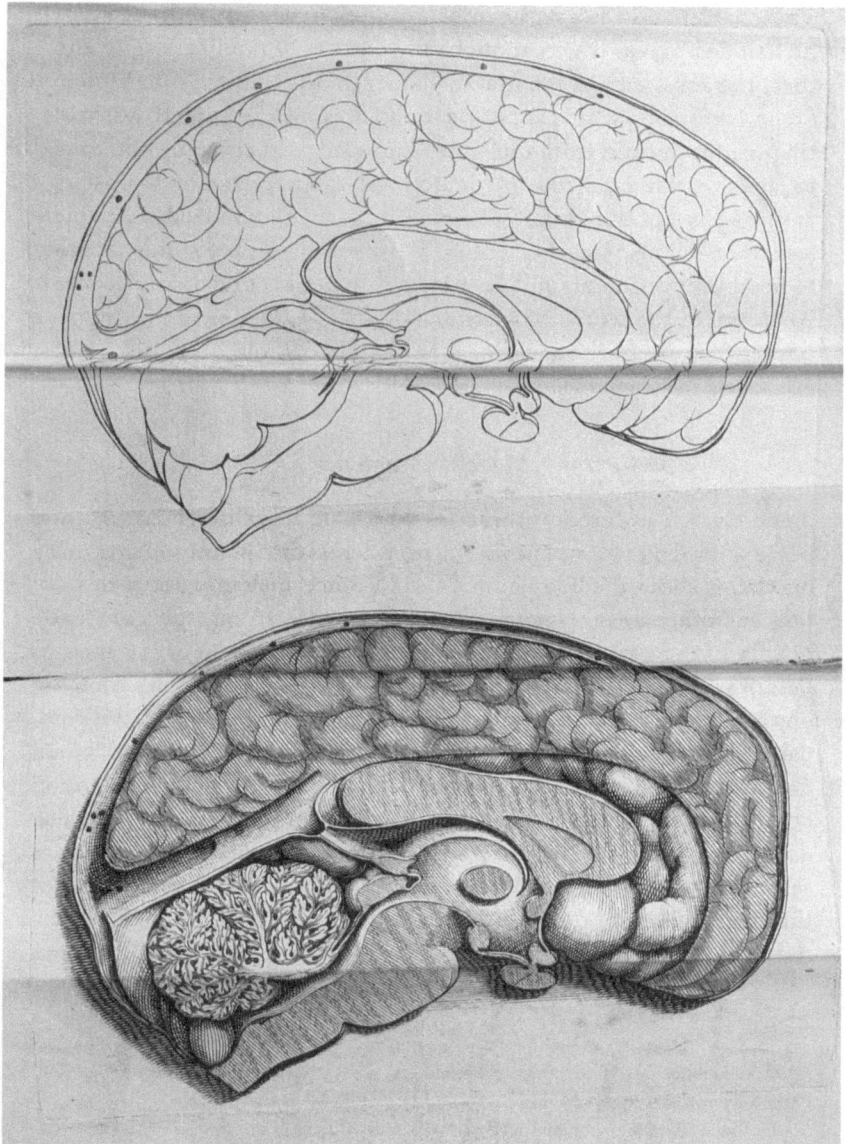

FIGURE 6.4. Cross-section of the brain, in Steno, *Discours sur l'anatomie du cerveau* (Paris, 1669). From the British Library Archive, General Collection 1477.bbb.32.

the brain. He accompanied his arguments with an illustrated cross-section of a human brain (see fig. 6.4), an approach that speaks to the cognitive benefits of seeing things through cross-sections, which also played a major role in his geological work. Above all, Steno did not add a new explanation as a replacement for the knowledge of the brain that he had destroyed. It was enough, and worthy of publication, to show that "the substance of the brain is [still] poorly known."[60] In Florence Steno had strong words about rigor and accuracy, writing that not distinguishing certain from uncertain knowledge in anatomy could lead to fatal errors in medicine.[61] If anatomists "did not hand on to posterity anything except things that were certain, then our knowledge would be less wide, but also less dangerous."[62] The solution was to limit anatomy to reliable experiments and demonstrations, and, as he put it, "to swear Anatomy in the words of Mathematics [*in Matheseos verba Anatome iuraret*]"[63]—in short, to use ideas and methods from other disciplines, especially mathematics.

Models in Anatomy

Steno's search for certainty forms the core of his interdisciplinary method because it led him to use knowledge from across disciplines and thus make his claims about the body more reliable. Since analogies between anatomy and other areas of knowledge also became a central aspect of Steno's history of the Earth, they deserve further exploration. For instance, as described in previous chapters, Steno often drew on examples from the physico-mathematical disciplines of mechanics and hydrostatics. He compared glands to blood sieves that functioned according to the rate of blood flow, described glandular secretions as lubricating fluids of the body, and compared veins to siphons and muscles to complex pulleys.[64] He also used examples from chymistry, such as when arguing that blood carries several substances within itself despite its uniform red color. He spoke of the chymical reaction of two non-red substances that also resulted in a red product (he mentioned the reaction between butter of antimony and

---

60. Steno, *Discours sur l'anatomie du cerveau*, 7.

61. On anatomical discoveries and medicine, see Bertoloni Meli, "Mechanistic Pathology and Therapy."

62. Steno, *Elementorum myologiæ specimen*, v. Steno made similar claims in *De musculis et glandulis*, 20–21. On the critique of medicine, see Stephen Pender, "Examples and Experience: On the Uncertainty of Medicine," *British Journal for the History of Science* 39, no. 1 (2006): 1–28.

63. Steno, *Elementorum myologiæ specimen*, v. Steno was probably using the line from Horace (65–8 BC) also used for Royal Society's motto: "not bound to swear in words to a master" (*Nullius addictus iurare in verba magistri*; Horace, *Epistles* I.1, 14).

64. Steno, *De musculis et glandulis*, 44; Steno, *Observationes anatomicæ*, 85; Steno to Bartholin, 30 April 1663, in EMC IV, 414–421; and Steno, *De musculis et glandulis*, 19–20.

spirit of niter).⁶⁵ This and other chymical analogies reflected an approach that was especially popular at the University of Leiden, where Steno had studied and worked.⁶⁶

But Steno's most sophisticated analogy was crafted in Florence. In the main part of the *Elementorum myologiæ specimen*, which he had completed before dissecting the shark's head, Steno argued that anatomy had to embrace mathematics to improve the epistemic strength of its claims. As he put it, there is no other reason for the innumerable errors of anatomy then its "disdain of the laws of mathematics until now [*Matheseos leges Anatome hactenus indignata fuerit*]."⁶⁷ As noted in chapter 5, this book was the first serious attempt to join mathematics with anatomy in the early modern period. It is thus a prime example of how these two disciplines intersected at this time.

The main argument of this book was that muscles should be understood as oblique parallelepipeds, with bases as rectangles and lateral surfaces as parallelograms.⁶⁸ This parallelepiped model mattered because it allowed Steno to argue against a very old anatomical theory. Since at least the time of Galen, anatomists used to explain muscle contraction by the theory of inflation of animal spirits. Everyone knows that if a muscle contracts, it looks inflated or larger. Premodern scholars explained this by saying that something entered the muscle to increase its size—namely, the so-called animal spirits.⁶⁹ A new version of this influx theory of spirits had just been proposed by Descartes in the posthumously published *De homine*. Steno's model of the parallelepiped was important because it threatened to shatter the role of animal spirits in muscle contraction.

His argument is easier to understand in two dimensions. Imagine two parallelograms, which are cross-sections of two parallelepipeds (see fig. 6.5). The parallelograms have different angles of inclination. The one on the left corresponds to a relaxed muscle, and the one on the right to

---

65. Steno, *Observationes anatomicæ*, §33, 30–31. As mentioned in chap. 5, Steno also compared the process of boiling a lobster to its chymical reaction with spirit of vitriol in Florence.

66. Ragland, "Chymistry and Taste in the Seventeenth Century"; and Ragland, "Experimenting with Chymical Bodies."

67. Steno, *Elementorum myologiæ specimen*, iv.

68. Steno, *Elementorum myologiæ specimen*, 3.

69. On the vast topic of animal spirits, see Maria Conforti, "Testes alterum cerebrum: Succo nerveo e succo seminale nella macchina del vivente di Giovanni Alfonso Borelli," *Medicina nei secoli arte e scienza*, n.s., 13, no. 3 (2001): 577–595; and Cobb, "Exorcizing the Animal Spirits."

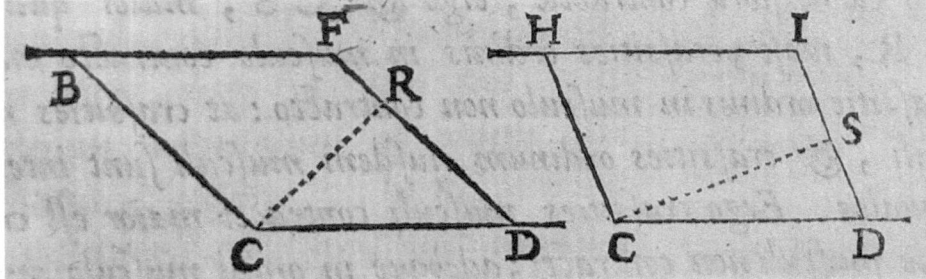

FIGURE 6.5. Relaxed (*left*) and contracted (*right*) muscles as parallelograms. Steno, *Elementorum myologiæ specimen*, 25. Courtesy of the Institute of the History of Medicine, Johns Hopkins University.

a contracted muscle. Steno explains that the thickness of the parallelogram increases from one to the other because line CS is longer than line CR. But since these parallelograms have the same base and height, their areas (base multiplied by height) remain the same. In short, if this model could be applied to muscle fibers, a muscle could change shape while keeping its volume.[70] No spirits were needed; geometry alone explained the inflation. Steno did not add anything else. He was happy just to show that "the idea of animal spirits is built on an uncertain foundation [*incerto fundamento superstructum*]."[71]

But how does this abstract model relate to Steno's emphasis on observations? Steno admitted that he did not see "all muscles of all animals to the extent that I believe with certainty [*adeoque certo crederem*]" that all planes were perpendicular.[72] This attitude displayed Steno's search for rigorous empirical knowledge and explains why he only made claims "with certainty in so far as I perceived it in many things [*id duntaxat certo hic affirmo me in plurimis talem deprehendisse*]."[73] Indeed, the link to observations in this case was Steno's explicit association of his mathematical insights with a novel cross-sectional cut of muscles along the length of their fibers. After first cutting muscles in this way in Leiden, Steno still took a while to decide on how to write and illustrate his

70. Steno, *Elementorum myologiæ specimen*, 30. See also Kardel, *Steno on Muscles*, 15–18; and Bertoloni Meli, "The Collaboration Between Anatomists and Mathematicians," 696–706.
71. Steno, *Elementorum myologiæ specimen*, v–vi.
72. Steno, *Elementorum myologiæ specimen*, 7.
73. Steno, *Elementorum myologiæ specimen*, 7.

geometrical ideas.[74] As demonstrated in the previous chapter, the solution only came to him in Florence, in collaboration with Viviani. In Florence Steno was confident enough to claim that the structure of muscles "requires almost by necessity that they must be explained mathematically [*musculos mathematice explicandos illorum fabrica quadam quasi necessitate postulat*]."[75] Therefore, although Steno did not exactly see a parallelepiped, he observed parallelograms in his cross-sections of muscles, which he then generalized to three dimensions. For the rest of the book he also relied on the format of Euclid's *Elements*, with a sequence of definitions, lemmas, and corollaries, many of which were illustrated with diagrams.

Comparisons were central to Steno's research in anatomy, and he used them in the awareness that they had varying degrees of certainty.[76] In his first work on muscles, Steno described muscle contraction as a pile driver "because an explanation by similar things [*per similia explicatio*] greatly pleases many people"; however, "this being only a comparison [*cum simile hoc tantum sit*]," he decided to pursue it no further.[77] In most cases, though, Steno used analogies for more than just illustrations. Mechanical metaphors of sieves and lubrication were central to his view on glands.[78] And the way he wrote about chymical reactions also carried enough epistemic strength for him to speak about the composite nature of blood. This reliance on analogies was not unusual, since they were also central to early modern mechanical worldviews.[79] Analogies as famous as those between macrocosm and microcosm were used by Johannes Kepler (1571–1630) or Johannes van Helmont (1580–1644) with great argumentative force.[80] Even Aristotle, who acknowledged the episte-

---

74. See Castel-Branco and Kardel, "Drawing Muscles with Diagrams."
75. Steno, *Elementorum myologiæ specimen*, iv.
76. I use the words "metaphor," "analogy," and "comparison" as loose synonyms. For a deeper discussion of these terms, see Thomas Kuhn, "Metaphor in Science," in *Metaphor and Thought*, ed. Andrew Ortony, 2nd ed. (Cambridge: Cambridge University Press, 1993), 533–542, esp. 538; and Peter Machamer, "The Nature of Metaphor and Scientific Description," in Hallyn, *Metaphor and Analogy in the Sciences*, 35–52.
77. Steno, *De musculis et glandulis*, 19–20.
78. See Bertoloni Meli, *Mechanism, Experiment, Disease*, 105–129, 150–169.
79. Dear, *The Intelligibility of Nature*, 15–38; and Bertoloni Meli, *Mechanism*, esp. 7–8.
80. Principe, *The Secrets of Alchemy*, 201–206, esp. 205; Gérard Simon, "Analogies and Metaphors in Kepler," in Hallyn, *Metaphor and Analogy in the Sciences*, 71–82. On natural history, see Paula Findlen, "Jokes of Nature and Jokes of Knowledge: The Playfulness of Scientific Discourse in Early Modern Europe," *Renaissance Quarterly* 43, no. 2 (1990): 292–331.

mological problems of analogies, used them often and consciously, such as when attributing the intentionality of a craftsman to nature ("nature does nothing in vain").[81] Therefore, analogies held significant cognitive benefits in Steno's anatomical project and would continue to do so in his subsequent research on fossils. Taking all this into consideration, this chapter now turns, one last time, to how Steno developed a history of the Earth.

The Search for Certainty in Earth History

When Steno started writing on fossils, he did not start from a blank slate. As a learned anatomist, he mastered typical works of natural history such as those by Guillaume Rondelet (1506–1566), Conrad Gesner (1516–1565), and Johannes Jonston (1603–1675), all of whom he quoted in his anatomical work.[82] Danish physicians even had a specific legacy in natural history. Ole Worm is the most famous example, owing to a widely read book on his cabinet of curiosities.[83] Thomas Bartholin also wrote specifically on tongue stones that he saw on the island of Malta, an observation that made Steno want to go there.[84] Even Borch, Steno's traveling companion in Europe, wrote a short essay "on the generation of stones in the macro and microcosm," commenting on similar formation processes of stones outside and inside the body (e.g., minerals and bladder stones).[85] Steno also mentioned chatting about seashells with Simon Paulli (1603–1680), a senior physician from Copenhagen whom he had met before starting university.[86] For these reasons, Steno was very familiar with the literature on fossils.

Yet, while many physicians wrote about fossils at the time, few of them wrote a history of the Earth's mountains and valleys before Steno. Indeed, Steno's familiarity with the history of the Earth went well

---

81. G. E. R. Lloyd, *The Revolutions of Wisdom: Studies in the Claims and Practices of Ancient Greek Science* (Berkeley and Los Angeles: University of California Press, 1989), 172–214, esp. 187–190, 200–203.

82. Steno, *De musculis et glandulis*, 60.

83. Worm, *Museum wormianum*, 67–68.

84. Bartholin, *Historiarum . . . centuria V et VI*, 193–201. Bartholin had a full manuscript on tongue stones, "De glossopetris melitensibus dissertatio," but it was lost in a 1670 fire before publication; see Thomas Bartholin, *De bibliothecæ incendio dissertatio* (Copenhagen: Matthius Godicchenius, 1670), 73–74. On Steno's comment, see *Elementorum myologiæ specimen*, 90.

85. Bartholin, *Acta medica et philosophica Hafniensia*, 5:184–196.

86. Steno, *De musculis et glandulis*, 59–60.

beyond natural history, for he delved into commentaries on Aristotle's *Meteorology*, the other discipline that dealt most with Earth history in the early modern period.[87] Steno defended a short dissertation on hot springs—a meteorological topic—in 1660 in Amsterdam.[88] Moreover, in the introduction to *Observationes anatomicæ*, he referred to scholars studying stars and planets alongside "the inside of the Earth" and its "concealed mysteries of minerals."[89] Steno mentioned them when describing the human mind's capacity to discern patterns between the micro- and macrocosm, but in the end the book was only about glands. In brief, Steno's knowledge of fossils and the Earth formed part of his intellectual background.

Accordingly, Steno knew that various other scholars also argued that fossils had an origin in sea animals. As Ivano Dal Prete has shown, the organic origin of fossils was the commonly accepted argument among ancient and medieval writers.[90] Scholars ranging from Aristotle through Isidore of Seville (c. 560–636) and Avicenna (980–1037) to Albert the Great (c. 1200–1280) all agreed that marine fossils showed that dry land had once been underwater. Steno therefore acknowledged that the true explanation of marine fossils had "become very uncertain in most recent times."[91] Indeed, according to a newly discovered letter to Dati, Steno was working on a list—now lost—of "all the ancients who believed that the ocean before was higher than in their time."[92] The problem that Steno faced in his time was that "many and great" were "the men" who did not agree with this opinion.[93] Authors whom Steno respected favored an origin of fossils completely detached from animals, such as Kircher, Ulisse

---

87. Ivano Dal Prete, *On the Edge of Eternity: The Antiquity of the Earth in Medieval and Early Modern Europe* (Oxford and New York: Oxford University Press, 2022), xi–xii, 45–47, 114–119, 122–125, 149.

88. Steno, *Disputatio physica de thermis*. On hot springs as part of the tradition of Aristotle's *Metereology*, see Craig Martin, *Renaissance Meteorology: Pomponazzi to Descartes* (Baltimore, MD: Johns Hopkins University Press, 2011), 89.

89. Steno, *Observationes anatomicæ*, preface.

90. Dal Prete, *On the Edge of Eternity*, 38–50.

91. Steno, *De solido*, 5.

92. Steno to Carlo Dati, 2 March 1669 in BNCF, Baldov.258.VII.31, fol. 105r: "fo una lista de' tutti quelli che fra gli antichi hanno creduto, che il mare altri tempi sia stato più alto di quello era ne' tempi loro."

93. Steno, *Elementorum myologiæ specimen*, 109. A few, like Andrea Cesalpino and Fabio Colonna, were with Steno; see Dal Prete, *On the Edge of Eternity*, 140; and Nicoletta Morello, *La nascita della paleontologia nel Seicento: Colonna, Stenone e Scilla* (Milan: FrancoAngeli, 1979).

Aldrovandi (1522–1605), and Mercati himself.⁹⁴ In their views, fossils were the result of incomplete processes of spontaneous generation, literally jokes of nature that displayed hidden links between the macro- and the microcosm. But from the perspective of a seventeenth-century anatomist, how could great scholars differ so much in their views if fossils and animal parts were so obviously similar? Steno had shaped his career around a search for certainty in anatomy. The time had come to apply to the Earth the same methods he had applied to the body.

Like his thoughts on muscles, Steno's ideas on the history of the Earth took a while to develop. He concluded his first text on fossils with six conjectures that offered "a glimpse of truth from the observations presented."⁹⁵ These conjectures were simple, but they dealt a significant blow to the idea that fossils were spontaneously generated on dry land. For instance, the first conjecture is that the soil where fossils are found does not seem to produce fossils today, whereas the third is that such soils were once covered with water.⁹⁶ Steno called these claims "conjectures [*conjecturæ*]" because he did not want to necessarily argue "that defenders of contrary views are wrong" and said that his opinion was only "similar to truth [*opinio vero simile*]."⁹⁷

Historians such as Paula Findlen have attributed this attitude to Steno's desire to please respectable authors with differing opinions, such as Kircher.⁹⁸ This is, in part, correct. Others have said that Steno was being careful to avoid the censorship of the Catholic Church.⁹⁹ In fact, scientific publications sponsored by the Medici at this time were famous for avoiding any interpretations of experimental results, and a few historians have associated this attitude with self-censorship.¹⁰⁰ The problem with

---

94. Ulisse Aldrovandi, *Musæum metallicum* (Bologna: Marcus Antonius Bernia, 1648), 820; and Gabriele Falloppio, *De medicatis aquis* (Venice, 1564), 109r–v. Steno mentioned Aldrovandi and Falloppio in Steno, "Ova viviparorum spectantes observationes aliae," 220, and *Elementorum myologiæ specimen*, 43. Kuang-Tai Hsu wrongly attributes the organic theory to Falloppio in Hsu, "The Path to Steno's Synthesis on the Animal Origin of Glossopetrae," in Rosenberg, *The Revolution in Geology from the Renaissance to the Enlightenment*, 93–106, esp. 96–97.

95. Steno, *Elementorum myologiæ specimen*, 93.

96. Steno, *Elementorum myologiæ specimen*, 93, 95.

97. Steno, *Elementorum myologiæ specimen*, 109.

98. Findlen, *Possessing Nature*, 235–237; and Francesco Luzzini, *Il miracolo inutile: Antonio Vallisneri e le scienze della terra in Europa tra XVII e XVIII secolo* (Florence: Leo S. Olschki, 2013), 21–23.

99. Cobb, *Generation*, 97–98.

100. Galluzzi, "L'Accademia del Cimento," 800–801, 823–832; Boschiero, *Experiment and Natural Philosophy in Seventeenth-Century Tuscany*, 192, 229–231; and Luzzini, *Il miracolo inutile*, 11, 17–18, 32.

this argument is that, *pace* the old historiography on censorship, making claims about rocks or the history of the Earth did not raise censorship problems in the seventeenth century.[101] Moreover, Steno's nondogmatic approach resonated with most scholarship on Aristotle's *Meteorology*, best exemplified by the probabilistic approach of Kircher's *Mundus subterraneus* (Amsterdam, 1665).[102] Instead, I argue that the best explanation for Steno's hesitations is the intellectual rigor of his search for certainty. Steno's careful approach derived from his interest in distinguishing certain from uncertain knowledge, an approach that he carried over from his research in anatomy. Steno himself said so: "Knowledge of [these] things is not yet there for me [*nondum ea mihi rerum cognitio est*], to the point that I would interpose here my judgment." Besides, since he had more travels ahead, he preferred not to claim that "that which I will observe in the rest of the journey is similar to what I have observed until now."[103]

Every new observation, however, seemed to confirm Steno's theory. Right before his first text on fossils came out of the press, Manfredo Settala (1600–1680), a prominent collector from Milan, visited Florence. After talking with him, Steno realized that "there were many things among the rarer pieces of his collection which quite clearly favor my conjectures."[104] Similarly, in a visit to the city of Lucca, a physician showed Steno "vertebrae

---

101. The works of Camilla Erculiani (fl. 1570), Isaac La Peyrère (1596–1676), and Antonio Vallisneri (1661–1730), which questioned different aspects of the biblical account of the flood, raised problems in ecclesiastical circles. But the problem with Erculiani and La Peyrère was that they challenged theological doctrines such as original sin. See Richard H. Popkin, *Isaac La Peyrère: His Life, Work, and Influence* (Leiden and New York: Brill, 1987), 1–3, 14–18; and Eleonora Carinci, "Una 'speziala' padovana: Lettere di philosophia naturale di Camilla Erculiani (1584)," *Italian Studies* 68, no. 2 (2013): 202–229, esp. 221–229. Yet, even Erculiani and La Peyrère were not censored, and because Vallisneri was a six-year-old child when Steno published in Florence, his work is not representative of censorship in the 1660s. See Dal Prete, *On the Edge of Eternity*, 132–137, 155. Censorship of another book published under Medici patronage in 1666 had also little to do with religion; see Bertoloni Meli, "Shadows and Deception," 384, 389.

102. Mark A. Waddell, "The World, as It Might Be: Iconography and Probabilism in the *Mundus subterraneus* of Athanasius Kircher," *Centaurus: An International Journal of the History of Science and Its Cultural Aspects* 48, no. 1 (2006): 3–22; and Craig Martin, "Conjecture, Probabilism, and Provisional Knowledge in Renaissance Meteorology," in "Evidence and Interpretation: Studies on Early Science and Medicine in Honor of John E. Murdoch" [special issue,] *Early Science and Medicine* 14, no. 3 (2009): 265–289.

103. Steno, *Elementorum myologiæ specimen*, 90.

104. Steno, *Elementorum myologiæ specimen*, 110.

found on the island of Malta most similar to the vertebrae of fish," one of which was found "clung to a lump of earth."[105] Steno still had time to add both notes to his text, but just barely: the last remark came so late in the process that it could only fit in the book's index.

In the following months, Steno dedicated more and more time to fossils and the Earth. One year later, he no longer hesitated.[106] He no longer feared the "objections of friends," reading other books, or even the "inspection of [new] sites."[107] Thus, he wrote another book, the *De solido intra solidum naturaliter contento dissertationis prodromus* (*Preliminary Dissertation on a Solid Naturally Contained Within a Solid*; Florence, 1669), which became his most important geological publication. The book was a preliminary one because Steno wanted only to show a few results to the grand duke, just as people in debt "pay what they have when they lack the means to pay it in full."[108] He still wanted to do more research for a final book on the topic, as he used to promise when writing anatomy, speaking of longer works on a given topic.[109] Unfortunately, the final work in geology is lost, but I found it listed in three seventeenth-century catalogs as the *Liber de solido in solidis, de glossopetris et aliis lapidibus qui in terra generantur vel alia quacunque re solida* (*Book on a Solid within Solids, on Tongue Stones, and on Other Stones that are Generated in the Earth and in Any Other Solid Things*; Florence, 1672), indicating that it was close to being printed.[110]

De solido was entirely framed around Steno's search for certainty. He regretted again that "things which cannot be determined with certainty are not kept separate from those that can be so determined."[111] For Steno, it was because this separation was rarely effected that there were absolute skeptics, "who are prevented by scruples from putting faith even in

---

105. Steno, *Elementorum myologiæ specimen*, 122.
106. Dating due to the December 1668 imprimatur in Steno, *De solido*, 77–78.
107. Steno, *De solido*, 8.
108. Steno, *De solido*, 4.
109. Steno, *De musculis et glandulis*, 24; and Steno, *Elementorum myologiæ specimen*, 47.
110. Martin Lipen, *Bibliotheca realis philosophica omnium materium* (Frankfurt: Johannis Frederici, 1682), 781; Cornelius à Beughem, *Bibliographia mathematica et artificiosa novissima* (Amsterdam: Johann Jansson van Waesberge, 1688), 129; and Johannes Mollerus, *Bibliotheca septentrionis eruditi sive syntagma tractatuum de scriptoribus illius* (Leipzig, 1699), 34. A folio at the end of Steno's papers in Florence, BNCF, Gal. 291, fol. 243v, also lists the book, following Mollerus. See also Alex Garboe, "Niels Stensen's Lost Geological Manuscript," *Meddelelser fra Dansk Geologisk Forening* 14, no. 3 (1960): 243–246.
111. Steno, *De solido*, 9. For a similar claim on muscle anatomy, see Steno, *Elementorum myologiæ specimen*, 66.

demonstrations," and fideists, who believed in "all things that seem admirable and ingenious."[112] This time, however, there were no more conjectures. Instead, Steno used "principles of nature ... [that are] in common use, widely accepted, and considered admissible by all from every school of thought."[113] He was looking for an explanation that forced assent across his wide range of readers. As with the main publication of the Accademia del Cimento, Steno detached his theory on the formation of fossils from specific philosophical worldviews.[114] Yet, nowhere in his book does he suggest that he did this in order to please his patrons or to avoid religious censorship. Instead, I suggest that he did so in order to reinforce the reliability of his claims. His arguments about the formation of rocks worked regardless of the worldview that one followed, be it corpuscular, chymical, or Aristotelian. Steno stated this explicitly: "Those things that I asserted on matter have a place ... regardless of whether matter has atoms, particles changing in a thousand ways, four elements, [or] chymical principles."[115] In short, Steno's philosophy-free approach allowed him to promote knowledge that was certain. The other method that he used to improve the epistemic strength of his claims was to apply the same analogies to the Earth that he had applied to the body.

Models in Earth History

Steno's history of the Earth hinged on the claim that a long time ago the soil containing fossils had been mixed with water. He explained that it was perfectly reasonable to believe this because, among other things, he had seen his Danish friend, the chymist Ole Borch, "reducing a very hard stone into water inside insipid water [*durissimum calculum insipida aqua in aquam redigentem*]."[116] But if solids and water had become one big mixture, how did they separate back again to form the Earth that we see today?[117] To answer this question, Steno spoke of various natural events in which fluids produce solids. First, he spoke of physiological phenomena, such as the appearance of stones inside the bladder and the cooling of blood. Blood only flows

---

112. Steno, *De solido*, 9. This quote is remarkably similar to Gassendi's desire to find "a middle way between skeptics ... and the dogmatics," as quoted in Osler, *Divine Will and the Mechanical Philosophy*, 106.
113. Steno, *De solido*, 10.
114. [Lorenzo Magalotti,] *Saggi di naturali esperienze fatte nell'Accademia del Cimento* (Florence: Giuseppe Cocchini, 1667), "Preface to the Readers."
115. Steno, *De solido*, 12.
116. Steno, *Elementorum myologiæ specimen*, 97, 99.
117. Steno, *Elementorum myologiæ specimen*, 100.

when warm; otherwise "it is separated in particles different in color and consistency."[118] Thus, by the same token, "warm juices" or "vapors from the Earth" separate "the more solid dust carried within them when the heat disappears."[119] Steno then turned to chymical examples to support his reasoning. He mentioned the chymical reaction in which "things dissolved in acids are precipitated by the arrival of salts." For the same reason, he continued, "extracted tinctures are separated from volatile spirit by the addition of water."[120] Finally, he saw something similar in Paris where "Borel[li], greatly versed in chymistry, joined two very limpid liquors, which immediately thickened, so that turning the glass upside down not even a drop fell."[121]

In addition to turning to anatomy and chymistry, Steno also used mechanical accounts of particles in motion to explain how liquids deposited solids in layers. As he put it, "once the water gradually returned to quietness, first the heavier bodies, then the less heavy settle down, whereas the lightest bodies float for longer . . . until they attach themselves" to the ground.[122] Steno concluded his list of examples with the common early modern comparison between macro- and microcosms, showing that this analogy survived well into the seventeenth century.[123] In his words, "what the diversity of a diet affects in the fluids of the microcosm [i.e., the body], so the vicissitudes of the sun and the moon and the changes of various other things may produce in the fluids of the Earth." Steno said that he read this comparison while reading Gassendi, who wrote "on the production of stones in his *philosophia*."[124]

These previous analogies were all written in 1667. By 1669, in *De solido*, Steno firmly concluded that "if a solid body is produced according to the laws of nature, it is produced from a fluid."[125] He supported this claim with

118. Steno, *Elementorum myologiæ specimen*, 101.
119. Steno, *Elementorum myologiæ specimen*, 101.
120. Steno, *Elementorum myologiæ specimen*, 102.
121. Steno, *Elementorum myologiæ specimen*, 102. "Borellum" was probably Jacques Borelly and not Pierre Borel; see Pierre Chabbert, "Jacques Borelly (16. .–1689): Membre de l'Académie royale des sciences," *Revue d'histoire des sciences et de leurs applications* 23, no. 3 (1970): 203–227.
122. Steno, *Elementorum myologiæ specimen*, 100.
123. For a different opinion, see Gary D. Rosenberg, "Introduction," in Rosenberg, *The Revolution in Geology from the Renaissance to the Enlightenment*, 1–11, esp. 3.
124. Steno, *Elementorum myologiæ specimen*, 102–103. See also Toshihiro Yamada, "Kircher and Steno on the 'Geocosm,' with a Reassessment of the Role of Gassendi's Works," in *The Origins of Geology in Italy*, ed. Gian Battista Vai, W. Glen Caldwell, and E. Caldwell (Boulder, CO: Geological Society of America, 2006), 65–80.
125. Steno, *De solido*, 18.

an analogy that reflects Steno's whole understanding of the body and how it relates to the Earth.[126] In this complex analogy, he spoke of external and internal bodily fluids. External fluids were those present in organs open to the outside, such as the mouth, stomach, intestines, uterus, or bladder. "Worms and stones generated inside our body are mostly produced in the external fluid," he concluded.[127] Steno then divided internal fluids into "common internal fluid [*fluidum internum commune*]" and "proper internal fluid [*fluidum internum proprium*]."[128] Common internal fluids were those "contained in veins, arteries and lymphatic vessels," whereas proper internal fluids were those located in specific parts of the body such as "around muscle fibers."[129] With these categories, Steno offered a sophisticated explanation for the formation of solids from bodily fluids: "Particles that are separated ... from external fluids are carried into the internal common fluid by an interceding sieving, from which they are also in several ways secreted and transmitted into the internal proper fluids through another sieving. [Then these particles] are added to solid parts either in the manner of fibers or organ tissue [*parenchyma*]."[130]

An example drawn from Steno's anatomy makes this description clearer. Steno understood that solid food entered the stomach and was then digested by means of external fluids produced therein. In the intestines, this new mixture of food with external fluids is transformed into one common internal fluid named chyle. This chyle flows via the thoracic duct to the subclavian vein, where it is mixed with blood (chyle and blood being common internal fluids). Next this composite bloodstream reaches the glands and muscles, where the blood is filtered into proper internal fluids. The latter's particles are then incorporated in muscle fibers. Steno concluded his analogy by saying that the external fluids of the Earth produced sediments, encrustations, or crystals, and that the internal fluids produced things such as "the hard substance joining broken bones."[131] Steno was not more specific on this latter analogy with internal fluids, perhaps because fossils were, after all, of organic origin.

---

126. Body-Earth analogies had been used since the thirteenth century; see Stefano Dominici, "The Volterra Cliff in the Mind of Philosophers, Savants, and Geologists (1282–1830)," *Geological Society, London, Special Publications* 543 (2023): 267–280.
    127. Steno, *De solido*, 20.
    128. Steno, *De solido*, 21.
    129. Steno, *De solido*, 21.
    130. Steno, *De solido*, 23.
    131. Steno, *De solido*, 24.

Having explained the mechanisms through which fluids secrete solids, Steno expanded his analysis to explain the formation of rocks over time, drawing on two propositions. First he concluded that "bodies entirely similar to one another are also produced in similar ways," then went on to say that "in two solids contiguous with each other, the first to harden is that which represents the properties of its own surface in the surface of the other."[132] Both propositions resonate with molding processes that Steno saw in his family's goldsmithing workshop—for example, wax hardening within a plaster mold from which it takes its shape.[133] Finally, Steno stated that "given a solid and its location, it will be easy to say something certain [*certum quid pronunciare*] about its place of production," even when that place is completely different from its current location.[134] This is still one of the main claims of modern geology.

But the strongest links between Steno's history of the Earth and the body are in his approaches to muscles—the topic that occupied him when he shifted his research to fossils. The first link has to do with format. In his mathematical treatise on muscles, Steno used a format similar to that of Euclid's *Elements*. He started with a list of definitions and then followed them with lemmas and corollaries that referred back to these definitions. In a striking parallel, he started his first text on fossils with a list of observations to which he had recourse in the following conjectures. He even printed in the book's margins the relevant observations used in each conjecture (see fig. 6.6). That is, just as he used a systematic model to understand muscle contraction, so too he developed a systematic model to explain fossilization.

More interestingly, Steno's visual arguments of the Earth's strata owe much to his muscle diagrams. In *De solido* Steno had to account not only for the aspect of sediments and strata (*historia*) but also for their formation in time (*actio*). He explained the origin of mountains by saying that "strata were themselves solids naturally contained within solids."[135]

---

132. Steno, *De solido*, 24. Steno supported these claims with observations in ibid., 15–18.
133. Kardel, "Prompters of Steno's Geological Principles," 131–132. Steno mentions wax modeling in *Elementorum myologiæ specimen*, 119. Medieval and Renaissance authors also described the formation of fossils with casting metaphors. See Ivano Dal Prete, "Ruins of the Earth: Learned Meteorology and Artisan Expertise in Fifteenth-Century Italian Landscapes," *Nuncius: Annali di storia della scienza* 33, no. 3 (2015): 415–441, esp. 428–429.
134. Steno, *De solido*, 24.
135. Steno, *De solido*, 26–37, esp. 37.

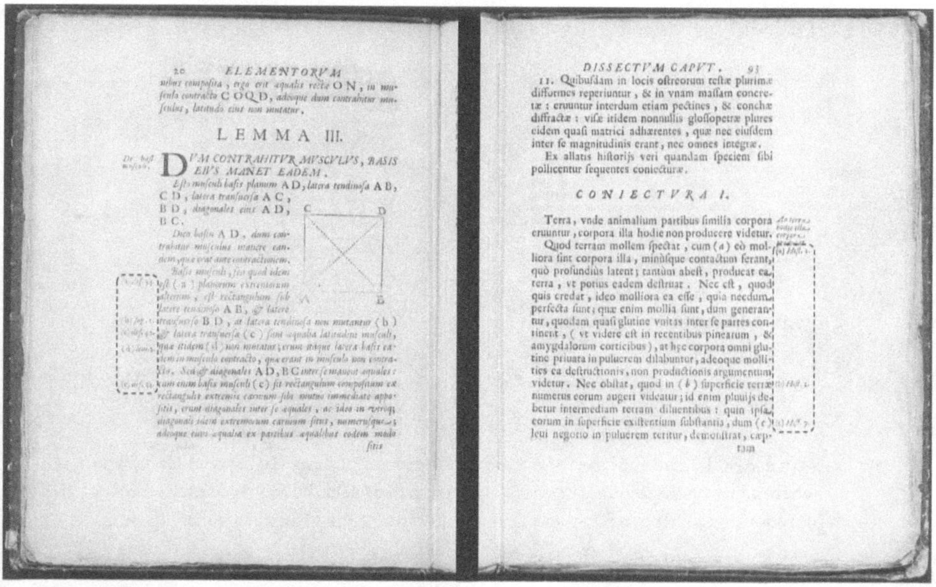

FIGURE 6.6. References to lemmas in the margins of the text on muscles (*left*) and to conjectures in the margins of the text on fossils (*right*). Steno, *Elementorum myologiæ specimen*, 20, 93. Courtesy of The Linda Hall Library of Science, Engineering & Technology.

Like fossils, he wrote, "whenever a certain stratum was formed, it was either surrounded by another solid body on the sides or it covered the entire globe of the Earth."[136] This is an early formulation of what is now known as Steno's principle of lateral continuity.[137] Steno then applied his ideas to the formation of the Tuscan landscape.[138] Strikingly, he relied on cross-sectional slices of these landscapes which he represented with geometrical diagrams (see fig. 6.7). Steno was neither the first nor the last to use diagrams in this context. Agostino Scilla (1629–1700), for example, used observational and painting techniques to depict fossils as animal parts.[139] And Descartes, in his *Principles of Philosophy* (Amsterdam, 1644),

---

136. Steno, *De solido*, 30.
137. Cutler, *The Seashell on the Mountaintop*, 112.
138. Steno, *De solido*, 67–76.
139. Paula Findlen, "Projecting Nature: Agostino Scilla's Seventeenth-Century Fossil Drawings," *Endeavour* 42, nos. 2–3 (2018): 99–132, esp. 122; and Domenico Laurenza, "Images and Theories: The Study of Fossils in Leonardo, Scilla, and Hooke," *Nuncius: Annali di storia della scienza* 33, no. 3 (2018): 442–463, esp. 449–456.

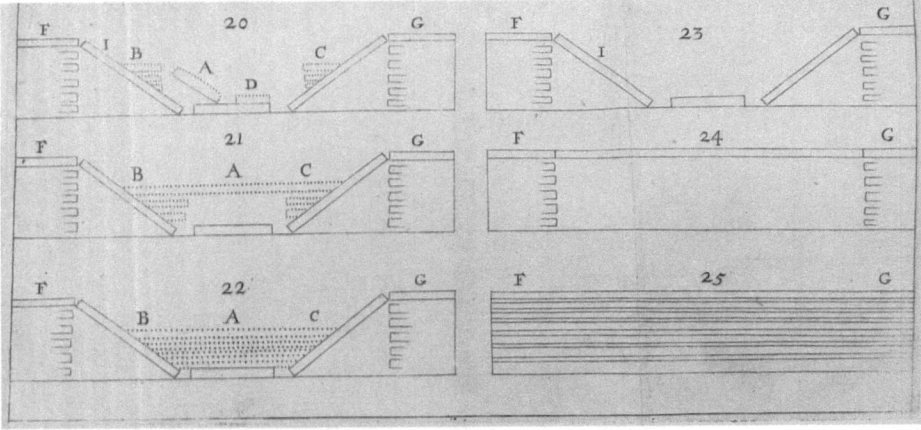

FIGURE 6.7. Diagram of the formation of Tuscan mountains. To be read from 25 to 20, where 25 is the original Earth and 20 is the Earth as seen by Steno. Steno, *De solido* (Florence, 1669). Reproduced with the permission of the Ministero della Cultura/Biblioteca Nazionale Centrale di Firenze.

which Steno knew well, drew cross-sections of the Earth years before Steno did them (see fig. 6.8).[140] Nonetheless, Steno's diagrams resemble Descartes's only superficially. First, Steno's diagrams represent a view in opposition to the Cartesian theory of the origin of mountains.[141] Descartes spoke of mountains growing through the Earth, with the destruction of the Earth's crust leading to a new mountain rising. But Steno spoke of waters drying up and strata breaking into valleys. More important, Steno's strata diagrams have a direct antecedent in his diagrams of muscle contraction. Just as his first diagrams were a result of observing cross-sections of muscles, his new diagrams represented a cross-sectional view of the Earth's strata. Indeed, cross-sectional diagrams became so useful to Steno that, even though he did not publish any more geology books, he used cross-sections again to depict a cave that he explored in the Alps two years later (see fig. 6.9).[142] In short, Steno's diagrams, alongside his comparative and

---

140. Steno mentions Descartes in *De solido*, 28. See also Daniel Garber, "Steno, Leibniz, and the History of the World," in Andrault and Lærke, *Steno and the Philosophers*, 201–229, esp. 202–204.

141. Garber, "Steno, Leibniz, and the History of the World," 202–204, 210.

142. Johannes Mattes, "Mapping the Invisible: Knowledge, Credibility and Visions of Earth in Early Modern Cave Maps," *The British Journal for the History of Science* 55, no. 1 (2022): 53–80, esp. 53–55, 72–74, 77, 79. Mattes claimed that Steno was the first to use cross-sections in geological maps.

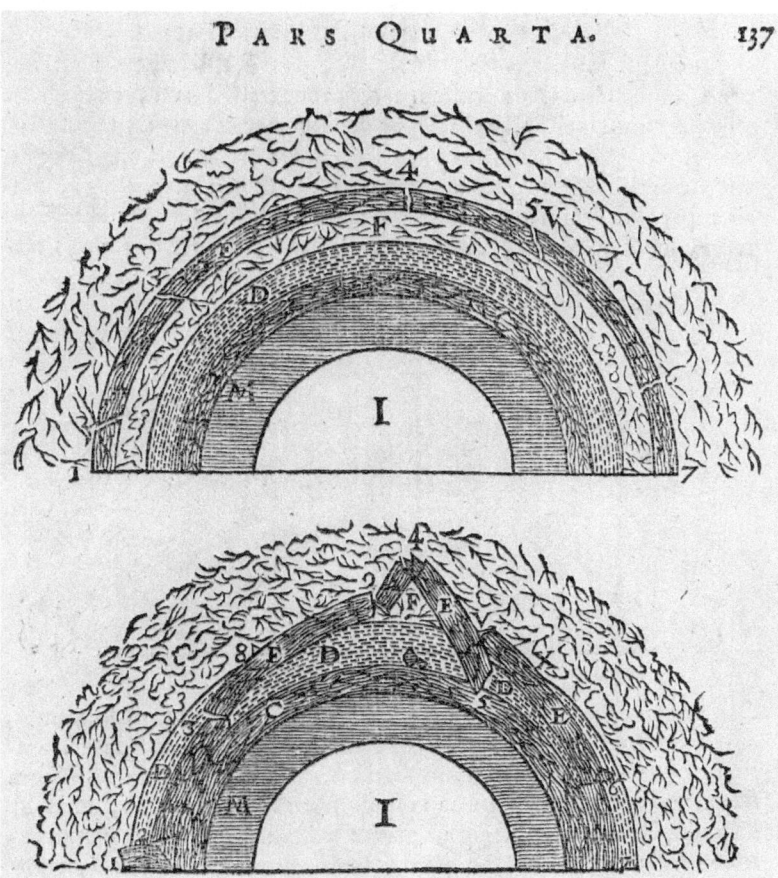

FIGURE 6.8. Descartes's diagram of the formation of the Earth. Descartes, *Principia philosophiæ* [sic]. Reproduced with the permission of the Wellcome Collection.

cross-disciplinary methods, confirmed to him the epistemic strength of his approach. "How well then everything fits! How [well] these things conspire between themselves in unanimous agreement!"[143]

✳

While crossing the roads of Europe from north to south, Nicolaus Steno, like all those who go hiking, probably marveled at the spectacular

---

143. Steno, *Elementorum myologiæ specimen*, 103–104. Exclamation marks in the original.

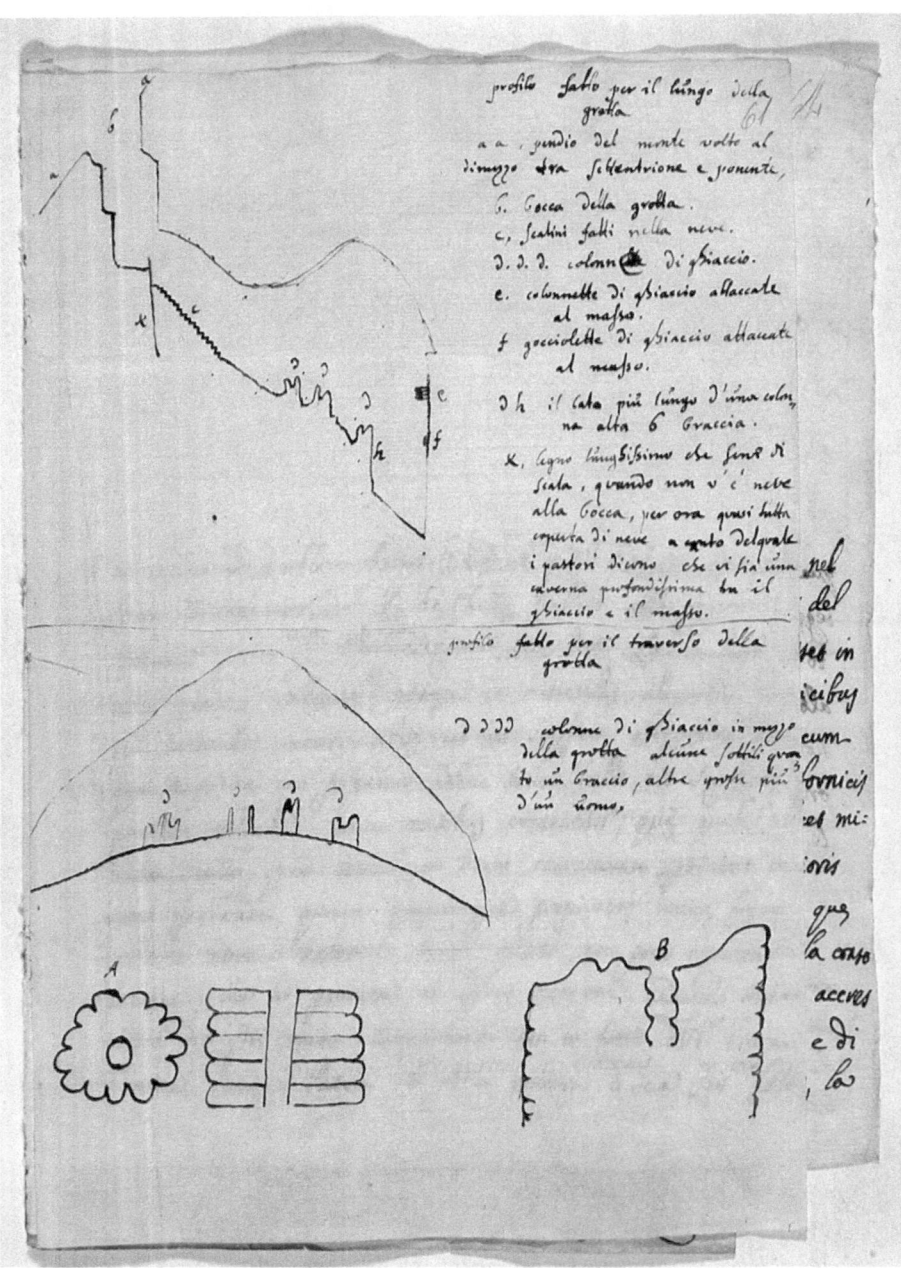

FIGURE 6.9. Cross-section by Nicolaus Steno of a cave in northern Italy. Florence, BNCF, Gal. 286, fol. 61r. Reproduced with the permission of the Ministero della Cultura/Biblioteca Nazionale Centrale di Firenze.

landscapes around him and wondered how they were formed. Through his work on fossils, Steno found a solution that in turn opened new fields of research that encompass more than just modern geology.[144] For instance, Charles Lyell's book *Principles of Geology* (London, 1830–1832), one of the main influences behind Darwin's theory of evolution, stated that Steno's geology was "the most remarkable work of that period."[145] But I suspect that neither Lyell nor Darwin knew that it all started with Steno's anatomy.

In this chapter, the disciplines of anatomy, Earth history, physicomathematics, and chymistry came together in Steno's hands as part of his search for certainty in science. Moreover, Steno's claims about the Earth were not entirely new. What was original and convincing was his empirical and analogical arguments, which he developed out of his anatomical training and research at the University of Leiden and further developed in Paris. That is, his writings in geology were as much a product of his European travels as of the context of the Accademia del Cimento. I am not diminishing the role of the Italian context in Steno's work. Rather, my analysis suggests that Florentine intellectual culture reinforced what Steno was already doing before he arrived in Italy, such as using mathematics and chymistry in anatomy.

Reading Steno's geology in the context of his anatomy also sheds new light on other aspects of seventeenth-century science. Steno's geological methods, such as using analogies and diagrams, had direct antecedents in his anatomy, which was often anti-Cartesian. Moreover, Steno's detachment of his work from philosophical commitments for epistemic reasons, and not for fear of religious censorship, confirms the recent reassessments of early modern censorship of science. This does not mean that religion did not have other roles in Steno's history of the Earth, but such a discussion falls outside the purview of this book. In short, Steno expanded his interests from anatomy to geology as a response to specific problems that he encountered in his historical context. But it is precisely because of this context that they were so significant to him, to his peers, and to historians of science today.

The *Elementorum myologiæ specimen* was a groundbreaking book in which Steno successfully developed a mathematical model of muscles, showed for the first time that women had ovaries, and explained the origin

---

144. William Poole, *The World Makers: Scientists of the Restoration and the Search for the Origins of the Earth* (Oxford: Peter Lang, 2010).

145. Charles Lyell, *Principles of Geology*, 3 vols. (London: John Murray, 1830), 1:40; and Janet Browne, *Darwin's "Origin of Species": A Biography* (New York: Atlantic Books, 2006), 22.

of fossils. In all these cases, his focused interdisciplinary method opened new areas of research in "mathematical anatomy," the female reproductive system, and the natural history of the Earth. By these lights, 1667 was Steno's most productive year. Strikingly, it was during this period of intensive production that Steno finally decided to convert to Catholicism on 2 November 1667—the Feast of All Souls. Yet again, the search for certainty in the sciences, travels, and friendships played a role in Steno's interior quest for spiritual certainty.

CHAPTER SEVEN

Anatomy of a Conversion

On 8 December 1667, the Feast of the Immaculate Conception, Nicolaus Steno knelt twice before the papal nuncio, the representative of the pope in Tuscany. The nuncio conferred on Steno the sacrament of confirmation and then gave him Holy Communion for the first time in his life.[1] Steno had been raised a devout Lutheran in Copenhagen, but from this moment on, a month before he turned thirty years old, he entered into full communion with the Catholic Church. In the words of this same nuncio, in a letter to the secretary of state of Rome, Steno wanted to receive these sacraments directly from him because of his "pious and ideal concept [*per suo pio ideale geroglifico*] of the doctrine of the infallibility of the Pope."[2] Wanting the nuncio to preside over these sacraments was no accident. Then as now, papal infallibility was the doctrine that distinguished Catholics from all other Christians. By choosing the papal nuncio to give him the sacraments of confirmation and the Eucharist, Steno publicly affirmed his new commitment to Catholicism. But why did the famous anatomist develop such a strong attachment to the Catholic Church? And why did his conversion happen at the end of such an intense research year in Italy?

Steno's religious conversion has been a topic of discussion among scholars for years.[3] Each study emphasizes distinct reasons for his conversion, to

---

1. Viviani to Magalotti, 13 December 1667, in Fabroni, *Delle lettere familiari*, 1:17–18.
2. Lorenzo Trotti to Decio Azzolino, 29 November 1667, in *Epistolae*, 2:925. On "geroglifico," see *Grande dizionario della lingua italiana*, ed. Salvatore Battaglia and Giorgio Squarotti, 21 vols. (Turin: Unione Tipografico Editrice Torinese, 1961–2009), 6:704–705. Although papal infallibility did not become dogma until 1870, papal authority was already a point of Protestant contention in the early modern period; see Carlos M. N. Eire, *Reformations: The Early Modern World, 1450–1650* (New Haven, CT: Yale University Press, 2016), 51–53, 151, 383–384.
3. The most detailed recent accounts are Frank Sobiech, *Herz, Gott, Kreuz: Die Spiritualität des Anatomen, Geologen und Bischofs Dr. med. Niels Stensen (1638–1686)*, Westfalia Sacra 13 (Munster: Aschendorff Verlag, 2004), 39–68; and Miniati, *Nicholas Steno's Challenge for Truth*, 185–190.

the point that one recent article even poses the question in its title, "Jesuits, Women, Money, or Natural Theology?"[4] Historians of science have also placed Steno's conversion in opposition to his scientific production by describing it as a "religious crisis" that brought his scientific career to an end.[5] These claims arose in part because of statements that Steno himself made. He, too, provided distinct reasons for his conversion when asked for them. And, in a book about his own conversion, he acknowledged that "the study of natural things in which I was wholly immersed did not allow me a serious study of religion." Things changed, he continued, only when "divine mercy unexpectedly dragged me from these other studies to the study of salvation [*a reliquis studiis ad studium salutis*]."[6] Yet, contrary to what Steno's claims may suggest, he continued to pursue his scientific interests for eight years after his conversion, up until he was ordained a priest. As a lay Catholic Steno traveled throughout Europe, dissected animals, observed rocks and mountains, and published new observations.

This chapter describes Steno's conversion by placing it in its historical context. It shows that the friendships, search for certainty, and humanistic skills that framed his itinerant career throughout Europe came together again in his turn to religion in Florence. Rather than asking *why* this conversion took place—which is a hard question to answer—I suggest that more can be learned about it by looking at *how* it happened.[7] However, understanding Steno's conversion is not a straightforward task because most narratives date from several years after it happened. The most detailed third-person account was written by the nun Maria Flavia del Nero (fl. 1660s), more than twenty years after Steno's conversion and two years after he died.[8] This means that all extant accounts are influenced by what happened to Steno after his conversion, such as the publication of a very successful book on geology in 1669; his elevation to the highest ranks of the church's hierarchy when he was ordained a bishop in 1677; and, most important, his publicly devout,

---

4. Sebastian Olden-Jørgensen, "Jesuits, Women, Money or Natural Theology? Nicolas Steno's Conversion to Catholicism in 1667," in Andrault and Lærke, *Steno and the Philosophers*, 45–62.

5. Findlen, *Possessing Nature*, 220; and Ole Peter Grell, "Between Anatomy and Religion: The Conversions to Catholicism of the Two Danish Anatomists Nicolaus Steno and Jacob Winsløw," in Grell and Cunningham, *Medicine and Religion in Enlightenment Europe*, 205–221, esp. 215.

6. Steno, *De propria conversione*, in OTH, 1:127.

7. For a similar approach see, e.g., Irene Fosi, *Inquisition, Conversion, and Foreigners in Baroque Rome* (Leiden: Brill, 2020), 12.

8. Sister Maria Flavia del Nero's account of Steno's conversion, 14 July 1688, in *Epistolae*, 2:987–990.

poor, and charitable life—which the Catholic Church officially recognized when Pope John Paul II beatified him in 1988. All these events make Steno's conversion look more significant later than when it actually happened. For these reasons, this chapter's anatomy of Steno's conversion focuses on sources closer to the event to locate it in its more immediate context.

I argue that Steno converted to Catholicism in Florence as the result of specific historical circumstances associated with the time Steno was in Italy. First, I explain that it was common to change religious denominations in early modern Europe, including from Protestantism to Catholicism. Moreover, although never mentioned in studies of Steno's conversion, the seventeenth-century papacy developed an increasingly open approach toward northern European scholars that also affected Steno. His firsthand experience of this openness toward Protestant intellectuals during his trip to Rome contributed to his developing a positive image of Catholicism. Second, Steno's intellectual collaboration and friendship with Francesco Redi was the main factor that put Steno on a track toward conversion, a social aspect that has rarely been emphasized until now.[9] Third, and more significant, I argue that, rather than an obstacle, Steno's scientific research implicitly supported his religious pursuits, especially because of his search for certainty. It is a fact that, when it came to studying theology, Steno did not rely on knowledge from other disciplines, as he did in science. Religion, therefore, was not part of his focused interdisciplinary work. At the same time, however, there were significant other resonances between his scientific and theological studies. Just like in anatomy and geology, Steno strongly relied on the humanistic education he acquired at the Copenhagen Cathedral School; he intentionally fitted theological studies into a busy research schedule; and benefited from friendships made in scientific circles, including with women. Indeed, Steno's conversion was in large part a response to conversations that he had with devout Catholic women whom he met in Europe's intellectual networks. Steno's openness to the advice of women thus extends the Renaissance phenomenon of spiritual motherhood described by Gabriella Zarri and others into the seventeenth century.[10] In a nutshell, this historical contextualization of Steno's conversion challenges the claim that the reasons that Steno gave for it are

---

9. Recent essays on Steno's conversion do not mention Redi; see Olden-Jørgensen, "Jesuits, Women, Money or Natural Theology?," and Grell, "Between Anatomy and Religion."

10. Gabriella Zarri, *Uomini e donne nella direzione spirituale (secc. XIII–XVI)* (Spoleto: Fondazione CISAM, 2016).

"contradictory"—something that Steno never said.[11] Instead, his conversion was a process as complex as human beings are and, for that reason, involved various social, intellectual, and emotional factors.

Religious Conversions in Seventeenth-Century Europe

Conversions to Catholicism were relatively new in seventeenth-century Europe. For over a millennium, despite political and cultural differences, Western Christendom experienced unity in faith and moral standards. The only conversions that Western medieval Europeans would have heard of were Muslims and Jews who, either freely or by force, became Christians, or Christians who lapsed in the opposite direction.[12] This scenario changed only when reformers such as Martin Luther and John Calvin rejected the intellectual and moral authority of the Catholic Church in the early sixteenth century.[13] By 1600, owing to the alignment of reformers with political leaders, Europe was splintered into multiple versions of Christianity, with massive consequences that changed the course of history and continue to influence world events to this day.[14] Crossing from one Christian denomination to another therefore constituted a novel practice for Western Europeans.[15] Yet, although novel, these religious conversions also affected the lives of early modern scientists.

As a consequence of these religious divisions, besides the various wars that began in Europe, nations worked to maintain religious unity within their borders. Most countries, Catholic and Protestant alike, adhered to the confession of their leaders—following the famous principle *cuius regio eius religio* (lit., whose kingdom, his religion)—and developed ways to keep other confessions away.[16] These efforts often affected higher

---

11. For the argument that they are contradictory, see Olden-Jørgensen, "Jesuits, Women, Money or Natural Theology?," 47–48.

12. David Nirenberg, *Communities of Violence: Persecution of Minorities in the Middle Ages* (Princeton, NJ: Princeton University Press, 1996), 127–128. I am assuming that, except in some parts of Eastern Europe, conversions to or from the Eastern Orthodox Church were rare.

13. Eire, *Reformations*, 19–42.

14. Brad S. Gregory, *The Unintended Reformation: How a Religious Revolution Secularized Society* (Cambridge, MA: Belknap Press of Harvard University Press, 2012), 82–96.

15. Benjamin J. Kaplan, *Divided by Faith: Religious Conflict and the Practice of Toleration in Early Modern Europe* (Cambridge, MA: Harvard University Press, 2007), 3–4, 266–276.

16. Kaplan, *Divided by Faith*, 99–124.

education.[17] Interestingly, Denmark was one of the most successful nations at keeping other faiths at bay.[18] Its newly Lutheran kings implemented a system of censorship and border control directed at stopping the circulation of heretical ideas, especially those that were Catholic.[19] After a few Jesuit attempts to regain influence in Denmark in the early seventeenth century, which succeeded in converting a number of Danes, the monarchy tightened its Lutheran grip even further. Every student at the University of Copenhagen had to sign a Lutheran profession of faith, and the city's few Catholic priests were not allowed to perform religious rites.[20] Theology university students were also forbidden to study in non-Lutheran schools abroad.[21] Therefore, the devout young Steno probably never met any Catholics in Copenhagen. Strikingly, medical students could visit and study at the best European universities, regardless of their faith. Steno's scientific training abroad would thus introduce him and other Danes to many Catholics.

Unlike Denmark, the Netherlands adopted Calvinism as its official religion and did not pursue the complete suppression of other confessions.[22] Besides Dutch Calvinists, there were also Dutch Catholics, Lutherans, Anabaptists, non-denominational Christians, and even a few deists, such as Spinoza—"deists" here meaning those who denied the intervening action of God in the world and criticized organized religion.[23] Ole Borch, Steno's traveling companion in the Netherlands, noted in his diary that he and Steno sometimes witnessed communal prayers of various confessions in Amsterdam and its surroundings.[24] They also met people who had converted from one confession to another. For instance, the alchemist Giuseppe Borri was a former Italian Catholic who held heterodox religious ideas.[25] The theologian Petrus Serrarius (1600–1669), whom Steno also

---

17. Gregory, *The Unintended Reformation*, 326–339; and Ole Peter Grell and Andrew Cunningham, eds., *Medicine and the Reformation* (New York: Routledge, 1993).

18. Eire, *Reformations*, 322.

19. Thorkild Lyby and Ole Grell, "The Consolidation of Lutheranism in Denmark and Norway," in *The Scandinavian Reformation from Evangelical Movement to Institutionalisation of Reform*, ed. Ole Peter Grell (Cambridge: Cambridge University Press, 1995), 114–143, esp. 117–119.

20. Lyby and Grell, "The Consolidation of Lutheranism in Denmark and Norway," 133–135.

21. Grell, "Caspar Bartholin and the Education of the Pious Physician," 93.

22. Eire, *Reformations*, 542–548.

23. Jorink, "'Outside God, There Is Nothing,'" 104.

24. They listened to an Armenian sermon and visited a Jewish temple; see OBI, 1:7, 25 December 1660. They also witnessed a Quaker ritual; OBI, 2:67–68, 26 February 1662.

25. OBI, 1:133, 24 May 1661.

met, was an English Calvinist turned millenarian—that is, one who believed that the end of the world was imminent.[26] And Franciscus van den Enden (1602–1674), whose name appeared several times in Borch's diary, was a former Jesuit who, after being expelled from the order, renounced his Catholic faith.[27] Van den Enden would become one of Spinoza's mentors in Amsterdam.[28]

Thanks to its plurality of religious ideas, the Netherlands became famous for religious freedom. Yet, as Benjamin Kaplan and Charles Parker have shown, early modern confessional coexistence was not the same as modern religious liberty.[29] Despite the Netherlands' claims of tolerance, Catholics were persecuted there. Catholic worship took place only in private homes, often at night.[30] "Papists," as they were called, could not hold public office or defend their faith in public, and priests were sometimes imprisoned or expelled from the country.[31] It seems that it was only through his scientific networks that Steno made Catholic friends, such as Regnier de Graaf, his colleague at the University of Leiden, and Catherine Questiers, mentioned in chapter 3. Their interactions, however, had little to do with religion, and there is no evidence that Steno attended Catholic events in the Netherlands. If anything, he would remember hearing "much evil about others who had converted to the Catholic faith [*quanta mala audiveram de aliis ad fidem catholicam conversis*]."[32] It was only after he crossed the border into Flanders that he visited Catholic churches and colleges, and it was probably while there that he began to develop a positive view of Catholicism.

Essentially, Steno's Dutch sojourn was an occasion for him to meet other kinds of believers for the first time, including those who rejected Christianity altogether or "declared it so dull that hardly anything remained except the name."[33] Such encounters were a shock to Steno's Lutheran faith. Years later, when writing about his conversion, Steno admitted that his years in the Netherlands "gradually diminished [*sensim debilitarent*]" his attachment to Lutheranism. The causes, he continued, were "the study of Cartesian philosophy," the multiplicity of Christian confessions, and "the political way of

---

26. OBI, 1:187, 9 August 1661; Jorink, "*Modus politicus vivendi*," 29–30.
27. Wim Klever, "Spinoza and van den Enden in Borch's Diary in 1661 and 1662," *Studia Spinozana* 5 (1989): 311–325.
28. On van den Enden and Spinoza, see Israel, *Spinoza*, 219–260.
29. Kaplan, *Divided by Faith*, 1–12.
30. Kaplan, *Divided by Faith*, 98, 122, 168–169, 172–174.
31. Charles H. Parker, *Faith on the Margins: Catholics and Catholicism in the Dutch Golden Age* (Cambridge, MA: Harvard University Press, 2008), 3, 12–13.
32. Steno, *Defensio . . . epistolae de propria conversione* (Hannover, 1680), in OTH, 1:392.
33. Steno, *Defensio epistolae*, in OTH, 1:389.

living of many [*modusque multorum vivendi politicus*]."[34] Historians agree that by this latter expression Steno meant a public detachment from religion, for instance changing one's confession for nonreligious reasons, perhaps political or financial.[35] In a later publication Steno, by that time a bishop, wrote of such people as having "no idea of religion other than a certain bond of human society."[36] Examples of this detachment abounded. One of Steno's colleagues at the University of Leiden, an Anabaptist, converted to Calvinism in order to accept a professorship at the university.[37] And Steno's friend Swammerdam once wrote that changing one's confession was not necessary because "only he who loves God and his neighbors like himself . . . may become blessed by Christ."[38] Many of these friends of Steno were radical followers of Cartesian philosophy. But unlike Swammerdam, who remained a Christian, Steno thought that "the perfection, or rather destruction of Cartesian philosophy" by Spinozists "made them all materialist [*totos reddidit materiales*]."[39]

In light of these radical Dutch influences, a few historians have claimed that the young Steno became a deist for a while, meaning that he rejected the basic tenets of Christianity.[40] The problem with this claim, however, is that no primary source shows that Steno's Dutch period destroyed his Lutheran faith.[41] On the contrary, when narrating his conversion, Steno explicitly wrote that even though some radical ideas were attractive, "I always adhered to the institutions of the father [Luther]."[42] It was not that Lutheranism was more persuasive than materialism, Steno continued, but rather that he was too focused on "natural studies and travels [*studia naturalia et peregrinatio*]" to think about these matters.[43] At the time of his conversion, no one in Florence associated Steno with heterodox beliefs (except for Lutheranism). Indeed, even though Steno flirted with radical ideas, his rejection of Cartesianism started early on in his scientific career, as I have been arguing throughout this book. His late claims of having been close to

---

34. Steno, *De propria conversione*, in OTH, 1:126; and *Defensio epistolae*, in OTH, 1:387.
35. Jorink, "*Modus politicus vivendi*," 26; and Miniati, *Nicholas Steno's Challenge to Truth*, 72.
36. Steno, *Defensio epistolae*, in OTH, 1:391.
37. Jorink, "*Modus politicus vivendi*," 33–34.
38. Swammerdam to Thévenot, January 1678, in G. A. Lindeboom, *The Letters of Jan Swammerdam to Melchisedec Thévenot* (Amsterdam: Swets & Zeitlinger, 1975), 84.
39. Steno, *Defensio epistolae*, in OTH, 1:389. He writes "*Spinoza et eius sequaces*."
40. Sobiech, *Herz, Gott, Kreuz*, 51–68; and Olden-Jørgensen, "Jesuits, Women, Money or Natural Theology?," 56–60.
41. This is also the opinion in Miniati, *Nicholas Steno's Challenge for Truth*, 76.
42. Steno, *De propria conversione*, in OTH, 1:126.
43. Steno, *De propria conversione*, in OTH, 1:126.

Cartesianism should be read as the statements of one who found proximity to Cartesian ideas very dangerous. Indeed, shortly before his episcopal ordination, Steno denounced Spinoza's works to the Holy Office in Rome as poisonous and as having done "much evil."[44] Regardless, the multiplicity of beliefs that Steno encountered in the Netherlands made him aware of early modern Christianity's crisis of certainty, discussed in further detail below. In addition, since in Florence he still thought of the Catholic faith as something to be "despised [*aborrita*]," the Dutch sojourn probably also reinforced the typical Protestant rejection of Catholicism with which he was raised.[45]

Nonetheless, despite the prevalence of anti-Catholic sentiments in Protestant lands, many northerners felt attracted to Catholicism in the early modern period. Swammerdam, who was officially a Calvinist, once told Thévenot that he was "more Catholic than Reformed," although he decided not to convert, for the reasons mentioned above.[46] In Italy the situation was different, and many crossed the confessional boundary into the Catholic Church.[47] Although there were laws against the free movement of non-Catholics in most of the peninsula, this prohibition was rarely enforced, and Protestants went there often for economic and intellectual reasons.[48] In fact, travel to Italy was so common that Protestant authors warned of its dangers, because it could lead to the traveler's conversion.[49] For example, when the sons of the royal physician of Denmark traveled to the Netherlands in the 1630s, they were explicitly warned not to proceed south. But, as a Capuchin friar reported, curiosity led the older son to Louvain, where he "began to doubt the verity of his sect," and to Paris, "where finally he abjured the sect of Luther and became a Catholic."[50] When the younger brother joined him in Genoa, where the elder had taken religious vows, the brother too became Catholic. In short, conversions to Catholicism were more common in the seventeenth century than an age of violent religious conflict may suggest.[51]

---

44. Steno to Holy Office, 4 September 1677, in Spruit and Totaro, *The Vatican Manuscript of Spinoza's "Ethica,"* 68.
45. Steno to Lavinia Arnolfini, undated (probably June 1668), in OTH, 1:8.
46. Swammerdam to Thévenot, undated (after September 1675, as it refers to Steno as "Bishop"), in Lindeboom, *The Letters of Jan Swammerdam,* 84.
47. Peter Mazur, *Conversion to Catholicism in Early Modern Italy* (New York: Routledge, 2016); and Fosi, *Inquisition, Conversion, and Foreigners.*
48. Mazur, *Conversion to Catholicism,* 43–65.
49. Fosi, *Inquisition, Conversion, and Foreigners,* 66–70, 126–127, 161.
50. As quoted in Mazur, *Conversion to Catholicism,* 57–58.
51. For a long, though not exhaustive, list of converts to Catholicism, see Andreas Räss, *Die Convertiten seit der Reformation nach ihrem Leben und aus ihren Schriften dargestellt,* 13 vols. (Freiburg, 1866–1880).

There are no sources from before Steno's conversion that suggest he was interested in becoming Catholic before he arrived in Florence. But a few letters written much later speak of important conversations that he had about Catholicism shortly after leaving Leiden. The first was in Cologne, where Steno sojourned on the way from Copenhagen to Paris in 1664. As mentioned in chapter 3, his mother had just died, and Steno was probably more anxious about his future in those days. Perhaps because of that, Steno engaged in a conversation about the merits of Lutheranism with a Jesuit priest in Cologne. Little is known about what they discussed, except that the Jesuit challenged Steno to identify exemplary Lutheran lives since the Reformation.[52] The goal was likely to argue that the Lutheran Reformation produced no saints, unlike the Catholic Church. In Paris Steno met Jean-Baptiste de La Barre (1609–1680), another Jesuit priest and the author of four works on doctrinal controversies with Protestants.[53] Steno was introduced to La Barre because of a discussion on "Christ's words about His body," that is, the Eucharist, which was one of the main points of contention across Christian denominations.[54] During this time Steno also learned of fellow scholars of natural studies who converted to Catholicism. The first was Sorbière. As already mentioned, Steno mastered Sorbière's arguments for the mathematization of medicine, while he was a young student in Copenhagen.[55] Sorbière had converted to Catholicism in 1653 and wrote a book about it.[56] In Paris Steno also met Graindorge, who had shifted from Calvinism to Catholicism.[57] Finally, in Rome, while still a Protestant, Steno joined the meetings of the academy of Queen Christina of Sweden.[58] Christina had abdicated the Swedish throne to become a Catholic in 1655, and her cultural patronage became part of the papal strategies to welcome Protestants to Rome.[59]

Since the late sixteenth century the papacy had invested in the spread of Catholicism across Europe as a response to the fragmentation of

---

52. Nicolaus Steno, *Occasio sermonum de religione* (Hannover, 1678), in OTH, 1:190.

53. See Augustin de Backer, Alois de Backer, and Carlos Sommervogel, eds. *Bibliothèque des écrivains de la Compagnie de Jésus*, 3 vols. (Paris, 1869–1876), 1:cols. 916–917.

54. Steno, *Occasio sermonum*, in OTH, 1:191.

55. Sorbière signed his mathematics essay with a pseudonym, but Steno at least knew of his prominence in Paris.

56. Sorbière, *Discours du sieur de Sorbière, sur sa conversion*.

57. Lux, *Patronage and Royal Science*, 10.

58. For Steno at Queen Christina's academy, see Benedetto Milino to Prince Leopoldo, 4 June 1667, in BNCF, Gal. 314, fols. 952r–953r.

59. Fosi, *Inquisition, Conversion, and Foreigners*, 153–154, 164–173.

Christianity. In the early 1600s Rome created a congregation specifically for "those who come to the faith spontaneously [*de iis qui sponte ad fidem veniunt*]," most of whom were Protestants.[60] An important agent in this program of conversions was Johannes Faber (1573–1629), a Vatican physician and member of the Accademia dei Lincei, who was involved in the conversion of Germans.[61] When Queen Christina converted, these efforts were amplified in new ways in Rome. The conversion of the Swedish queen coincided with the election of Pope Alexander VII, who had previously been a vicar in northern Germany and thus had a special interest in the conversion of Germans.[62] A document from those years on the congregation for Catholic converts reports that the primary concern of the congregation was "to not leave the northerners [*oltramontani*], and particularly heretics, . . . [without] direction from authoritative and respected persons." Rather, they should depart Rome "satisfied by the kindness and prudence of the pope's ministers."[63] As part of this strategy, Protestant scholars and princes arriving in Rome were invited to visit the Apostolic Palace, the Vatican Library, and other Roman libraries.[64] These tours expanded the range of experiences already available in Rome, such as Kircher's museum at the Jesuit Roman College.[65] Steno visited all these institutions when he arrived in Rome in May 1666. His conversations were mostly about anatomical research and mathematics, as discussed earlier. But in a report on Steno's suitability to the episcopacy, written more than ten years later, the archbishop of Florence suggested that Steno "started doubting his religion" in Rome, where he had "various and long discourses with Fr. Honoré Fabri."[66] However, even if Steno spoke about religion with Fabri, it is unlikely that he was considering conversion by that time, as explained below.

Above all, this welcoming approach shows that the Catholic Church combined the patronage of science with efforts to disseminate a positive view of its culture and thus bring Protestants back into the fold. This attitude was not entirely new in Rome: Galileo wrote his work to show "that we in Italy, and especially in Rome, know as much about this subject

60. Fosi, *Inquisition, Conversion, and Foreigners*, 7–8, 129.
61. Fosi, *Inquisition, Conversion, and Foreigners*, 85–106.
62. Fosi, *Inquisition, Conversion, and Foreigners*, 150–151.
63. Quoted in Fosi, *Inquisition, Conversion, and Foreigners*, 61 n. 51; 164. By this time the congregation had changed its name to *de conversis ad fidem adiuvandis*.
64. Fosi, *Inquisition, Conversion, and Foreigners*, 166–177.
65. Waddell, *Jesuit Science and the End of Nature's Secrets*, 91–116.
66. Cardinal Francesco Nerli, Report on Steno's suitability for the episcopacy, 1 June 1677, in *Epistolae*, 2:932.

[Copernicanism] as transalpine diligence can have ever imagined."[67] More important, making Catholic culture more attractive to foreigners explicitly circumvented any attempt to force Protestants to convert. When the Benedictine monk and future cardinal Giovanni Bona (1609–1674) wrote to Prince Leopoldo to acknowledge the receipt of Steno's mathematical work, he praised the Medici court's hospitality to all scholars. Bona wrote that the Tuscan court had all the best intellectuals, "even if [they were] foreigners or from the most remote countries."[68] Indeed, various Protestant scholars worked for the Medici family, such as the Anglican physician John Finch (1626–1682), professor of anatomy at the University of Pisa, and the Calvinist anatomist Tilmann Trutwin (d. 1677).[69] Steno himself benefited from Medici patronage for more than a year while still a Lutheran, which shows that he did not need to convert for financial reasons or due to pressure at the court. A few years later Steno explicitly associated his first days in Italy with freedom of religion and ideas: "I lived in places where the Inquisition exists, in Rome and Florence as well as in Pisa and Livorno, and I saw everywhere that the non-Catholics enjoyed the greatest freedom of life if they did not do anything objectionable. Yes, even at the papal court I heard people freely dispute with prelates against the faith, the ecclesiastical hierarchy, or the monarchical regime."[70] In light of his conversations and religious tours in Paris and Rome, Steno was already familiar with the doctrines and institutions of Catholicism by the time he arrived in Florence in the spring of 1666. But the conversion would only come after he completed his work on muscle motion, when his mind turned to the history of the Earth.

Friendships, from Collaboration to Conversion

Steno started working for the Medici court in the summer of 1666. In February 1667 he moved to Pisa and Livorno to accompany the court to their winter residences. There Steno lodged with Redi. This was a logical

67. Galileo Galilei, *Dialogue on the Two Chief World Systems*, preface, translated in *The Essential Galileo*, ed. Maurice Finocchiaro (Indianapolis: Hackett, 2008), 190.
68. Bona to Leopoldo, 4 June 1667, in BNCF, Gal. 312, fol. 98v: "benchi sia straniero, e de' più rimoti paesi."
69. See Archibald Malloch, *Finch and Baines: A Seventeenth Century Friendship* (Cambridge: Cambridge University Press, 1917), 24–25. It was because of Steno that Trutwin converted to Catholicism, shortly before dying; see Jacopo del Lapo to Mario Fiorentini, 30 January 1676, in Domenico Maria Manni, *Vita del letteratissimo Monsig. Niccolò Stenone di Danimarca* (Florence: Giuseppe Vanni, 1775), 158–161.
70. Steno, *Defensio et plenior elucidatio scrutinii reformatorum* (Hannover: Wolfgang Schwendimann, 1679), in OTH, 1:286.

choice because Redi was the Chief Physician of the Medici court, and, as I have described elsewhere, Steno and Redi quickly noticed that they shared similar opinions on their anatomical research.[71] They both acknowledged the importance of the circulation of the blood and incorporated it into their work. Steno also understood the action of poison through the circulatory system, in a way that was similar to how Redi explained the action of the poison of vipers in his own book. More important, both Steno and Redi accorded great epistemic relevance to observations and conducted experiments together, sometimes in entertaining ways. Once, having gone hunting with the court, they finished the day by dissecting a boar that they had just killed in the Tuscan woods.[72] Another time they went fishing in the Ligurian Sea and examined the anatomical details of each fish they captured, as Redi wrote in his notebook that day.[73] Anatomy was so central to their social life that they discussed it constantly, including "by the fire, after eating dinner."[74] Redi also shared Steno's emerging interest on fossils, lent him a relevant manuscript to read, and would later review Steno's *De solido* at the request of the Holy Office.[75] All these factors sustained what became a close friendship between the two scholars.[76]

However, science was not all the good friends talked about. At the end of March 1667 Redi confessed in a letter to a priest friend from Prato that he "certainly hoped that Mr. Nicolaus Steno would convert to the Catholic religion, abandoning Lutheranism. I have so much [evidence] on hand [*io ho tanto in mano*] that I can tell you this with confidence. Thanks to the Blessed Lord! Believe me, . . . Mr. Nicolaus is truly an angel in his manners [*il Sig. Niccolò è veramente un angiolo di costume*], besides being that great philosopher, and great anatomist, and great mathematician that he is."[77] This is the earliest known reference to Steno's possible conversion to Catholicism. It shows that conversations with Steno revealed the Dane's proximity to the Catholic faith, leading Redi to think that his friend was close to conversion. Most important, it demonstrates how a scientific collaboration developed into a friendly conversation about confessional differences.

Ironically, Steno did not feel as close to the Catholic Church as Redi (and probably everyone else) thought. In later accounts of his conversion,

---

71. Castel-Branco, "Friendship Fostered by Poison."
72. Francesco Redi, *Opere*, 9 vols. (Milan, 1809–1811), 6:187–189.
73. Redi, 11 April 1667, in Florence, Biblioteca Marucelliana, Redi 32, fol. 290r–v.
74. Redi, *Opere*, 6:70.
75. Steno, *Elementorum myologiæ specimen*, 109; and Steno, *De solido*, 77–78.
76. Castel-Branco, "Friendship Fostered by Poison."
77. Redi to Valerio Inghirami, 30 March 1667, in Redi, *Opere*, 7:407.

Steno mentions that he only started considering religion seriously weeks later, after the Medici court returned to Florence. Rather than following the court, Steno traveled to Lucca and Pistoia to continue his new research on fossils and to meet scholars there.[78] Redi introduced Steno to two friends in Lucca. The first was Francesco Maria Fiorentini (1603–1673), one of Lucca's leading physicians. Redi asked whether Fiorentini could host Steno, "the most famous of the anatomists of our time," who was traveling there to meet him.[79] As described in previous chapters, Fiorentini provided Steno with important accounts that confirmed his new ideas about fossils.

More relevant to his conversion was Lavinia Cenami Arnolfini (1631–1710), a devout noblewoman and the wife of Lucca's ambassador in Florence. In his introductory letter, Redi praised Steno, subtly writing that "if Steno did not have impressed in his soul the dogmas of Luther, she would judge him as a man of no ordinary perfection."[80] He also asked whether she and the ambassador could receive Steno in their palace and perhaps "satisfy there that native curiosity that made him a Pilgrim of the World."[81] It is not clear whether Redi wanted Lavinia to bring Steno closer to conversion. The Arnolfini family had lived in Paris for several years and overlapped with Steno there, probably sharing some acquaintances.[82] Given Steno's mastery of courtly manners and the French language—then as now essential to Western diplomacy—there would have been some resonance between Steno and the family.[83] Most likely Redi was only looking for suitable friends for Steno to meet in Lucca.

Perhaps because Redi mentioned Steno's faith, when Steno visited the Arnolfinis, Lavinia steered the conversation to discuss "religious matters [*de religione varia*]."[84] The sources do not tell exactly what they talked

---

78. Molara to Viviani, 11 April 1667, in Florence, BNCF, Gal. 163, fol. 110v: "[lui] va a Lucca e poi a Pisttoia."

79. Redi to Francesco Maria Fiorentini, 11 April 1667, in Eugenio Lazzareschi, *Lettere di Nicola Stenone a Lavinia Cenami Arnolfini* (Lucca: Scuola Tip. Artigianelli, 1936), 5.

80. Redi to Lavinia Arnolfini, 11 April 1667, Florence, Biblioteca Marucelliana, Redi 8, fol. 119r: "se egli non avesse impressi nell'anima i dogmi di Lutero che fosse giudicato da VS per un uomo di non ordinaria perfezione."

81. Redi to Lavinia Arnolfini, 11 April 1667, Florence, Biblioteca Marucelliana, Redi 8, fol. 119r: "appagare costì quella nativa curiosità che lo ha fatto Pellegrino del Mondo."

82. Cesare Bambacari, *Descrizione delle azioni e virtù dell'illustrissimma signora Lavinia Felice Cenami Arnolfini* (Lucca: Pellegrino Frediani, 1715), 12–29.

83. Jean-Benoît Nadeau and Julie Barlow, *The Story of French* (New York: St. Martin's Press, 2006), 129, 448.

84. Steno, *De propria conversione*, in OTH, 1:127.

about, but at some point Lavinia made an intense and emotional plea for Steno to embrace the Catholic faith. In a letter he wrote some five to ten years later, Steno quoted Lavinia as saying that "if my blood were enough to demonstrate to you the necessity of it [the Catholic faith], God is my witness that I would give my life in this very moment for your salvation." In response, "moved by this unexpected argument of Christian charity [*commotus hoc non expectato Christianæ caritatis argumento*]," Steno acknowledged that until then he had given "greater attention to other studies than to my own salvation." He promised on the spot to undertake "a serious examination of religion [*serium religionis examen*]."[85] Steno mentioned this encounter with Lavinia in various letters after his conversion, confirming that this moment triggered something in Steno's heart and mind.[86] For the first time in many years, religion (not necessarily Catholicism) appeared important to him.

However, unless this conversation took place two months later at the Arnolfini palace in Florence, Steno did not turn his attention to religion right away.[87] He returned to Florence to finalize the publication of his book on muscles. And a few weeks later he was again on the move. Steno departed Florence in late May. Vincenzo Viviani asked a friend in Rome to support Steno on his arrival, "despite the bond of friendship that has been constructed between him and me."[88] This contrastive clause probably speaks of Viviani's disappointment with his friend's new travels after working so hard to find new lodgings for him in Florence.[89] More than anyone else, Viviani was aware of Steno's new interest on the origin of fossils, since he had to deal with the new material Steno wanted to include in the manuscript last minute. Thus, in his letter to Rome, Viviani also spoke of "the wandering of his [Steno's] doctrine, well-known to the entire world of scholars."[90] In short, Viviani became slowly aware that

---

85. Steno, *De propria conversione*, in OTH, 1:127.

86. Lazzareschi, *Lettere di Nicola Stenone a Lavinia Cenami Arnolfini*.

87. Steno's conversation with Lavinia could have happened later at the Arnolfini palace in Florence. Such a visit, however, is not documented in the sources, whereas the Lucca visit is.

88. Viviani to Ricci, 25 May 1667, in Florence, BNCF, Gal. 158, fol. 186v: "ma non ostante il legame dell'amicizia che tra esso e me si è costrutta, mi stringe di raccomandarlo al favorevole patronato di V.S. per qualunque occorrenza che gli possa sopra avvenire."

89. Viviani to Molara, 1 April 1667, in BNCF, Gal. 158, fol. 179r–v.

90. Viviani to Ricci, 25 May 1667, in Florence, BNCF, Gal. 158, fol. 186v: "si fa largo col proprio merito per la vagolaria della sua dottrina, troppo nova alla... università tutta de' letterati."

Steno's intellectual itineracy was intrinsically tied to his traveling. Steno spent a few days in Rome and by early June was in Livorno, the port city of Tuscany. From there he wrote to Florence with news of shells that he had collected.[91] Steno was traveling to make new observations about rocks and strata.

Yet again, in the midst of this research trip, Steno found himself grappling with what he called "the uncertainty of my soul [*l'incertezza del l'animo mio*]."[92] Steno's time in Livorno coincided with the Feast of Corpus Christi, during which Catholics process through the streets with the Eucharist. Then as now, Catholics believed that during Mass the communion bread ceases to be bread and becomes the Body, Blood, Soul, and Divinity of Christ, thenceforth known as the Eucharistic Host. The Council of Trent further added that this "Real Presence" of Christ remained in the bread *after* Mass, an idea that all Protestant confessions rejected.[93] According to this belief, Catholics could process with and adore the Eucharistic Host, as Steno witnessed in Livorno. As he later wrote, when the Eucharistic Host passed in procession "with such pomp throughout the city ... I felt waking up in my mind this argument [*sentii svegliarmisi nella mente*]: either that Host is a simple piece of bread, and silly are those who give so much honor to it, or here is the true body of Christ, and why don't I honor it too?" This thought seemed to contain a deeper dilemma for Steno's mind, one that touched on the problem of uncertainty. Either the entire Catholic world, "numerous in educated and learned men [*numeroso d'uomini svegliati e dotti*]," was wrong, or Lutherans, with whom "I was born and raised [*in cui ero nato ed allevato*]," were at fault. In Steno's words, there is no way "to reconcile two propositions that contradict each other [*non vi è modo di conciliare insieme due proposizioni che si contradicono*]."[94] His reaction seemed to have been triggered by witnessing the devotional gestures of pious Catholics toward the consecrated Host. However, emotions were likely not enough to create for Steno such an intellectual dilemma. As a learned scholar, he had been aware of the different Eucharistic doctrines between Catholics and Protestants for years. Instead, these emotions arose

---

91. Steno to Molara, 11 June 1667, as mentioned in Molara to Steno, 14 June 1667, in *Epistolae*, 1:192–193.

92. Steno to Lavinia, undated, in OTH, 1:9.

93. *Canons and Decrees of the Sacred and Ecumenical Council of Trent*, trans. James Waterworth (London, 1848), 76–80 (Council of Trent, session 13). Except for Lutherans, all Protestants denied the Real Presence entirely. For Luther's explicit rejection of Corpus Christi processions, see Martin Brecht, *Martin Luther*, trans. James L. Schaaf, 3 vols. (Minneapolis, MN: Fortress Press, 1985–1999), 2:122, 3:295.

94. Steno to Lavinia, undated, in OTH, 1:9.

from his life circumstances at the time, namely the encounter with Lavinia and his own scientific pursuits. Somehow, the question of whether the Eucharist was God must have resonated with the questions of whether the heart was a muscle, and, even more timely that month, whether tongue stones were shark's teeth.

All in all, the conversation with Lavinia Arnolfini and seeing the Eucharistic procession in Livorno compelled Steno to dive deeper into the truth of each of the Christian confessions. From then on his research included not just fossils and the bodies of animals at the dissection table, but also what happened on the altar, where the body of Christ was supposed to become present during Mass. Back in Florence, Steno took "almost all the morning hours [*matutinas horas fere omnes*]" to study theology. Booking the morning hours to work on a specific task is reminiscent of his efforts to avoid distractions and study medicine while writing the "Chaos" manuscript in Copenhagen. With the same scholarly urge that led him to read everything about the anatomy of an organ, he read books "borrowed from friends, [and] not only Catholic books but also others [*ab amicis petitos libros, non catholicos modo, sed et alios*] like the *Centuriatores Magdeburgenses*," the most famous Lutheran church history at the time.[95] Lavinia also introduced Steno to her confessor, the Jesuit Emilio Savignani (d. 1678), who would later become Steno's confessor.[96] Yet, conversations with Savignani and other Catholics were not satisfying his search for certainty. As he wrote to Lavinia, "not satisfied with talking about this matter with learned people . . . I wanted to apply all diligence in learning about the original texts of Sacred Scripture and the most ancient authors. And [I wanted to do] this in various ways, particularly in a famous library with the oldest Greek and Hebrew manuscripts, so that I would not trust the Latin versions without further examination but rather compare them with the original texts in the two languages, since I already knew them through my previous studies [*giacché per lo studio già fattone le possedevo*]."[97] In these accounts Steno describes how his friends and patrons came to the rescue of his religious anxieties by lending him books and granting access to the Medici's Laurentian Library. But perhaps more significant is Steno's knowledge of ancient Greek and Hebrew, which he had acquired at the Copenhagen Cathedral School. The same institution that planted in Steno the mathematical interests that accompanied him throughout his itinerant career was now contributing to his religious studies and therefore to his

---

95. For this and the previous quote, see Steno, *De propria conversione*, in OTH, 1:127.
96. Cardinal Francesco Nerli's report, 1 June 1677, in *Epistolae*, 2:932.
97. Steno to Lavinia, undated, in OTH, 1:9.

conversion. Thus was the breadth, unity, and value of the formation of an early modern humanist.

The Search for Certainty in Religion

Even after his conversion, Steno continued to search for further arguments and evidence to consolidate his conclusion that the Catholic Church was the true church. His ideas were published in apologetic works that he wrote during the last years of his scientific career as a lay anatomist and then, after 1675, during his ecclesiastical career as priest and bishop. Unsurprisingly, most of Steno's theological works speak of his conversion and contain many arguments in favor of the Catholic Church.[98] But which arguments mattered to him at the time of his conversion? And what relation did they have with his scientific work, if any?

In 1667 Steno witnessed the maximum efficacy of his comparative methods to achieve reliable knowledge in science. The anatomical similarities of muscles and the tissues of the heart, and between fish ovaries and "female testicles," led him to conclude that each pair was of the same substance. Steno's analogy between muscle fibers and parallelograms also supported his claim that animal spirits were unnecessary for muscle motion. Steno published all these claims in the *Elementorum myologiæ specimen* around the end of April 1667, only a few days after his life-changing meeting with Lavinia Arnolfini in Lucca. It was also in Lucca that Steno heard of a new fossil that supported his theory on the origin of fossils, which further confirmed the epistemic success of his comparative method outside of anatomy.

These comparative methods helped Steno resolve uncertainties about the body and the Earth that he discovered in his research, while traveling through Europe. In the medical pilgrimage that he was undertaking, however, Steno also became increasingly aware of uncertainties about religion, especially during the years he spent in the Netherlands. Therefore, when Lavinia Arnolfini challenged him to grapple with Catholicism, he understood the intellectual depth of the challenge. A serious engagement with the Catholic faith meant that he had to tackle the uncertainties of early modern religion, especially answering the difficult question of which church was the true church. A few years after his conversion, Lavinia asked Steno to write the reasons that led him "to abandon the Lutheran beliefs . . . and embrace the Roman Catholic

---

98. Miniati, *Nicholas Steno's Challenge for Truth*, 213–249.

faith." In that letter Steno described his reaction to the Corpus Christi procession in Livorno in terms of his search for certainty, writing that he "took all possible diligence in searching for the truth [*adoperai ogni possibile diligenza nel cercare la verità*]." The main purpose of his studies, he continued, was satisfying "the uncertainty of my soul [*per soddisfare all'incertezza del l'animo mio*], agitated in this aforementioned mystery of the Eucharist."[99]

Steno's search for certainty is essential to understanding his conversion and, more specifically, his perception of the church's magisterial authority as the key to solving the uncertainties of early modern Christianity. In his studies Steno realized that it was impossible to resolve differences between Christian confessions through Scripture alone (a doctrine known as *sola scriptura*), as Protestants wanted.[100] As he later explained, one of the main things that led him away from Lutheranism was that Protestants and Catholics held "contradictory doctrines based on the same principle [*ex eodem principio contradictorias doctrinas probantium*]," namely the Scriptures.[101] The consequence of relying on *sola scriptura*, Steno continued, was that each confession was "carried about by every wind of doctrine [*omni vento doctrinae circumferar*]."[102] This was the case with early modern Lutheranism, where "so many winds seemed to arise and still arise, inspiring new doctrines [*novas doctrinas spirantes*]."[103] Steno probably had in mind the rise of Calvinist ideas in Lutheran churches in seventeenth-century Europe, including in Denmark.[104] Instead, the solution for Steno was the "word of God, both written *and* handed down [*omne verbum Dei, et scriptum et traditum*]," that is, Scripture *and* tradition, as taught by the Catholic Church.[105] In other words, among all the Christian confessions, the Catholic Church alone held a special epistemic space for magisterial authority. Steno also argued that this epistemic authority was acknowledged by the church fathers. When Steno returned to the Netherlands as a Catholic in 1670, someone presented him a quote from Augustine that seemed to favor the *sola scriptura* principle that the only epistemic religious source was the Sacred Scriptures. In response, Steno said that Augustine had also written that "I would not believe the Gospel, unless the authority of the

99. Steno to Lavinia, undated, in OTH, 1:9.
100. Miniati, *Nicholas Steno's Challenge for Truth*, 219 n. 22.
101. Steno, *Defensio epistolae*, in OTH, 1:387.
102. Steno, *Defensio epistolae*, in OTH, 1:381.
103. Steno, *Defensio epistolae*, in OTH, 1:381.
104. Lyby and Grell, "The Consolidation of Lutheranism in Denmark and Norway."
105. Steno, *Defensio epistolae*, in OTH, 1:381. My italics.

Church moved me to it."[106] In a nutshell, Steno's reasoning led him to the teachings, or magisterial authority, of the Catholic Church because it gave him "the true tranquility of the soul [*vera animae tranquilitas*]."[107]

But how could Steno accept doctrines such as "the matter of the sacraments, Purgatory, and the authority of the pope" while at the same time being so rigorous in his scientific research?[108] Or, to put it in another way, how could Steno embrace Catholic dogmas after rejecting Cartesian tenets because of their apparent dogmatism?[109] First, it is important to acknowledge that as a Lutheran Steno already believed in such religious dogmas as the Holy Trinity and the divinity of Christ. There were, however, other practical and observational reasons that somewhat resemble his attachment to empiricism in scientific research. As he wrote in a public letter on his conversion, he was deeply impressed by "the life of some Catholic friends," whose excellence he had not observed "in friends of other religions."[110] Steno would also comment later that although Catholic countries were accused of corruption, such failures were "the fault of men, not of the doctrine." "There is uniformity of vices everywhere," he continued, including in Protestant countries. On the other hand, when it came to those who lived holy lives, Catholic countries were more diverse than others: "The confessors of our religion, martyrs, virgins, celibates, the poor, missionaries, and countless other examples of true Christian life, present in people of all conditions and sexes, testify that there is great diversity in virtues."[111] This final reference to examples of both sexes living good lives is remarkable and speaks to the role that women had in his own conversion. Specifically, Steno's fascination for morally good people explains his reaction to the words of Lavinia Arnolfini, whom he described as "a noble lady of holy and distinguished life [*vitae sanctimonia conspicua nobilis matrona*]."[112] Steno was particularly attentive to good moral lives because he too was making a special effort to lead one, including when he was still a Lutheran. As a response to the religious indifferentism he found in the Netherlands in the early 1660s, he later wrote, "I began years before my conversion to abandon everything" to the providence of God.

---

106. Steno, *Occasio sermonum*, in OTH, 1:195.
107. Steno, *De propria conversione*, in OTH, 1:129.
108. Trotti to Azzolino, in *Epistolae*, 2:925.
109. In seventeenth-century France, Cartesians were sometimes compared to dogmatists. See Lennon, "Pierre-Daniel Huet, Skeptic Critic of Cartesianism."
110. Steno, *De propria conversione*, in OTH, 1:126.
111. Steno, *Occasio sermonum*, in OTH, 1:190.
112. Steno, *De propria conversione*, in OTH, 1:127.

In particular, he adopted a "rule of life" that helped him each day "avoid everything that could be thought unwise in light of the Gospel."[113] This ascetic decision coincided with the beginning of his tour of Catholic France and Italy and may have made devout Catholics more attractive to him, and Steno to them. Therefore, what he perceived to be the exemplary lives of his Catholic friends became something like the observational component of his search for religious truth.[114]

Finally, and to return to his comparative methods, Steno argued that it was only when Luther and other reformers appeared in the sixteenth century that the divine origins of the Roman church were questioned. Since "the spirit of truth cannot be the author of opposed churches," he asked, which of the reformed confessions was the true one?[115] The answer to this question necessarily entails a comparison between confessions and the supposed marks of the true church. This method therefore resonated especially well with Steno at a time when his comparative scientific methods were proving to be remarkably successful. After weeks searching for "the marks [*characteres*]" of the true church among the various confessions, Steno concluded that "all [reformers] boasted purity of doctrine, all appealed to Sacred Scriptures, and all said that their church is apostolic.... [But] none of them has anything which proves itself to be truer than the others." Therefore, "none of the reformers was to be followed," since none had a way to prove their divine origin. In the end, Steno concluded that it was best to follow that "which our ancestors held as divine before..., confirmed as much by the witness of the holy fathers as by the blood of the martyrs."[116] This was the church "from which all other churches departed," namely the Catholic Church.[117]

However, Steno acknowledged that reason and evidence alone did not make Catholicism truly certain for him. Instead, his arguments only worked "by reason of the falsity of the others [*ratione falsitatis aliorum*]."[118] That is, Steno could only conclude with certainty that the reformed churches *were not* the true church. This approach resembles the one he used when showing what was not certain about the brain and when arguing that animal spirits were not needed for muscle motion. This negative approach continued to have significant epistemic value for Steno, including in religion.

---

113. Steno, *Defensio epistolae*, in OTH, 1:393.
114. Miniati, *Nicholas Steno's Challenge for Truth*, 216–217, 222–226.
115. Steno, *De propria conversione*, in OTH, 1:128.
116. Steno, *De propria conversione*, in OTH, 1:128.
117. Steno, *De propria conversione*, in OTH, 1:129.
118. Steno, *De propria conversione*, in OTH, 1:129.

At the same time, he admitted, the arguments for "the Catholic truth ... are only probable."[119] Not surprisingly, his claims about the history of the Earth, which he was developing at the time, were also increasingly probabilistic. As I explained in chapter 6, Steno could not know for certain which mechanism led to the formation of the Earth's strata because he did not have direct access to the Earth's past. Instead, he presented various possible mechanisms for sedimentation and said that at least "it is certain that they [i.e., the strata] could have accumulated in such ways."[120] But whereas Steno considered probabilistic reasoning useful in geology, that was not the case in religion. By October 1667 this lack of intellectual certainty was not enough for Steno to make the leap into the Catholic Church. What, then, catalyzed his decision to become a Catholic?

Steno and the Devout Women

Early in November 1667 Steno's mind was "distracted by so great and so many varied worries that, as if almost out of control, it did not find an end to its unhappiness."[121] This restlessness, however, lasted only "until, on the Feast of All Souls, around evening, all at once so many arguments and circumstances came together for me [*tot simul et argumenta concurrerent et circumstantiae*]."[122] On that evening of 2 November 1667, Steno finally embraced the Catholic faith. As he described in the public letter on his conversion, it was "the divine certitude of divine grace [*divina certitudo gratiae divinae*]" that gave him the certainty that he lacked.[123] Looking back at his story, Steno understood that the missing piece in his search for religious truth was the grace of God, "which blessed me so abundantly," as he often wrote.[124] Reason alone could not give him the complete certainty of Catholicism, but God's grace could.

But how exactly did divine grace operate in Steno's soul, to continue using his terms? The answer lies in events that took place just before his conversion. On that day Steno went to the Arnolfini palace in Florence to speak again with Lavinia.[125] According to her biographer—the only

---

119. Steno, *De propria conversione*, in OTH, 1:129.
120. Steno, *Elementorum myologiæ specimen*, 103.
121. Steno, *De propria conversione*, in OTH, 1:127.
122. Steno, *De propria conversione*, in OTH, 1:127.
123. Steno, *De propria conversione*, in OTH, 1:129.
124. Steno, *De propria conversione*, in OTH, 1:129.
125. See Steno to Lavinia Arnolfini, 1 November 1684, in Lazzareschi, *Lettere di Nicola Stenone a Lavinia Cenami Arnolfini*, 20–21; and Maria Flavia's account, 14 July 1688, in *Epistolae*, 2:987–990.

source on the exact words of that conversation—Lavinia said that their meetings, which were "against my style," occurred with the only goal of obtaining Steno's "eternal salvation." However, if Steno "did not want to surrender to what is true [*se voi non volete arrendervi alla cognizione del vero*]," that is, if he did not want to become a Catholic, he should not return to talk with her, since she did not want to waste her time.[126] That evening, after an intense conversation with Savignani, Steno finally decided to enter the Catholic Church.[127] Two days later, according to a report in Rome, Steno "appeared spontaneously at the Holy Office of Florence" to abjure his Lutheran faith, which was the norm for early modern converts.[128] In short, Lavinia's words appear to have been the catalyst that finally led Steno to become a Catholic. Many years later Steno explicitly acknowledged Lavinia's influence, writing to her of "the part that God wanted you to have in my conversion."[129] So deep was her role in this conversion that Steno spoke of the "obedience that I owe, as spiritual son [*comme fils spirituelle*], to a person that has done so much for me to be born in God and the holy Church."[130] Also for this reason, Steno addressed her as "my most honorable mother in God" or "in Jesus Christ," even after he became a bishop.[131]

Strikingly, almost no one who wrote about Steno's conversion at the time mentioned Lavinia's role in it.[132] In his comments to the nuncio of Florence, the pope's secretary of state referred only to "the works of the local fathers of the Society [of Jesus]" and the "pious generosity" with which the Medici "enticed this gentleman to stay for so long" in Florence.[133] These two factors speak to the typical agents of conversions in the early modern period: Jesuit preachers and patronage.[134] However, despite Lavinia's absence from the official sources, Steno was not the

---

126. Bambacari, *Descrizione delle azioni*, 37–38.

127. Bambacari, *Descrizione delle azioni*, 38.

128. Report on Steno's abjuration, Rome, Archivum S. Congregationis de Propaganda Fide, SRCG 465, fol. 290r, as quoted in *Epistolae*, 1:363 n. 4. The original abjuration was lost.

129. Steno to Lavinia, 1 November 1684, in Lucca, Archivio di Stato di Lucca, Arnolfini, Reg 97, fasc. 16, lett. 5.

130. Steno to Lavinia, 19 March 1683, in Lucca, Archivio di Stato di Lucca, Arnolfini, Reg 97, fasc. 16, lett. 1.

131. See Lazzareschi, *Lettere di Nicola Stenone a Lavinia Cenami Arnolfini*.

132. One of the few exceptions was Maria Flavia's account, 14 July 1688, in *Epistolae*, 2:989.

133. Azzolino to Trotti, in *Epistolae*, 2:925.

134. Mazur, *Conversion to Catholicism*, 27–28, 36–42.

only one to benefit from her spiritual influence. Her life of devotion was publicly known at the Medici court, and various women in Florence wrote to her "to direct their souls."[135] Even the grand duke spoke "with pleasure of her virtues" and asked for her prayers.[136] Historians know well that Lavinia was far from being the only noblewoman to influence early modern religious culture.[137] But perhaps what is not as well-known is that she was not the first to influence Steno. Indeed, Lavinia is only the last in a series of various women who functioned as spiritual mentors throughout Steno's life. Even if their appearances in the sources are brief, these meetings with a few Catholic women place Lavinia's final touch in Steno's conversion within a wider context of female religious leadership.[138]

Steno's susceptibility to female influence probably started with his mother at home. As explained in chapter 1, Steno's father died early in his life, and so too did his subsequent stepfather. With the lack of a father at home, his mother became the most important adult figure in Steno's childhood and thus played a crucial role in Steno's education.[139] Unsurprisingly, Steno interrupted his European travels to return to Copenhagen in 1664 and be with his mother in her final days. These family circumstances probably also helped Steno become closer to his sister, with whom he stayed whenever he returned to Copenhagen, even after he had become a Catholic.[140] Thus Steno developed an intellectual affinity with women from early on at home. This respect may have been the impetus behind the poem he wrote on the occasion of the burial of Regina Kallenbach, the mother of one of his Danish professors.[141] It appeared in 1657 and was Steno's first published work.

---

135. Bambacari, *Descrizione delle azioni*, 44.

136. Bambacari, *Descrizione delle azioni*, 44.

137. Elisja Schulte van Kessel, "Gender and Spirit, *pietas et contemptus mundi*: Matron-Patrons in Early Modern Rome," in *Women and Men in Spiritual Culture, XIV–XVII Centuries: A Meeting of South and North*, ed. Elisja Schulte van Kessel (The Hague: Netherlands Government Publishing Office, 1986), 47–68; and Sharon Strocchia, *Forgotten Healers: Women and the Pursuit of Health in Late Renaissance Italy* (Cambridge, MA: Harvard University Press, 2019).

138. This has nothing to do with the "*cherchez la femme* principle" mentioned in Olden-Jørgensen, "Jesuits, Women, Money or Natural Theology?," 48–49.

139. Viktor Hermansen, "Niels Stensens Mutter," *Stenoniana Catholica* 6, no. 3 (1960): 71–81.

140. Steno to Maria Flavia, 30 August 1672, in *Epistolae*, 1:271.

141. See Gustav Scherz, "Supplement: 2. Two Poems by Nicolaus Steno," in Scherz, *Nicolaus Steno and His Indice*, 290 n. 1.

As Steno traveled across Europe, he continued to interact with women he met through his intellectual networks. In Amsterdam he became friends with the Catholic artist and alchemist Catharina Questiers. Catharina wrote a poem in Steno's *album amicorum* on the unity between science and the virtues, mentioned in chapter 3. Little is known about her religious beliefs. But considering that she never married, she may have been one of the many lay celibate women who contributed to the development of the underground Catholic Church in the Netherlands.[142] A few years later, when he moved to Paris, Steno continued his (perhaps unintentional) practice of meeting Catholic women in his research environment. After his conversion, Steno thanked Thévenot for contributing to it by introducing him "to the friendship of the servant of God Perriquet."[143] Marie Perriquet, Thévenot's cousin, was accustomed to meeting scholars from the Parisian scientific academies, such as Pascal and Christiaan Huygens.[144] What Steno and Perriquet discussed is unclear, but since Steno associated meeting her with his conversion, they clearly talked about religion. Thus, a friendship with a woman that Steno cultivated in a scientific context developed into one that had significant ramifications on his views about religion.

Most important, Steno met the Danish noblewoman and Catholic convert Maria Elisabeth von Rantzau (1625–1706) in Paris. Elisabeth had moved to Paris to accompany her husband, a Danish officer who shifted to the service of the French king during the Thirty Years War.[145] As a result of interactions with a Catholic nun and the priest Vincent de Paul (1581–1660) during a time described as a "contemplative revival" in France, the Rantzau family converted to Catholicism.[146] After her husband died in 1650, Elisabeth joined the Order of the Annunciation, whose Italian founder was also a widow.[147] In Paris Rantzau introduced

---

142. Parker, *Faith on the Margins*, 155–158. In the Netherlands there were also non-Catholic female religious leaders such as Anna Maria von Schurmann and Antoinette Bourignon.

143. Steno to Thévenot, 4 February 1678, in *Epistolae*, 1:371–372.

144. Cornelis Andriesse, *Huygens: The Man Behind the Principle* (Cambridge: Cambridge University Press, 2005), 135–137; and Ch.-H. Boudhors, "'Une amie de Pascal?' Marie Perriquet et sa soeur Geneviève (suite et fin)," *Révue d'histoire littéraire de la France* 36, no. 3 (1929): 355–387.

145. Gisela Nowak, *Maria Elisabeth von Rantzau: Ein Leben für Caritas und Einheit im Glauben* (Hildesheim: Bernward, 1984), 20–27.

146. Nowak, *Maria Elisabeth von Rantzau*, 27–43; see also Barbara Diefendorf, *From Penitence to Charity: Pious Women and the Catholic Reformation in Paris* (Oxford and New York: Oxford University Press, 2004), 135–172, esp. 158.

147. Nowak, *Maria Elisabeth von Rantzau*, 57.

Steno to Jean-Baptiste de La Barre, her confessor, with whom he talked about the Eucharist before his conversion.[148] She was also likely the "most noble woman" who told Steno to search in Luther's writings for the real motivation "that [his] reform was once enlivened by."[149] This suggestion speaks to how well-read Elizabeth was in theology, a feature that was becoming increasingly common among devout noblewomen.[150] Regardless, Steno must have been impressed to meet a fellow Dane and a learned noblewoman who abandoned Lutheranism to become a Catholic and, later, a nun.

In Florence Steno developed a friendship with another nun who, like Lavinia Arnolfini, played an important role in his conversion. Steno met Sister Maria Flavia del Nero at the Annalena house, a Dominican convent close to the Palazzo Pitti.[151] The Annalena was one of the main convent pharmacies in Florence and for that reason was a hub for the circulation of medical knowledge.[152] Steno went there "to buy ointments [*manteche*] and other similar things," but as soon as Sister Maria Flavia realized that he was a Protestant, she "felt inspired" to convince him to become a Catholic. Steno seemed to enjoy the conversation "at the grille [*alle grate*]," the convents' physical separation between nuns and visitors, and they quickly became friends.[153] The sister wrote an extremely detailed account of Steno's conversion to Catholicism that is still the one most frequently used for modern accounts of it.[154] However, her narrative is problematic because of dating incongruencies and because she may have exaggerated her role in Steno's conversion, since Steno rarely spoke of her in his own conversion accounts. Regardless, Sister Maria Flavia's account mentions specific moments at which Steno was moved by her suggestions. Once she invited Steno to pray the Hail Mary with her. Although still a Protestant, Steno was open to the suggestion, but only "until *fructus ventris tui*"—that is, the part of the prayer taken directly from the Bible. He told her that he was not comfortable to pray the rest because of problems with the "intercession of the Most Blessed Virgin and the Saints." Still, on the nun's advice, Steno visited the famous

---

148. Steno, *Occasio sermonum*, in OTH, 1:191.
149. Steno, *Defensio et plenior elucidatio scrutinii reformatorum*, in OTH, 1:267–268.
150. Keith Luria, *Sacred Boundaries: Religious Coexistence and Conflict in Early-Modern France* (Washington, DC: Catholic University of America Press, 2005), 197–205.
151. Maria Flavia del Nero's account, 14 July 1688, in *Epistolae*, 2:987.
152. Strocchia, *Forgotten Healers*, 1–13, 95, 100–102, 135–140.
153. Maria Flavia del Nero's account, 14 July 1688, in *Epistolae*, 2:988.
154. See Scherz's biography in Kardel and Maquet, *Nicolaus Steno: Biography and Original Papers*, 227–230, 236–238; and Sobiech, *Herz, Gott, Kreuz*, 39–46.

miraculous image of the Virgin Mary at the church of the Santissima Annunziata—the place where he would celebrate his first Mass as a Catholic priest.[155] After entering the Catholic Church, Steno donated "fifty pounds to make a pair of silver chandeliers for a miraculous Virgin" as an act of thanksgiving for the nuns' prayers for him.[156]

Steno's admiration for devout women continued after his conversion and grew in depth, even if it did not increase in quantity. Three days before his episcopal ordination, Steno wrote to Sister Maria Flavia offering "part of this office" for her and "the other servants of God [i.e., the other nuns] who for a long time" prayed for his conversion.[157] That is, Steno wanted to entrust his new ministry to the nuns' prayers. Around that time Steno also became a leading supporter of Elizabeth von Rantzau, who had left Paris to become the superior of a monastery in Hildesheim, in northern Germany.[158] Even before becoming a priest, when returning from his position as royal anatomist in Denmark, Steno sent and paid for two young Danish women, converts to Catholicism, to enter Rantzau's monastery in Hildesheim in 1674.[159] Years later, as a bishop in Hannover, Steno supported the conversion of Dorothea Hedwig (1636–1692), daughter of a Danish duke, who also found refuge in Hildesheim later on.[160] But most important, Steno continued to ask for spiritual guidance from his old friend Lavinia Arnolfini, whose correspondence became one of Steno's consolations in the final years of his life. Steno shared with Lavinia such serious concerns as his regret for having ordained priests who he thought were grave sinners and the hard time he had finding a good confessor.[161] On these and other matters, Steno asked her for prayers and advice. This impressive epistolary exchange between a Catholic bishop and a devout noblewoman shows that Steno understood well Lavinia's wisdom with regard to specific spiritual matters. In that sense, Bishop Steno still resembled his young

---

155. Maria Flavia del Nero's account, 14 July 1688, in *Epistolae*, 2:988. On Steno's first Mass, see Steno to Cosimo III, 4 April 1685, in *Epistolae*, 2:768.

156. Maria Flavia del Nero's account, 14 July 1688, in *Epistolae*, 2:989. Domenico Maria Manni saw these silver candlesticks one century later, with the inscription "Di Niccolò Stenone"; see Manni, *Vita del letteratissimo Monsig. Niccolò Stenone*, 84–85.

157. Steno to Maria Flavia, 15 September 1677, in *Epistolae*, 1:360.

158. Nowak, *Maria Elisabeth von Rantzau*, 77–85.

159. See Scherz's biography in Kardel and Maquet, *Nicolaus Steno: Biography and Original Papers*, 384.

160. See Scherz, *Niels Stensen: Eine Biographie*, 2:77–79.

161. Letters from Steno to Lavinia, 19 March 1683 and 24 September 1684, in Lazzareschi, *Lettere di Nicola Stenone a Lavinia Cenami Arnolfini*, 14–16, 18–20.

anatomist self, who used to learn religion from devout ladies across Europe.

❋

News of Steno's conversion to Catholicism spread rapidly among his scientific peers. Only a few days later Redi, with whom Steno had dissected three deer five days before converting, was already corresponding about it with a friend.[162] By the end of the month Orazio Rucellai (1604–1673), an atomistic philosopher at the Medici court, wrote to a cardinal friend that "Mr. Nicolaus Steno from Denmark, who presented you the book on muscles ... felt himself led by an invisible hand to the holy resolution [of becoming Catholic]."[163] And Viviani, the grand duke's mathematician, wrote that Steno, "who lacked nothing else to make himself, so to speak, adorable [cui altro non mancava per rendersi, per così dire, adorabile], resurrected precisely on the day of the dead [i.e., on All Souls' Day] by declaring himself Catholic." He also mentioned that after the Mass where Steno received his first Catholic sacraments, all his friends, including the grand duke and Prince Leopoldo, were filled with "an uncommon joy."[164]

Interestingly, Steno's conversion was not the only thing that these accounts referred to. They also mentioned how virtuous Steno seemed to be. Less than a month after he converted, the nuncio of Florence reported to Rome that "[Steno's] conversion occurred with such edification" that it deserved a separate account of "the Holy Spirit in guiding to the truth he who searched for it with so much effort ... and with moral goodness."[165] A year later in Rome, the mathematician Michelangelo Ricci wrote to Prince Leopoldo about how impressed he was with Steno's "modesty and sincerity and his intellect so clear and rich in sciences."[166] When Steno finally returned home to Denmark to take up the position of royal anatomist, five years after his conversion, the new nuncio in Florence wrote to Rome that Steno "made such progress that he may be counted among the perfect. His

---

162. Cosimo Brunetti to Redi, 19 November 1667, in Florence, BML, Redi 220, fol. 89r. On the dissection, see Florence, Biblioteca Marucelliana, Redi 30, fols. 148r–149r.

163. Orazio Rucellai to Cardinal Giovanni Delfino, 26 November 1667, in Domenico Moreni, *Saggio di lettere d'Orazio Rucellai e di testimonianze autorevoli in lode e difesa dell'Accademia della Crusca* (Florence: Magheri, 1826), 49. On Rucellai, see Federica Favino, "Ricasoli Rucellai, Orazio," in DBI, 87:143.

164. Viviani to Magalotti, 13 December 1667, in Fabroni, *Delle lettere familiari*, 1:17–18.

165. Trotti to Azzolino, in *Epistolae*, 2:925.

166. Ricci to Leopoldo, 17 November 1668, in Angelo Fabroni, *Lettere inedite di uomini illustri*, 2 vols. (Florence: Francesco Moücke, 1773–1775), 2:161–163.

goodness and virtues flourish in such a way that I compare him to the holiest people."[167] By striving to lead a good life, Steno practiced a way of living that, when he saw it in others, convinced him of the truth of Catholicism. More important, by adopting a life of asceticism he also confirmed himself in his conversion. As he wrote, "divine certitude cannot be demonstrated to anyone except to the person who is experiencing it."[168]

Interestingly, Steno was still a layman at this time. He was regularly performing dissections and observations of rocks at the same time he became involved in religious polemics. In truth, Steno's life did not change much after his conversion until his priestly ordination in 1675. The intellectual concerns that gradually grew inside him—anatomical, geological, and religious—continued to shape his travels and conversations. But his religious interests slowly started to take up the better part of his life. His commitment to the Catholic faith, which started the same year he published his geometry of muscles and his first account on fossils, led him to a priestly vocation in the Catholic Church.[169] In 1675 Steno was ordained a priest at the Cathedral of Santa Maria del Fiore in Florence, and in 1677 he was ordained a bishop in Rome and sent by the pope to northern Germany as apostolic vicar. From then on his life was focused on his pastoral duties, with some exceptions shown by a few personal notes related to the soul and the brain.[170] Nonetheless, he recalled where his Christian zeal had come from. In Hannover, Steno told Leibniz, the court's librarian, about "how God, through anatomical revelations [*comment Dieu par les découverts anatomiques*] . . . gradually brought me to embrace the love of Christian humility [*me reduis à peu a peu a recevoir l'amour de l'humilité chrétienne*]."[171] As a solid within a solid, his intellectual and geographical travels had surrounded his entire person, in science as in religion.

167. Oppizio Pallavicini to Federico Baldeschi, 16 August 1672, in *Epistolae*, 2:926.
168. Steno, *De propria conversione*, in OTH, 1:129.
169. On Steno's vocation for the priesthood, see Miniati, *Nicholas Steno's Challenge for Truth*, 200–203.
170. Gal. 291, fols. 88r–90v; see also Andrault, "Human Brain and Human Mind," 106–109.
171. Steno to Wilhelm Gottfried Leibniz, undated (possibly November 1677), in *Epistolae*, 1:366–369.

Epilogue

The decade between the composition of the "Chaos" manuscript in 1659 and the publication of *De solido* in 1669 was transformative for the history of science. Boyle published his first experiments on the vacuum with the air pump, the *Treatise on Man* divulged Descartes's mechanical understanding of the body, and Huygens dedicated his discovery of Saturn's rings to Prince Leopoldo de' Medici.[1] During these years the French Académie des Sciences and the English Royal Society were also founded, and the *Journal des sçavans* and *Philosophical Transactions*—two still-ongoing academic publications—released their first issues. All these events speak to the increasing role of observations in creating new knowledge, the stimulus mechanical philosophies gave to scientific research, and the scholarly demand for patronage by royal courts and academies at this time. Through his itinerant life, Nicolaus Steno contributed significantly to establishing these trends as he crossed the intellectual and social networks of early modern Europe.

To early modern eyes, Steno was an anatomist with diverse skills and a broad scope of knowledge. As this book has argued, it was in his anatomical work that Steno applied most of his eclectic knowledge. His interdisciplinary responses to anatomical problems therefore had an impact that extended well beyond anatomy and led his research from the body to the Earth and finally to theology. However, his identity as an anatomist remained with him for years, partly because he continued working on it. Shortly after he was ordained a Catholic priest in June 1675, a friend asked Steno about his "physical and anatomical works, which you had thought of

---

1. Robert Boyle, *New Experiments Physico-Mechanical: Touching the Spring of the Air, and Its Effects* (Oxford: The University, 1660); Descartes, *De homine*; and Huygens, *Systema saturnium*.

publishing."[2] This friend was Rudolf Christian von Bodenhausen (d. 1698), a German mathematician living in Florence who would contribute greatly to Leibniz's work on dynamics in the 1690s.[3] Bodenhausen converted to Catholicism under the influence of Steno in Florence less than a year after Steno's own conversion.[4] This influence of the anatomist-interested-in-mathematics on Bodenhausen's conversion shows that religion rapidly became intertwined with Steno's scientific friendships. More significant for his ongoing anatomical prestige was that Steno almost received the crown jewel for an early modern anatomist: the chair of anatomy at the University of Padua, a position previously held by Andreas Vesalius (1514–1564) and Fabricius d'Acquapendente. In early March 1676, while visiting the Habsburg court in Vienna, the Venetian ambassador approached Lorenzo Magalotti to ask him about "the qualities of Mr. Nicolaus Steno."[5] The reason was that the University of Padua was thinking of Steno to "fill the chair, now vacant, of anatomy."[6] Magalotti, perhaps fearing that Steno might accept such an offer, responded that the Dane was unavailable, since he served the prince of Tuscany as "master of philosophy and mathematics."[7]

Describing Steno as a master of subjects other than anatomy shows that Steno's broad intellectual opinions and expertise mattered greatly for his successful navigation of Europe's social networks. In Paris, for example, Steno's dissections had been central to the informal academy of Thévenot, to the point that attendees seemed to lose interest in the meetings when Steno was away. The intrinsic connection between Steno's ideas and social life also cannot be fully understood unless one considers the specific content of his work. In Florence, Steno's dissections at the Medici court

---

2. Rudolf Christian von Bodenhausen to Nicolaus Steno, 8 June 1675, in Gal. 291, fol. 224v: "delle sue opere Fisiche ed' Anatomiche, le quali V.S. Ill.ma aveva pensato di pubblicare."

3. Antognazza, *Leibniz*, 306–307. See also Charlotte Wahl, "Between Cosmopolitanism and Nationalism: The Role of Expatriates in the Dissemination of Leibniz's Differential Calculus," *Almagest: International Journal for the History of Scientific Ideas* 5, no. 2 (2014): 40–68, esp. 44–50.

4. Florence, Archivio Storico Arcivescovile di Firenze, TIN 34.13, fols. 74r–76v, esp. fol. 76v: "pro testibus ad hoc vocatis . . . Dom[i]no Nicolao Stennone."

5. Lorenzo Magalotti to Francesco Marucelli, 8 March 1676, in Florence, ASF, MdP, f. 4412, fol. 23r: "informazione delle qualita del Sig. Niccolo Stenone."

6. Magalotti to Marucelli, 8 March 1676, in Florence, ASF, MdP, f. 4412, fol. 23r: "per provedere la Catedra quivi vacante di Notomia."

7. Magalotti to Marucelli, 8 March 1676, in Florence, ASF, MdP, f. 4412, fol. 23r: "maestro di filosofia e di matematiche."

were a success not simply because he demonstrated new organs that were hard to dissect, such as the lymphatics, but also because he claimed that anatomy had to be studied with the help of mathematics—a claim whose impact I could only discern in rarely studied Italian manuscripts about the reception of Steno's *Elementorum myologiæ specimen*. The first Italian readers of this book praised the mathematical parts of his work, not its opinions about skin glands, ovaries, and fossils. Steno's mathematization of anatomy was so central to his positive reception in Italy that almost a hundred years later, Angelo Maria Bandini (1726–1803), the librarian of the Laurentian Library, entitled a historical report on Steno "On the Death and Actions of Bishop Steno, a Most Outstanding Mathematician [*mathematicus praestantissimus*]."[8]

Yet, as I have shown throughout this book, mathematics was only one among his other interests: chymistry, mechanics, hydrostatics, and epistemology. Interestingly, unlike the work of encyclopedists or polymaths, Steno did not aim to display a mastery of all these subjects. Instead, he crossed this interdisciplinary knowledge with focused problems in anatomy and, later, in the history of the Earth. This focused interdisciplinary work, as I have called it here, was not unique to Steno. But his search for certainty and the networks he formed in his itinerant life made this work methodology more visible in ways that illuminate the work of those that he met along his way, and which had a significant impact in early modern science. To put it another way, Steno's itinerant career unveils a history of early modern science that is highly complex.

From a social perspective, historians of science have often demonstrated the importance of artisanal and domestic backgrounds, rather than solely university training, to early modern science.[9] Steno's career, however, provides a fresh view of why these contexts mattered, since they were strongly tied to his university studies and research. Historians have named "artisanal epistemology" the specific knowledge that artisans had about the world. Although applicable in some instances, this categorization does not really apply to Steno, whose work in Danish goldsmith workshops was about doing and not knowing.[10] It was only when Steno returned to his study room that he adopted artisanal practices, such as measuring weights

---

8. Angelo Maria Bandini, *Collectio veterum aliquot monimentorum ad historiam præcipue litterariam pertinentium* (Arezzo: Michael Bellotti, 1752), 78.

9. See the section "Sites of Natural Knowledge" (chaps. 8–16) in Park and Daston, *Early Modern Science*, 206–364.

10. Interestingly, even most chymists in the second half of the seventeenth century were more engaged in producing pharmaceuticals than in knowing nature, so Robert

and building devices, to understand the inner structure of matter or the heart's motion. Steno's artisanal background was crucial to the success of his scientific career, but it was because of skills and practices rather than epistemologies that he acquired in his family workshop.[11] Therefore, Steno's career invites historians of science to expand the role of artisanal practices to include what they were: handicraft skills that greatly benefited natural research in the early modern period.

Steno's anatomy classes at the university further contributed to his skills with the knife and other instruments as he cut through delicate organs, inserted probes in glands, and ligated veins. Like goldsmith workshops, a dissection table was a space where research questions in anatomy were closely tied to craftsmanship. In short, the world of practices and ideas met in the interdisciplinary research of an anatomist who started developing quantitative and chymical interests in his home workshop and the university.[12] Steno's career therefore shows that academic environments were as crucial to early modern science as artisanal workshops, princely courts, private houses, and hospitals.[13] In the northern universities, especially in Leiden, Steno learned the new practices and methods of experimentation, met inspiring professors who encouraged students to produce new research, and befriended like-minded peers who shared an interest in post-Harveian anatomy. Even at older and more conservative institutions such as the universities of Paris and Pisa, Steno freely exchanged ideas with the local faculty. However, it is true that as he traveled south, Steno found the informal academies of Paris and the princely courts of Italy more attractive places for his research.

This study of Steno's scientific career also points to a more interesting history of the circulation and mathematization of knowledge than has hitherto been acknowledged. As I have shown, to understand the factors

---

Boyle and Wilhelm Homberg were an exception; see Principe, *The Transmutations of Chymistry*, 140, 404–405.

11. On ways of knowing, see Pamela Smith, *The Body of the Artisan: Art and Experience in the Scientific Revolution* (Chicago: University of Chicago Press, 2004).

12. This is therefore a different scenario from Pamela Long's trading zones, which were places of interaction between sixteenth-century scholars and practitioners. See Pamela O. Long, *Artisan/Practitioners and the Rise of the New Sciences, 1400–1600* (Corvallis: Oregon State University Press, 2011).

13. For the varied environments in which seventeenth-century anatomists worked, see Bertoloni Meli, *Mechanism, Experiment, Disease*, 6–12. On universities, see Roy Porter, "The Scientific Revolution and Universities," in *Universities in Early Modern Europe*, ed. Hilde De Ridder-Symoens, vol. 2 of *A History of the University in Europe*, ed. Walter Rüegg (Cambridge: Cambridge University Press, 1996), 531–564.

that led Steno and other anatomists to mathematics, it is helpful to look away from the shadow of great names such as Galileo and Descartes and focus more on broader intellectual trends. When Steno arrived in Italy, he already knew mathematics, which he had learned as a teenager at the Cathedral School of Copenhagen. Rather than a break with his previous anatomical work, Steno's mathematical model of muscle motion and his theory on the origin of fossils were a progression of his earlier research. The environment of Florence simply brought the knowledge and methods that Steno had learned in northern Europe to another level, especially the uses of mathematics and comparisons in anatomy. Thus, Steno's itinerant career highlights the essentially transnational elements of the history of science that a narrow focus on specific European geographies often misses.

Furthermore, contrary to what has always been claimed, Steno was never a Cartesian in his active research life. Neither was he probably a full Gassendist, although his mechanical philosophy and epistemology seem closer to Gassendi's than to Descartes's. As Steno's case illustrates, early modern mechanical philosophies could be as varied as the number of scholars that adopted them. The work of Steno and his colleagues thus casts a wide view on seventeenth-century mechanical philosophies. This breadth would lead Richard Westfall to reconsider his claim that the uses of mechanics in biology, or iatromechanics, led to "no significant discovery whatever."[14] Even if not always acknowledged, the study of glands as filters and the muscles of respiration as mechanical agents in blood motion, as promoted by Steno and others, undergirds modern principles of glandular and cardiovascular science.[15] Troels Kardel has also shown that Steno's geometrical model of muscle fibers has been used with success in modern computer models simulating animal locomotion.[16] Above all, the focused interdisciplinary work method that Steno developed as an anatomist became essential for his description of glands and muscles and his history of the Earth.

Studying Steno on his own terms, as I have done in this book, has also brought greater continuity to a period traditionally understood as one of revolution. Steno knew that he was opening new paths in the

---

14. Richard S. Westfall, *The Construction of Modern Science: Mechanisms and Mechanics* (Cambridge: Cambridge University Press, 1971), 104.

15. Harald Moe and Finn Bojsen-Møller, "The Fine Structure of the Lateral Nasal Gland (Steno's Gland) of the Rat," *Journal Ultra Structure Research* 36, no. 1 (1971): 127–148; and Furst, *The Heart and Circulation*, 199–202.

16. Troels Kardel, "Steno's Myology: The Right Theory at the Wrong Time," in Andrault and Lærke, *Steno and the Philosophers*, 138–173.

understanding of the human body, but he was also convinced that the ancients would most likely agree with him. Whenever possible, he introduced his discoveries as if they were already known by the ancients, such as when he wrote that Hippocrates thought of the heart as a muscle.[17] As he wrote, the ancients "kindled the light. Our task is to keep it shining and, by going forward, make it burn more luminously."[18]

My claim that Steno's choice of publication genre was as significant for the history of science as his discoveries also brings further continuity to his work. The genre of *observationes anatomicæ* in which Steno chose to publish gave him the freedom to include various kinds of studies in his books in the order in which he performed them, not unlike a journal editor compiling the articles he receives into a journal issue. Steno's story illustrates Gianna Pomata's argument that the rise of *observationes medicæ* was vital in establishing scientific journals in the seventeenth century.[19] When Steno handed Thomas Bartholin his unpublished dissection notes, Bartholin turned them into nine "articles" in the *Acta medica Hafniensia*. More important, this book's revelation that Steno intended to publish all his anatomical works as books of *observationes anatomicæ* sheds a unifying light on his diverse intellectual career. The discovery that even the *Elementorum myologiæ specimen* was planned as a book of observations reveals how significant this genre was for Steno. I suggest that this way of publication continued to influence Steno even as he started to spend more time on theology. Steno published most of his theology in short booklets made up of letters addressed to Protestant adversaries.[20] The epistolary format was unusual for a theological treatise, but it was strikingly similar to the genre of anatomical observations, which were sometimes published in the form of letters. The problem is that short accounts of empirical results were valuable in anatomy but not in theology, which may have been what led Leibniz to comment that Steno "from being a great physician became a mediocre theologian."[21] Mogens Lærke has remarked that one of the things that Leibniz disliked about Steno's theology was his controversial attitude—an attitude that his short, *observationes*-like publications further highlighted.[22]

17. Steno, *De musculis et glandulis*, 4.
18. Steno, *Observationes anatomicæ*, §4, 4.
19. Pomata, "Sharing Cases," 225.
20. See also Miniati, *Nicholas Steno's Challenge for Truth*, 227 n. 58.
21. Leibniz, *Essais de theodicée*, trans. in Mogens Lærke, "Leibniz and Steno, 1675–1680," in Andrault and Lærke, *Steno and the Philosophers*, 63.
22. Lærke, "Leibniz and Steno, 1675–1680," 72–75.

In science, this epistemic genre allowed Steno to explore new textual and research techniques in ways that fit his broad interests. In Steno's hands, the genre appeared as a hybrid between a commonplace book, with observations from other authors' books, and a laboratory notebook, with notes of dissections and different kinds of anatomical observations. Steno improved on this blending of empirical and textual traditions by describing his dissections with vivid narratives, persuading the reader that the vessels he discovered presented themselves to his hands. The intrinsic flexibility of the genre of *observationes anatomicæ* also gave Steno the space to explore interdisciplinary interests within an anatomical publication. By looking at Steno's anatomical research with an eye to interdisciplinarity, this book thus explains his ability to swiftly shift his work from the body to the Earth. As an anatomist who routinely applied mechanical and chymical analogies to the body in *historiæ anatomicæ*, it is not surprising that he would also use them in his history of the Earth.

Unsurprisingly, others also pursued mathematical, mechanical, and chymical approaches to anatomical research. As this book has shown, Steno learned to use ideas from other disciplines in a focused way by studying the medical works of Borel and Pecquet. Other physicians and mathematicians whom Steno met in his travels, such as Deusing and Fabri, are also part of this story of mathematization.[23] Since Steno's focused interdisciplinary work benefited from the scholars he met along the way, it also casts further light on seventeenth-century collaborations between mathematicians and anatomists recently explored by Domenico Bertoloni Meli. In my account, the relationship between Steno and Viviani emerges with greater depth than previously acknowledged. Viviani's impact on Steno's Florentine research project and stay makes sense, considering that it was the last mathematician with whom Steno collaborated since his youth. He learned mathematics with Hilarius, his mentor at the Cathedral School in Copenhagen, and befriended the Leiden mathematics professor Gollius. In Amsterdam Steno also became especially aware of the collaboration between the physician Willem Piso and the mathematician Marcgrave. Early modern scientific collaborations are challenging to study owing to their oral nature and the period's complex authorship practices. However, biographical studies such as the present one are especially helpful by tracing their appearance in Steno's itinerant career and their positive impact on his mind.

---

23. Anton Deusing, *Disquisitio physico-mathematica, gemina, de vacuo* (Amsterdam: Pieter van den Berge, 1661); and Fabri, *Tractatus duo*.

Steno interacted with so many European scholars that his travels reveal new networks of significance across the northern and southern European divide in the 1660s. In particular, less well-known scholars whom Steno met also made essential contributions to the history of science, such as the Amsterdam physicians Blasius and Kerckring, the Leiden anatomists de Graaf and Swammerdam, the Parisian surgeon Morel, the anatomist Bellini in Pisa, the mathematicians Ricci and Gradi in Rome, and so on.[24] Among the actors whom Steno's trajectory illuminates were also several women that Steno met in the intellectual circles of Europe and who played a leading role in his religious conversion.

Steno converted to Catholicism on All Souls' Day, 2 November 1667, the same year he published his mathematical theory of muscle contraction and his first history of the Earth. This religious conversion was firmly tied to his travels and friendships with other intellectuals. His desire to visit Europe, typical of early modern Danes in a *peregrinatio medica*, led him to visit Catholic institutions that also contributed to his conversion. In particular, he unwittingly participated in the efforts of the early modern papacy to transform its sponsorship of science into a subtle tool of evangelization. Steno's tour of the leading scholarly institutions of Rome, such as the Vatican Library, the Collegio Romano, and Queen Christina's academy in the year before his conversion, gave him a positive image of the Catholic Church that was necessary for his conversion. However, the most crucial link between Steno's scientific and religious lives was Francesco Redi, who spoke with Steno about religion. During a research trip to Lucca, Redi introduced Steno to Lavinia Arnolfini, whom Steno later called a spiritual mother. Through a serious study of Steno's religious conversion, this book has thus brought to light the influence of Catholic women in the learned circles of Amsterdam, Paris, and Florence, which has been hitherto ignored. Even the nun Maria Flavia, who seemed more pious and solely interested in Steno's conversion, was no stranger to medical knowledge, as suggested by the recent work of Sharon Strocchia.[25]

These women and Redi led Steno to turn his studies to theology, also because the search for certainty in Christianity resonated with Steno's successful searches for certainty in anatomy and the origin of fossils. The language skills that Steno acquired in the Cathedral School of Copenhagen also facilitated his study of theology, showing the ultimate impact of his pre-university education. Interestingly, Steno's epistemic certainty in

---

24. A quick search on Isis Current Bibliography (www.isiscb.org) confirms that much has yet to be done on these scholars.

25. Strocchia, *Forgotten Healers*.

religion helped him only to rule out churches that he thought were not an authentic version of Christianity, just as he ruled out Cartesian claims about the brain. His conversion, in the end, was not exclusively an intellectual decision, as he would explain in print. Nonetheless, this epistemic and methodological bridge between Steno's scientific and religious research expands the scope of productive interactions between early modern science and religion.[26]

In this book I have shown how the Steno who wrote about the history of the Earth gradually evolved from the Steno who was distracted by various topics when writing the "Chaos" manuscript. As part of his quest for certainty, Steno's interest in the underlying causes of things led him to drop old projects and pursue new ones.[27] His research on respiration, he admitted, could not progress until he first studied the muscles, which needed a focused understanding of the motion of muscle fibers. As he said in *De solido*, "when I investigated more carefully both each place and body, these gave rise to a succession of doubts, indissolubly connected, which assailed me day by day." Steno compared these doubts to the heads of a "Hydra, since whenever one disappears, countless others are born from beneath."[28] Fortunately for Steno, these heads grew out of the same body—a world united by the same natural laws, which he could tackle with observations and analogical reasonings.

Nicolaus Steno was one of the most famous and respected anatomists of his time. Therefore, it is no wonder that his choice of a demanding ecclesiastical vocation, eight years after his conversion, disappointed some of his friends. A few months after Steno became a bishop, Swammerdam complained to Thévenot that Steno was too busy with "many pursuits," wishing that Steno was "still like he was when he sought God in the bible of nature."[29] Leibniz wrote that Steno "was a great anatomist and deeply versed in natural science, but unfortunately gave up research."[30] These complaints created a modern narrative in which "Steno did not seem capable of reconciling his interests in natural philosophy . . . with his new

26. Bernard Lightman, ed., *Rethinking History, Science, and Religion: An Exploration of Conflict and the Complexity Principle* (Pittsburgh: University of Pittsburgh Press, 2019).

27. Steno wrote he was planning a longer work on myology; see *Elementorum myologiæ specimen*, 47.

28. Steno, *De solido*, 3.

29. Swammerdam to Thévenot, undated (probably January 1678), in Lindeboom, *The Letters of Jan Swammerdam*, 84.

30. Leibniz, *Essais de theodicée* (Amsterdam, 1710), trans. in Lærke, "Leibniz and Steno, 1675–1680," 63–84, esp. 63.

commitment to religion."[31] However, stopping scientific research does not mean rejecting it. On the contrary, Steno continued to foster his scientific interests in compelling ways, albeit not with new publications. Therefore, it is fitting to conclude this book with a brief account, based on new archival research, of Steno's scientific interests after his religious conversion and ordination.

In the eight years between his conversion and priestly ordination, Steno continued to produce cutting-edge research in geology and anatomy, often in a focused way. Among Steno's papers at the National Library in Florence that he produced after the "Chaos" manuscript, I discovered that the few mysterious notes in English, previously dated to March 1665, were reading notes of the *Philosophical Transactions* taken years later.[32] Steno copied passages from the first issue, published on 6 March 1665, to the forty-fourth, dated 15 February 1669. Given the vast interests of contributors to this journal, each issue of the *Philosophical Transactions* covers an extensive range of topics. Yet, this did not stop Steno from reading it even more selectively than when he wrote the "Chaos" manuscript. When reading the May 1667 issue of the *Philosophical Transactions*, Steno did not excerpt from "a letter of M. Pecquet" on anatomy.[33] Instead, he took notes of observations about the organic origin of minerals in Hungary and Egypt and submerging waters in Guinea.[34] And when reading the September 1667 issue, Steno did not excerpt from accounts of injections "of liquors into the veins of animals" by Carlo Fracassati, or Malpighi's discoveries on the brain and tongue, even though these topics were right up Steno's anatomical-research alley.[35] Instead, Steno took notes about waters, minerals, and encrustations from the "Caribe-Islands" and the River Thames.[36] These notes point toward observational accounts of minerals and plants in places that Steno had not visited, such as Great Britain and Hungary, but most especially the New World, Africa, and Asia. This means that Steno wanted to widen the scope of his geological claims from Tuscany—as he had written in *De solido*—to the entire world.[37] In

---

31. Pina Totaro, "Steno in Italy: From Florence to Rome," in Andrault and Lærke, *Steno and the Philosophers*, 270–287, esp. 272.

32. Gal. 291, fols. 76r–83r.

33. *Philosophical Transactions* 2, no. 25 (May 1667): 461–464.

34. BNCF, Gal. 291, fols. 77ar–7av. *Philosophical Transactions* 2, no. 25 (May 1667): 467–472.

35. *Philosophical Transactions* 2, no. 27 (September 1667): 490–493.

36. BNCF, Gal. 291, fols. 77av, 77v.

37. For Steno's geological research after writing the *De solido*, see Cutler, *The Seashell on the Mountaintop*, 125–126; and Scherz's biography as translated in Kardel and Maquet, *Nicolaus Steno: Biography and Original Papers*, 295–296.

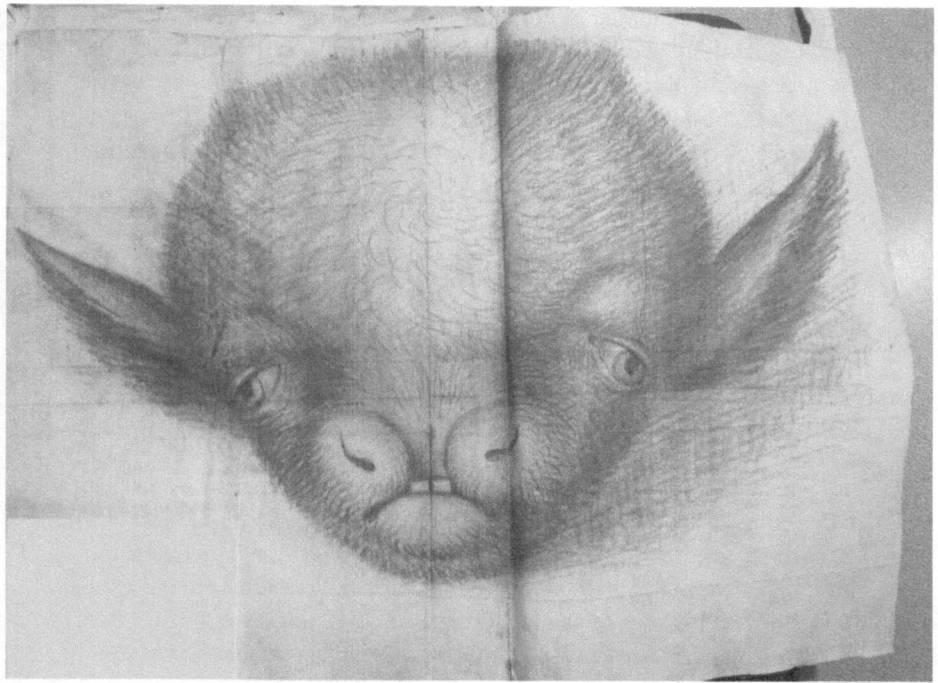

FIGURE 8.1. Drawing in sanguine or red chalk of a large-headed calf. Florence, ASF, MdP, f. 1020, fol. 665r. Courtesy of the Ministero della Cultura/Archivio di Stato di Firenze.

short, Steno was taking these notes for his second book on the history of the Earth, the now-lost *Liber de solido*.

As for anatomy, Steno's dissection account of a large-headed calf in Innsbruck in 1669 remains a groundbreaking yet rarely noted contribution to the history of brain disease.[38] The condition that Steno studied, known as hydrocephalus since antiquity, is still poorly understood today.[39] Steno applied his usual interdisciplinary research tools to study this water-filled brain: he delicately cut thin membranes and vessels, weighed the excess liquid, analyzed the effects of water pressure on the brain, and compared

---

38. Troels Kardel, "Steno on Hydrocephalus."

39. Isabella Bonati, "Hydrocephalus in Context: A History from Graeco-Roman Sources," *Early Science and Medicine* 27, no. 4 (2022): 333–350. For its modern complexity, see Gurjit Nagra and Marc R. Del Bigio, "Pathology of Pediatric Hydrocephalus," in *Pediatric Hydrocephalus*, ed. Giuseppe Cinalli, M. Memet Ozek, and Christian Sainte-Rose (Cham: Springer, 2018), 1–25, esp. 2.

the junction of extended bones with geometrical figures.[40] A remarkable drawing made in red chalk of a large-headed calf, not included in the printed version, has remained in the state archives of Florence ever since the archduchess of Austria sent it to her brother Ferdinand II after Steno's dissection (see fig. 8.1).[41] This is one of the few seventeenth-century illustrations of brain pathology.[42]

In 1675, after working in Copenhagen for two years as royal anatomist, Steno returned to Florence to be ordained a Catholic priest. From then on he did what most clergymen do. He celebrated his first Mass at the church of the Santissima Annunziata, where Sister Maria Flavia had recommended that he go years before, while still a Lutheran, and listened to confessions at the baroque church of San Gaetano.[43] Yet, even as a priest, he did not abandon his scientific interests. It is known that Steno continued to attend dissections at the Medici court, although he probably did not perform them.[44] More revealing of Steno's lasting commitment to anatomy was his use of anatomical skills after becoming a bishop in 1677. On occasion he returned to the dissection table for reasons of apologetics, an unusual activity for a bishop. At the end of a religious debate at the court of Hannover, Steno dissected a calf's head and heart to show the beauty of God's wisdom.[45] Years later, in Hamburg, when trying to convert a lapsed Franciscan who had rejected his religious vows and was approaching the "beginnings of atheism," Steno also dissected an animal heart to speak again about the "wisdom of God."[46] Finally, as recent research has suggested, notes taken by Bishop Steno in the years before he died show that he kept working on the relationship between body and soul.[47]

Searching for the unknown fueled Steno's travels from his early years in Copenhagen to his time in Florence. On a historic occasion on 28 January 1673 Steno, still a layman, was invited by his former professor Thomas Bartholin to perform a public dissection at the anatomical theater of the

40. Bartholin, *Acta medica et philosophica Hafniensia*, 1:249–262.
41. Anna de' Medici to Ferdinand II, 16 June 1669, in Florence, ASF, MdP, 1020, fol. 665r.
42. Domenico Bertoloni Meli, *Visualizing Disease: The Art and History of Pathological Illustrations* (Chicago: University of Chicago Press, 2018) 1–22.
43. *Epistolae*, 2:768–770, 933; OTH, 2:364. On Steno's priestly years, see Miniati, *Nicholas Steno's Challenge for Truth*, 254–270.
44. Holger Jakobsen speaks of Steno attending his dissections in Pisa; see Scherz's biography in Kardel and Maquet, *Nicolaus Steno: Biography and Original Papers*, 406.
45. Sobiech, *Herz, Gott, Kreuz*, 61–65, 82.
46. Steno to Cosimo III de' Medici, 11 October 1684, in *Epistolae*, 2:729.
47. Andrault, "Human Brain and Human Mind," 106–109.

University of Copenhagen. Steno had learned to dissect in this theater more than a decade earlier. Now he had returned home to lecture his former teachers on anatomy. He famously framed his anatomical lecture around the search for certainty, the primary motivation of his research throughout his career. This time Steno spoke of it in terms of beauty. Perhaps historians, too, can use a lesson from this lecture to think of early modern science: "Beautiful are the things that are seen, more beautiful the ones that are known, by far the most beautiful are those that are ignored [*pulchra sunt, quae videntur, pulchriora, quae sciuntur, longe pulcherrima, quae ignorantur*]."[48]

48. Nicolaus Steno, "Prooemium demonstrationum anatomicarum in Theatro Hafniensi anni 1673," in Bartholin, *Acta medica et philosophica Hafniensia* 2:359–366, esp. 363.

Acknowledgments

When I started studying the life and work of Nicolaus Steno, I never imagined that I too would become a committed traveling scholar. This book was written on the road from Baltimore to Berlin, Florence, Lisbon, and Oxford. Like many early modern enterprises, this book would not have been possible without the friends, colleagues, and patrons who supported me along the way.

I started writing this book as a graduate student in the history of science at Johns Hopkins University. I am especially thankful to Lawrence Principe and María Portuondo, who read this book in the various forms it has taken up until now. Larry's unfailing guidance showed me how to walk in the footsteps of giants in the history of science. María continues to teach me, better than anyone else, how to make a topic speak to broader audiences. The same gratitude is due to Gianna Pomata, in whose seminar the idea to write this book appeared, and who also read various versions of this manuscript. Her positive influence and scholarly rigor are easily discernible throughout these pages. All errors and failures, however, are my own. It was also during a Johns Hopkins colloquium that I first met Evan Ragland, who has been sharing his vast knowledge with me ever since. Evan too read various versions of many chapters. I am grateful to Ryan Hearty for fostering my enthusiasm for history and science. Finally, I thank Marta Hanson, whose insightful conversations about this book started in Baltimore and continued when we both took positions in Berlin.

I am especially appreciative of the support and encouragement of Katja Krause, at the Max Planck Institute for the History of Science in Berlin, in my academic pursuits. It was in Berlin that I first met Sophie Roux, whose advice and clear ideas about mechanical philosophy strongly influenced my own. Many scholars and friends at Harvard University's Villa I Tatti and at Oxford's All Souls College have also been kind enough to comment and offer criticism of earlier versions of some chapters. I am especially thankful to Philippa Ovenden in Florence and Noel Malcolm in Oxford. I am also grateful to audiences at various seminars, conferences, and colloquiums because they gave me opportunities to deepen and rethink my ideas. Various other scholars—too many to mention here—generously shared unpublished work, novel ideas, and crucial comments with me. In Lisbon, I am grateful to Inês Bénard da Costa, who provided editorial support. At the University of Chicago Press, I am indebted to Karen Darling for believing in this project, and Fabiola Enríquez Flores for guiding me through the adventures of production. I also thank the three anonymous reviewers, whose suggestions significantly improved this manuscript.

This book's research benefited from various generous institutions. The Harry Woolf Fund and the Charles Singleton Center for the Study of Premodern Europe at Johns Hopkins University allowed me to travel to Italy for an initial survey of the archives and then to spend a few months there. The Santorio Fellowship for Medical Humanities and Science also contributed to that research. The Huntington Library supported a month of research in Oxford and London. The Max Planck Institute for the History of Science gave me thirteen months to develop my arguments and pursue other, similar projects. The Harvard University Center for Italian Renaissance Studies at Villa I Tatti awarded me another year for similar pursuits, this time in Steno's beloved Florence. Finally, the Warden and Fellows of All Souls College provided me with the ideal place to focus and complete the final version of the manuscript at Oxford.

Studying Steno came with the benefit of meeting the world's leading authors on the topic. I am especially blessed to have met and befriended Alan Cutler, Stefano Dominici, and Troels Kardel. They all read and commented on various parts of this book. Troels Kardel, who was my fact-checker for all of it, left us too soon, while this book was in production. May he rest in peace. I also thank Raphaële Andrault for her work on Steno and her comments on my ideas. There is also no way around acknowledging the tremendous work of those who transcribed, edited, and translated Steno's writings in decades past. I am especially thinking of Gustav Scherz and August Ziggelaar, S.J., whose foundational work made my research possible. All translations in this book are mine unless otherwise noted, but such translations were made quicker thanks to their work and to the complete English edition of Steno's scientific writings, translated and edited by Kardel and Paul Macquet.

Finally, various parts of this book were written at (and inspired by) the views of the highest mountains of mainland Portugal and the strata of the cliffs of its western coast. I was able to stay in both places due to my parents- and siblings-in-law, who kindly endured my constant writing, sometimes amid family vacations. Nor would this book be possible without my parents and siblings, in whose home I learned to love science, history, and truth. I also have in Portugal the roots of my intellectual formation, especially in my mentor and friend Henrique Leitão, whose search for knowledge and ability to spread it is unparalleled. Conversations with my friend J. P. also transformed my productivity habits—and therefore this book—for the better.

Above all, the greatest joy in writing this book was to have as my traveling companions Leonor, Catarina, Tomás, and little Nuno. They are the main reasons on Earth why I wrote this book. Leonor is present in all pages as my better half. There are not enough words to thank for all her support across the roads of Europe and Maryland. In many ways, writing this manuscript was a joint project. This book is dedicated to her.

※

This book was awarded and is partly supported by a Lila Wallace—Reader's Digest Publications Subsidy from Harvard University's Villa I Tatti.

Parts of chapter 3 appeared in "Dissecting with Numbers: Mathematics in Nicolaus Steno's Early Anatomical Writings, 1661–64," *Substantia: An International Journal on the History of Chemistry* 5, no. 1 (2021): 29–42. An earlier version of chapter 6 was published in "Thinking the Earth with the Body: How the Anatomist Nicolaus Steno (1638–1686) Read History in the Earth's Strata," *Isis* 115, no. 2 (2024): 312–334.

This book is complemented by two appendices available online at https://press.uchicago.edu/dam/ucp/books/pdf/Castel-Branco _appendices.pdf. Appendix 1 is the list of books excerpted in the "Chaos" manuscript. Appendix 2 is a table listing all the known letters mentioning Nicolaus Steno between 1659 and 1669.

Bibliography

Abbreviations

ASF	Florence, Archivio di Stato di Firenze
BML	Florence, Biblioteca Medicea Laurenziana
BNCF	Florence, Biblioteca Nazionale Centrale di Firenze
DBI	*Dizionario biografico degli italiani*, 100 vols. Rome: Treccani, 1960–2020.
EMC I	Thomas Bartholin, *Epistolarum medicinalium... centuria I et II*. Copenhagen: Matthia Godicchenni, 1663.
EMC III	Thomas Bartholin, *Epistolarum medicinalium centuria III*. Copenhagen: Matthia Godicchenni, 1667.
EMC IV	Thomas Bartholin, *Epistolarum medicinalium centuria IV*. Copenhagen: Matthia Godicchenni, 1667.
Epistolae	Nicolaus Steno, *Epistolae et epistolae ad eum datae*. Edited by Gustav Scherz. 2 vols. Copenhagen: Nyt Nordisk Forlag Arnold Busck, 1952.
MdP	Mediceo del Principato
MsElem	Copenhagen, Royal Library, NKS 4019 kvart
OBI	*Olai Borrichii Itinerarium 1660–1665: The Journal of the Danish Polyhistor Ole Borch*. Edited and with an introduction by H. D. Schepelern. 4 vols. Copenhagen: C. A. Reitzels Forlag, and London: E. J. Brill, 1983.
OPH	*Nicolai Stenonis Opera philosophica*. Edited by Vilhelm Maar. 2 vols. Copenhagen: Vilhelm Tryde, 1910.
OTH	*Nicolai Stenonis Opera theologica*. Edited by Knud Larsen and Gustav Scherz. 2 vols. Copenhagen: Nyt Nordisk Forlag, 1941–1947.

Manuscripts

Berlin, Staatsbibliothek
  Darm 3a 1650
Copenhagen, Danish National Archives
  Metropolitanskolen (1646–1737), "Liber scholae Hafniensis Rectore scholae Georgio Hilario 1666,"
Copenhagen, Royal Library of Denmark
  Acc. 2019/54, KBs arkiv (indtil 1943) E 2: Catalogi Bibliothecae Gerstorffianae, NKS 4019 kvart, NKS 4660 kvart, Thott 1926 kvart.

Florence, Archivio di Stato di Firenze
   MdP, filza 1020, filza 2662, filza 4412, filza 4495
Florence, Archivio Storico Arcivescovile di Firenze
   TIN 34.13
Florence, Biblioteca dell'Accademia della Crusca
   Ms. 33 (BCF-Cat.1a), as quoted in
   www.accademicidellacrusca.org/scheda?IDN=2148.
Florence, Biblioteca Marucelliana
   Redi 7, Redi 8, Redi 30, Redi 32
Florence, Biblioteca Medicea Laurenziana (BML)
   Acq. e doni. 859, Ashburnham 1866, Redi 203, Redi 206, Redi 214, Redi 220, Redi 222.
Florence, Biblioteca Nazionale Centrale di Firenze (BNCF)
   Aut. Palat. IV. 49, Baldov.258.VII.31, Fondo Nazionale II.IV.539, Gal. 130, Gal. 158, Gal. 163, Gal. 243, Gal. 251, Gal. 252, Gal. 254, Gal. 255, Gal. 269, Gal. 277, Gal. 278, Gal. 280, Gal. 281, Gal. 282, Gal. 291, Gal. 312, Gal. 313, Gal. 314, Gal. 315, Magl.VIII.432, Magl.VIII.518, Magl.VIII.643, Magl.VIII.718
Leiden, Leiden University Library
   ASF 10.
London, Royal Society of London
   EL/S1/92
Lucca, Archivio di Stato di Lucca
   Arnolfini, Reg 97, fasc 16
Oxford, Bodleian Library
   MS Lister 5
Paris, Bibliothèque numérique Medica
   "Liste funèbre des chirurgiens de Paris, qui sont morts depuis l'année 1315 . . . ,"
   www.biusante.parisdescartes.fr/histmed/medica/page?ms02118&p=201
Rome, Archivio della Pontificia Università Gregoriana
   Ms. 563
Rome, Archivum S. Congreagationis de Propaganda Fide
   SRCG 465
Siena
   Biblioteca Communale di Siena, D. V. 7

Printed Sources

Anon. "An Account of Some Books . . . II. Joh. Swammerdam, M.D. Amsterodamentis *De respiratione et usu pulmonum*." *Philosophical Transactions* 2, no. 28 (1667): 532–538.

Anon. *Kjøbenhavns Universitets Matrikel*. Vol. 1, *1611–1667*. Edited by S. Birket Smith. Copenhagen: Gyldendalske Bøghandels Forlag, 1890.

Anon. "Inquiries for Hungary, Transylvania, Egypt and Guiny." *Philosophical Transactions* 2, no. 25 (1667): 467–472.

Anon. "Review of Steno's *De musculis et glandulis*." *Journal des sçavans* (23 March 1665): 139–142.

Albanese, Massimiliano. "Magliabechi, Antonio." In DBI, 67:422–427.

Album studiosorum Academiæ Ludguno Batavæ. The Hague: Martin Nijhoff, 1875.
Aldrovandi, Ulisse. *Musæum metallicum*. Bologna: Marcus Antonius Bernia, 1648.
Ambrose, Charles. "Immunology's First Priority Dispute: An Account of the 17th-Century Rudbeck-Bartholin Feud." *Cellular Immunology* 242, no. 1 (2006): 1–8.
Andersen, Jesper. *Thomas Bartholin: Laegen & anatomen; Fra enhjørninger til lymfekar*. Copenhagen: FADL's Forlag, 2017.
Andersen, Kirsti. "An Impression of Mathematics in Denmark in the Period 1600–1800." *Centaurus* 24, no. 1 (1980): 316–334.
Andersen, Kirsti, and Henk Bos. "Pure Mathematics." In Park and Daston, *Early Modern Science*, 696–723.
Andrault, Raphaële. "Anatomy, Mechanism and Anthropology: Nicolas Steno's Reading of *L'Homme*." In *Descartes' Treatise of Man and Its Reception*, edited by Delphine Antoine-Mahut and Stephen Gaukroger, 175–192. Cham: Springer, 2016.
Andrault, Raphaële. "The Chain of Motions and the Chain of Thoughts: The Diachronic Mechanism of Spinoza's Friends." In *Mechanism, Life and Mind in Modern Natural Philosophy*, edited by Antonio Clericuzio, Paolo Pecere, and Charles Wolfe, 119–135. Cham: Springer, 2022.
Andrault, Raphaële. "Divine Organs? Leibniz's 'Hymn to Galen' and the Best of All Possible Bodies." In *Galen and the Early Moderns*, edited by Matteo Favaretti and Emanuela Scribano, 135–153. Cham: Springer, 2022.
Andrault, Raphaële. "Human Brain and Human Mind: The *Discourse on the Anatomy of the Brain* and Its Philosophical Reception." In Andrault and Laerke, *Steno and the Philosophers*, 87–112.
Andrault, Raphaële. "Introduction." In Stensen, *Discours sur l'anatomie du cerveau*, ed. Andrault, 7–74. Paris: Éditions Classiques Garnier, 2009.
Andrault, Raphaële. "Mathématiser l'anatomie: La myologie de Stensen (1667)." *Early Science and Medicine* 15, nos. 4–5 (2010): 505–536.
Andrault, Raphaële. "Spinoza's Missing Physiology." *Perspectives on Science* 27, no. 2 (2019): 214–243.
Andrault, Raphaële. "Tirer sur la corde: Douleur et vivisection animale à l'âge classique." In *L'homme et la brute au XVIIe siècle: Une éthique animale à l'âge classique*, edited by Marine Bedon and Jacques-Louis Lantoine, 83–104. Lyon: ENS Éditions, 2022.
Andrault, Raphaële, and Mogens Lærke, eds. *Steno and the Philosophers*. Leiden: Brill, 2018.
Andriesse, Cornelis. *Huygens: The Man Behind the Principle*. Cambridge: Cambridge University Press, 2005.
Angeli, Roberto. *Niels Stensen: Il beato Niccolò Stenone, uno scienzato innamorato del vangelo e dell'Italia*. 2nd ed. Milan: San Paolo, 1996. Originally published as *Niels Stensen: Anatomico, fondatore della geologia, servo di Dio* (Florence: Libreria Editrice Fiorentina, 1968).
Antognazza, Maria Rosa. *Leibniz: An Intellectual Biography*. Cambridge: Cambridge University Press, 2009.
Backer, Augustin de, Alois de Backer, and Carlos Sommervogel, eds. *Bibliothèque des écrivains de la Compagnie de Jésus*. 3 vols. Paris: L. Grandmont-Donders, 1869–1876.
Bambacari, Cesare. *Descrizione delle azioni e virtù dell'illustrissima signora Lavinia Felice Cenami Arnolfini*. Lucca: Pellegrino Frediani, 1715.

Bandini, Angelo Maria. *Collectio veterum aliquot monimentorum ad historiam praecipue litterariam pertinentium*. Arezzo: Michael Bellotti, 1752.

Banks, David. *The Birth of the Academic Article: "Le journal des sçavans" and the "Philosophical Transactions," 1665–1700*. Sheffield: Equinox, 2017.

Bartholin, Caspar. *Anatomicae institutiones corporis humani*. Wittenberg: Bechtoldum Raaben, 1611.

Bartholin, Caspar. *De studio medico*. Copenhagen: Georg Hantzsch, 1628.

Bartholin, Caspar. *Opusucula quatuor singularia*. 4 vols. Copenhagen: Georg Hantzsch, 1628.

Bartholin, Erasmus. *De naturæ mirabilibus quaestiones academicæ*. Copenhagen: Georg Gödian, 1674.

Bartholin, Thomas. *Anatomia ex Caspari Bartholini parentis institutionibus . . . reformata*. Leiden: Franciscus Hack, 1651.

Bartholin, Thomas. *Anatomia, ex Caspari Bartholini . . . reformata*. The Hague: Adriaan Vlacq, 1655.

Bartholin, Thomas. *Anatomia, ex Caspari Bartholini . . . reformata*. The Hague: Adriaan Vlacq, 1663.

Bartholin, Thomas. *Anatomia parentis Caspari Bartholini*. Leiden: Franciscus Hack, 1641.

Bartholin, Thomas. *Anatomica aneurysmatis dissecti historia*. Palermo: Alfonso dell'Isola, 1644.

Bartholin, Thomas. *Cista medica Hafniensia*. Copenhagen: Mattheus Godicchenius, 1662.

Bartholin, Thomas. *De bibliothecæ incendio dissertatio*. Copenhagen: Matthius Godicchenius, 1670.

Bartholin, Thomas. *De lacteis thoracicis . . . historia anatomica*. Copenhagen: Melchior Martzan, 1652.

Bartholin, Thomas. *De libris legendis*. Copenhagen: Daniel Paull, 1676.

Bartholin, Thomas. *De peregrinatione medica*. Copenhagen: Daniel Paull, 1674.

Bartholin, Thomas. *Dissertatio anatomica de hepate defuncto*. Copenhagen: Christian Wering, 1661.

Bartholin, Thomas. *Historiarum anatomicarum rariorum centuria III & IV*. The Hague: Adriaan Vlacq, 1657.

Bartholin, Thomas. *Historiarum anatomicarum rariorum centuria V et VI*. Copenhagen: Henric Gödian, 1661.

Bartholin, Thomas. *Responsio de experimentis bilsinianis*. Copenhagen: Petrus Haubold, 1661.

Bartholin, Thomas. *Spicilegia bina ex vasis lymphaticis*. Amsterdam: Pieter van den Berge, 1661.

Bartholin, Thomas. *Vasa lymphatica . . . et hepatis exequiæ*. Copenhagen: Petrus Hakius, 1653.

Bartholin, Thomas, ed. *Acta medica et philosophica Hafniensia*. 5 vols. Copenhagen, 1673–1680.

Bartholin, Thomas. *Institutiones anatomicæ*. Edited by Caspar Bartholin. Leiden: Franciscus Hack, 1641.

Battaglia, Salvatore, and Giorgio Squarotti, eds. *Grande dizionario della lingua italiana*. 21 vols. Turin: Unione tipografico editrice torinese, 1961–2009.

Beaune, Florimond de. *De æquationum natura, constitutione, et limitibus opuscula duo . . . edita ab Erasmio Bartholino, Medicinæ et Mathematum in Regia Academia Hafniensi Professore publico.* Amsterdam: Ludovicus et Danielis Elsevier, 1659.

Begum, Shamima, Michele Cianci, Bo Durbeei, et al. "On the Origin and Variation of Colors in Lobster Carapace." *Physical Chemistry, Chemical Physics* 17, no. 26 (2015): 16723–16732.

Bek-Thomsen, Jakob. "From Flesh to Fossils: Nicolaus Steno's Anatomy of the Earth." In Duffin, Moody, and Gardner-Thorpe, *A History of Geology and Medicine*, 289–305.

Bek-Thomsen, Jakob. "Nicolaus Steno and the Making of an Early Modern Career: Nature, Knowledge and Networks at the Court of the Medici, c. 1650–1675." Ph.D. diss., Aarhus University, 2012.

Bek-Thomsen, Jakob. "Steno's *Historia*: Methods and Practices at the Court of Ferdinand II." In Andrault and Laerke, *Steno and the Philosophers*, 233–258.

Beretta, Marco. "At the Source of Western Science: The Organization of Experimentalism at the Accademia del Cimento (1657–1667)." *Notes and Records: The Royal Society Journal of the History of Science* 54, no. 2 (2000): 131–151.

Beretta, Marco, Antonio Clericuzio, and Lawrence Principe, eds. *The Accademia del Cimento and Its European Context*. Sagamore Beach, MA: Science History Publications, 2009.

Bernardi, Walter. "Uno scienziato Aretino protagonista della nascita della modernità." In *Francesco Redi Aretino: Atti del convegno*, edited by Lorella Mangani and Giuseppe Martini, 17–36. Arezzo: Accademia Petrarca di lettere arti e scienze, 1990.

Berryman, Sylvia. "Galen and the Mechanical Philosophy." *Apeiron* 35, no. 3 (2011): 235–253.

Bertoloni Meli, Domenico. "Authorship and Teamwork Around the Cimento Academy: Mathematics, Anatomy, Experimental Philosophy." *Early Science and Medicine* 6, no. 2 (2001): 65–95.

Bertoloni Meli, Domenico. "The Collaboration Between Anatomists and Mathematicians in the Mid-Seventeenth Century with a Study of Images as Experiments and Galileo's Role in Steno's *Myology*." *Early Science and Medicine* 13, no. 6 (2008): 665–709.

Bertoloni Meli, Domenico. "The Color of Blood: Between Sensory Experience and Epistemic Significance." In Daston and Lunbeck, *Histories of Scientific Observation*, 117–134.

Bertoloni Meli, Domenico. "Early Modern Experimentation in Live Animals." *Journal of the History of Biology* 46 (2013): 199–226.

Bertoloni Meli, Domenico. "Machines of the Body in the Seventeenth Century." In Distelzweig, Goldberg, and Ragland, *Early Modern Medicine and Natural Philosophy*, 91–116.

Bertoloni Meli, Domenico. *Mechanism: A Visual, Lexical, and Conceptual History*. Pittsburgh: University of Pittsburgh Press, 2019.

Bertoloni Meli, Domenico. *Mechanism, Experiment, Disease: Marcello Malpighi and Seventeenth-Century Anatomy*. Baltimore, MD: Johns Hopkins University Press, 2011.

Bertoloni Meli, Domenico. "Mechanistic Pathology and Therapy in the Medical *Assayer* of Marcello Malpighi." *Medical History* 51, no. 2 (2007): 165–180.

Bertoloni Meli, Domenico. "Reliability and Generalization in Early Modern Anatomy." In *La tradizione Galileiana e lo sperimentalismo naturalistico d'età moderna: Atti del seminario internazionale di Studi (Milan, 15–16 ottobre 2010)*, edited by Maria Monti, 1–26. Biblioteca dell'edizione nazionale delle opere di Antonio Vallisneri, 8. Florence: Leo Olschki, 2011.

Bertoloni Meli, Domenico. "Shadows and Deception: From Borelli's *Theoricae* to the *Saggi* of the Cimento." *The British Journal for the History of Science* 31, no. 4 (1998): 383–402.

Bertoloni Meli, Domenico. *Visualizing Disease: The Art and History of Pathological Illustrations*. Chicago: University of Chicago Press, 2018.

Beughem, Cornelius à. *Bibliographia mathematica et artificiosa novissima*. Amsterdam: Johann Jansson van Waesberge, 1688.

Biagioli, Mario. "Etiquette, Interdependence, and Sociability in Seventeenth-Century Science." *Critical Inquiry* 22, no. 2 (1996): 193–238.

Biagioli, Mario. *Galileo, Courtier: The Practice of Science in the Culture of Absolutism*. Chicago: University of Chicago Press, 1993.

Biagioli, Mario, and Peter Galison, eds. *Scientific Authorship: Credit and Intellectual Property in Science*. London: Routledge, 2003.

Birkenhead, Lord Cohen of. "On the Motion of Blood in the Veins." *British Medical Journal* 3, no. 5774 (1971): 551–557.

Blair, Ann. "An Early Modernist's Perspective." *Isis* 95, no. 3 (2004): 420–430.

Blair, Ann. "Natural Philosophy." In Park and Daston, *Early Modern Science*, 365–406.

Blair, Ann. *The Theater of Nature: Jean Bodin and Renaissance Science*. Princeton, NJ: Princeton University Press, 1997.

Blair, Ann. *Too Much to Know: Managing Scholarly Information Before the Modern Age*. New Haven, CT: Yale University Press, 2010.

Blasius, Gerard. *Anatome animalium*. Amsterdam: Johan van Someren, 1681.

Blasius, Gerard. *Medicina generalis, nova accurataque methodo fundamenta exhibens*. Amsterdam: Pieter van den Berge, 1661.

Blasius, Gerard. *Medicina universa*. Amsterdam: Petrus van den Berge, 1665.

[Blasius, Gerard.] *Observationes anatomicae selectiores collegii privati Amstelodamensis* Amsterdam: Caspar Commelin, 1667.

Bloch, Olivier René. *La philosophie de Gassendi: Nominalism, matérialisme et métaphysique*. Archives internationales d'histoire des idées/International Archive of the History of Ideas 38. The Hague: Martinus Nijhoff, 1971.

Bonati, Isabella. "Hydrocephalus in Context: A History from Graeco-Roman Sources." *Early Science and Medicine* 27, no. 4 (2022): 333–350.

Borel, Pierre. *De vero telescopii inventore, cum brevi omnium conspiciliorum historia ... accessit etiam centuria observationum microcospicarum*. The Hague: Adriaan Vlacq, 1655.

Borel, Pierre. *Historiarum, et observationum medicophysicarum centuria prima et secunda*. Toulouse: Arnald Colomeri, 1653.

Borel, Pierre. *Historiarum et observationum medicophysicarum centuriæ IV*. Paris: Jean Billaine & Veuve Mathurin Dupuis, 1656.

Borel, Pierre. *Vitæ Renati Cartesii summi philosophi compendium*. Paris: Jean Billaine & Veuve Mathurin Dupuis, 1656.

Borelli, Giovanni, *De motu animalium*. 2 vols. Rome: Angeli Bernabo, 1680-1681.

Borrichii, Olai [Ole Borch]. *Olai Borrichii Itinerarium 1660–1665: The Journal of the Danish Polyhistor Ole Borch*. Edited and with an introduction by H. D.

Schepelern. 4 vols. Copenhagen: C. A. Reitzels Forlag, and London: E. J. Brill, 1983. [Hereafter cited as OBI.]
Boschiero, Luciano. *Experiment and Natural Philosophy in Seventeenth-Century Tuscany: The History of the Accademia del Cimento*. Dordrecht: Springer, 2007.
Boudhors, Ch.-H. "'Une amie de Pascal?' Marie Perriquet et sa soeur Geneviève (suite et fin)." *Révue d'histoire littéraire de la France* 36, no. 3 (1929): 355–387.
Boyer, Marjorie Nice. "Resistance to Technological Innovation: The History of the Pile Driver Through the 18th Century." *Technology and Culture* 26, no. 1 (1985): 56–68.
Boyle, Robert. *New Experiments Physico-Mechanical, Touching the Spring of the Air, and Its Effects*. Oxford: The University, 1660.
Brahe, Tycho. *Epistolarum astronomicarum liber*. Uraniborg, 1596.
Brecht, Martin. *Martin Luther*. 3 vols. Translated by James L. Schaaf. Minneapolis, MN: Fortress Press, 1985–1999.
Brockliss, Laurence, and Colin Jones. *The Medical World of Early Modern France*. Oxford and New York: Oxford University Press, 1997.
Broecke, Peter van den. *Epistolarum libri tres*. Lucca: Hyacinthus Pacius, 1684.
Brookfield, Michael E. *Principles of Stratigraphy*. Malden, MA: Blackwell, 2004.
Brown, Harcourt. *Scientific Organizations in Seventeenth Century France (1620–1680)*. Baltimore, MD: Williams & Wilkins, 1934.
Brown, Harcourt. "Martin Fogel e l'idea accademica Lincea." *Rendiconti della R. Accademia nazionale dei Lincei* 11, nos. 11–12 (1935): 814–833.
Browne, Janet. *Darwin's "Origin of Species": A Biography*. New York: Atlantic Books, 2006.
Buchwald, Jed Z. "Descartes's Experimental Journey Past the Prism and Through the Invisible World to the Rainbow." *Annals of Science* 65, no. 1 (2008): 1–46.
Bunge, Wiep van. "The Early Dutch Reception of Cartesianism." In Nadler, Schmaltz, and Antoine-Mahut, *The Oxford Handbook of Descartes and Cartesianism*, 417–433.
Burke, Peter. *The Polymath: A Cultural History from Leonardo da Vinci to Susan Sontag*. New Haven, CT: Yale University Press, 2020.
Bustaffa, Francesco. "Michelangelo Ricci (1619–1682): Biografia di un cardinale innocenziano." Ph.D. diss., Università di San Marino, 2010.
Bylebyl, Jerome J. "Nutrition, Quantification and Circulation." In "Owsei Temkin at 75," [special issue,] *Bulletin of the History of Medicine* 51, no. 3 (1977): 369–385.
Canons and Decrees of the Sacred and Ecumenical Council of Trent. Translated by Waterworth. London: Burns & Oates, 1848.
Carinci, Eleonora. "Una 'speziala' padovana: *Lettere di philosophia naturale* di Camilla Erculiani (1584)." *Italian Studies* 68, no. 2 (2013): 202–229.
Carré, Marie-Rose. "A Man Between Two Worlds: Pierre Borel and His *Discours nouveau prouvant la pluralité des mondes* of 1657." *Isis* 65, no. 3 (1974): 322–335.
Carruccio, Ettore. "Cavalieri, Bonaventura." In *Complete Dictionary of Scientific Biography*, vol. 3, 149–153. Detroit, MI: Charles Scribner's Sons, 2008.
Castel-Branco, Nuno. "Friendship Fostered by Poison: The Collaboration of Nicolaus Steno and Francesco Redi." In *Poison: Knowledge, Uses, Practices*, edited by Caterina Mordeglia and Agostino Paravicini Bagliani, 325–346. Florence: SISMEL—Edizioni del Galluzzo, 2022.

Castel-Branco, Nuno. "Physico-Mathematics and the Life Sciences: Experiencing the Mechanism of Venous Return, 1650s–1680s." *Annals of Science* 79, no. 4 (2022): 442–467.

Castel-Branco, Nuno. "Who Was Borelli Responding To? Nicolaus Steno in *De motu animalium* (Rome, 1680–1681)." *Physis* 57, no. 2 (2022): 431–448.

Castel-Branco, Nuno, and Troels Kardel. "Drawing Muscles with Diagrams: How a Novel Dissection Cut Inspired Nicolaus Steno's Mathematical Myology (1667)." *Notes and Records: The Royal Society Journal of the History of Science* (2022). https://doi.org/10.1098/rsnr.2022.0005.

Cavallo, Sandra. *Artisans of the Body in Early Modern Italy: Identities, Families and Masculinities.* Manchester: Manchester University Press, 2010.

Chabbert, Pierre. "Jacques Borelly (16. .–1689): Membre de l'Académie royale des sciences." *Revue d'histoire des sciences et de leurs applications* 23 (1970): 203–227.

Chang, Ku-ming (Kevin). "From Oral Disputation to Written Text: The Transformation of the Dissertation in Early Modern Europe." *History of Universities* 19, no. 2 (2004): 129–187.

Chen, Thomas S. N., and Peter S. Y. Chen. "William Harvey as Hepatologist." *The American Journal of Gastroenterology* 83, no. 11 (1988): 1274–1277.

Chene, Dennis Des. "Mechanisms of Life in the Seventeenth Century: Borelli, Perrault, Régis." *Studies in History and Philosophy of Science Part C: Studies in History and Philosophy of Biological and Biomedical Sciences* 36, no. 2 (2005): 245–260.

Cheung, Tobias. "Omnis Fibra ex Fibra: Fibre Œconomies in Bonnet's and Diderot's Models of Organic Order." In "Transitions and Borders Between Animals, Humans and Machines, 1600–1800," [special issue,] *Early Science and Medicine* 15 (2010): 66–104.

Clericuzio, Antonio. "Meccanismo ed empirismo nell'opera di Stensen." In Vitoria and Gómez, *Scienza, filosofia e religione nell'opera di Niels Steensen*, 123–138.

Coari, Giulio, and Claudio Mutini. "Bellini, Lorenzo." In DBI, 7:713–716.

Cobb, Matthew. "Exorcizing the Animal Spirits: Jan Swammerdam on Nerve Function." *Nature Reviews: Neuroscience* 3, no. 5 (2002): 395–400.

Cobb, Matthew. *Generation: The Seventeenth-Century Scientists Who Unraveled the Secrets of Sex, Life, and Growth.* New York: Bloomsbury, 2006.

Cochrane, Eric. *Florence in the Forgotten Centuries, 1527–1800: A History of Florence and the Florentines in the Age of the Grand Dukes.* Chicago: University of Chicago Press, 1973.

Cohen, Bernard. "Roemer and the First Determination of the Velocity of Light (1676)." *Isis* 31, no. 2 (1940): 327–379.

Collegium privatum Amstelodamense. *Observationes anatomicae selectiores.* Amsterdam: Caspar Commelinus, 1667, 1673.

Conforti, Maria. "'Se fusse meno cartesiano lo stimarei molto': Anti-Cartesian Motifs in Italian Medicine." In *Descartes and Medicine: Problems, Responses and Survival of a Cartesian Discipline*, edited by Fabrizio Baldassarri, 437–449. Turnhout: Brepols, 2023.

Conforti, Maria. "Testes alterum cerebrum: Succo nerveo e succo seminale nella macchina del vivente di Giovanni Alfonso Borelli." *Medicina nei secoli arte e scienza: Giornale di storia della medicina*, n.s. 13, no. 3 (2001): 577–595.

Cook, Harold J. *Matters of Exchange: Commerce, Medicine, and Science in the Dutch Golden Age.* New Haven, CT: Yale University Press, 2007.

Cooper, Alix. "The Museum and the Book: The *Metallotheca* and the History of an Encyclopaedic Natural History in Early Modern Italy." *Journal of the History of Collections* 7, no. 1 (1995): 1–23.
Craik, Elizabeth. "The Reception of the Hippocratic Treatise *On Glands*." In Horstmanshoff, King, and Zittel, *Blood, Sweat and Tears*, 65–82.
Cunningham, Andrew. *The Anatomical Renaissance: The Resurrection of Anatomical Projects of the Ancients*. 2nd ed. London: Routledge, 2016.
Cunningham, Andrew. "The Bartholins, the Platters and Laurentius Gryllus: The *peregrinatio medica* in the Sixteenth and Seventeenth Centuries." In Grell, Cunningham, and Arrizabalaga, *Centers of Medical Excellence?*, 3–16.
Cunningham, Andrew. *"I Follow Aristotle": How William Harvey Discovered the Circulation of the Blood*. London: Routledge, 2022.
Cutler, Alan. *The Seashell on the Mountaintop: A Story of Science, Sainthood, and the Humble Genius Who Discovered a New History of the Earth*. New York: Dutton, 2003.
Dal Prete, Ivano. *On the Edge of Eternity: The Antiquity of the Earth in Medieval and Early Modern Europe*. Oxford and New York: Oxford University Press, 2022.
Dal Prete, Ivano. "The Ruins of the Earth: Learned Meteorology and Artisan Expertise in Fifteenth-Century Italian Landscapes." *Nuncius: Annali di storia della scienza* 33, no. 3 (2018): 415–441.
Danneskiold-Samsøe, J. F. C. *Muses and Patrons: Cultures of Natural Philosophy in Seventeenth-Century Scandinavia*. Ugglan, Minervaserien 10. Lund: Lund University Press, 2004.
Daston, Lorraine, and H. Otto Sibum. "Introduction: Scientific Personae and Their Histories." *Science in Context* 16, nos. 1–2 (2003): 1–8.
Daston, Lorraine, and Elizabeth Lunbeck, eds. *Histories of Scientific Observation*. Chicago: University of Chicago Press, 2011.
Dear, Peter. *Discipline and Experience: The Mathematical Way in the Scientific Revolution*. Chicago: University of Chicago Press, 1995.
Dear, Peter. *The Intelligibility of Nature: How Science Makes Sense of the World*. Chicago: University of Chicago Press, 2006.
Dear, Peter. *Mersenne and the Learning of the Schools*. Ithaca, NY: Cornell University Press, 1988.
Dear, Peter. "Mixed Mathematics." In *Wrestling with Nature: From Omens to Science*, edited by Peter Harrison, Ronald L. Numbers, and Michael H. Shank, 149–172. Chicago: University of Chicago Press, 2011.
De Beer, E. S., ed. *The Correspondence of John Locke*. 8 vols. Oxford and New York: Oxford University Press, 1976–1989.
de Graaf, Regnier. *De mulierum organis generationi inservientibus*. Leiden: Hack, 1672.
de Graaf, Regnier. *Disputatio medica de natura et usu succi pancreatici*. Leiden: Hack, 1664.
Deitz, Luc. "Ioannes Wower of Hamburg, Philologist and Polymath: A Preliminary Sketch of His Life and Works." *Journal of the Warburg and Courtauld Institutes* 58, no. 1 (1995): 132–151.
Delorme, Suzanne. "Montmor, Henri." In *Complete Dictionary of Scientific Biography*, 9:497–499. Charles Scribner's Sons, 2008.
Del Prete, Antonella. "Teaching Cartesian Philosophy in Leiden: Adriaan Heereboord (1613–1661) and Johannes De Raey (1622–1702)." In *Descartes in the Classroom: Teaching Cartesian Philosophy in the Early Modern Age*, edited by

David Cellamare and Mattia Mantovani, 60–78. Medieval and Early Modern Philosophy and Science 35. Leiden: Brill, 2023.

Denis, Jean-Baptiste. "A Letter Concerning a New Way of Curing Sundry Diseases by Transfusion of Blood." *Philosophical Transactions* 2, no. 27 (1667): 490–493.

Descartes, René. *De homine*. Leiden: Petrus Leffen & Franciscus Moyardus, 1662.

Descartes, René. *A Discourse on the Method*. Translated by Ian Maclean. Oxford and New York: Oxford University Press, 2006.

Descartes, René. *L'homme*. Paris: Charles Angot, 1664.

Descartes, René. *Principia philosophiæ*. https://wellcomecollection.org/works/wduad66n/items.

Descartes, René. *Principles of Philosophy*. In *The Philosophical Writings of Descartes*, translated by John Cottingham, Robert Stoothoff, and Dugald Murdoch, 1:193–266. 2 vols. Cambridge: Cambridge University Press, 1985.

Descartes, René. *Treatise of Man*. Edited and translated by Thomas Steele Hall. Cambridge, MA: Harvard University Press, 1972.

Deusing, Anton. *Disquisitio physico-mathematica, gemina, de vacuo*. Amsterdam: Pieter van den Berge, 1661.

Dew, Nicholas. *Orientalism in Louis XIV's France*. Oxford and New York: Oxford University Press, 2009.

Diefendorf, Barbara. *From Penitence to Charity: Pious Women and the Catholic Reformation in Paris*. Oxford and New York: Oxford University Press, 2004.

Distelzweig, Peter M. "Fabricius' Galeno-Aristotelian Teleomechanics of Muscle." In *The Life Sciences in Early Modern Philosophy*, edited by Ohad Nachtomy and Justin E. H. Smith, 65–84. Oxford and New York: Oxford University Press, 2014.

Distelzweig, Peter M. "The Use of *Usus* and the Function of *Functio*: Teleology and Its Limits in Descartes's Physiology." *Journal of the History of Philosophy* 53, no. 3 (2015): 377–399.

Distelzweig, Peter, Benjamin Goldberg, and Evan R. Ragland, eds. *Early Modern Medicine and Natural Philosophy*. History, Philosophy and Theory of the Life Sciences 14. New York: Springer, 2016.

Dobre, Mihnea. "Jacques Rohault and Cartesian Experimentalism." In Nadler, Schmaltz, and Antoine-Mahut, *The Oxford Handbook of Descartes and Cartesianism*, 388–401.

Dominici, Stefano. "A Man with a Master Plan: Steno's Observations on Earth's History." *Substantia* 5, no. 1, suppl. 1 (2021): 59–75.

Dominici, Stefano. "The Volterra Cliff in the Mind of Philosophers, Savants, and Geologists (1282–1830)." *Geological Society, London, Special Publications* 543, no. 1 (2023): 267–280.

Donders, Franciscus. *Physiologie des Menschen*. Leipzig, 1856.

Dopper, Jantien. "A Life of Learning in Leiden: The Mathematician Frans van Schooten (1615–1660)." Ph.D. diss., Utrecht University, 2014.

Drexel, Jeremias. *Joseph aegypti prorex*. Antwerp: Cornelius Leysser, 1641.

Duffin, C., R. T. J. Moody, and C. Gardner-Thorpe, eds. *A History of Geology and Medicine*. London: Geological Society of London, 2013.

Eire, Carlos M. N. *Reformations: The Early Modern World, 1450–1650*. New Haven, CT: Yale University Press, 2016.

Eloy, N. F. J. *Dictionnaire historique de la médecine ancienne et moderne*. 4 vols. Paris: H. Hoyos, 1778.

Fabri, Honoré. *Tractatus duo: Quorum prior est de plantis et de generatione animalium et posterior de homine*. Paris: François Muguet, 1666.

Fabricius ab Acquapendente, Hieronymus. *De visione. De voce. De auditu*. Venice, 1600.

Fabroni, Angelo. *Delle lettere familiari del Conte Lorenzo Magalotti e di altri insigni uomini a lui scritte*. 2 vols. Florence: S.A.R., 1769–1775.

Fabroni, Angelo. *Lettere inedite di uomini illustri*. 2 vols. Florence: Francesco Moücke, 1773–1775.

Falloppio, Gabriele. *De medicatis aquis*. Venice: Lodovico Avans, 1564.

Falloppio, Gabriele. *Observationes anatomicæ*. Venice: Marcus Antonius Ulmus, 1561.

Favino, Federica. "Ricasoli Rucellai, Orazio." In DBI, 87:143.

Favino, Federica, and Giulia Giannini. "Borelli Reloaded: Contexts and Networks in 17th-Century Italy." *Physis* 57, no. 2 (2022): 289–300.

Findlen, Paula. "Academies, Networks, and Projects: The Accademia del Cimento and Its Legacy." *Galilaena: Journal of Galilelean Studies* 7, no. 7 (2010): 277–298.

Findlen, Paula. "Controlling the Experiment: Rhetoric, Court Patronage and the Experimental Method of Francesco Redi." *History of Science* 31, no. 1 (1993): 35–64.

Findlen, Paula. "Introduction: 'The Last Man Who Knew Everything . . . or Did He? Athanasius Kircher, S.J. (1602–80) and His World.'" In *Athanasius Kircher: The Last Man Who Knew Everything*, edited by Paula Findlen, 1–50. New York: Routledge, 2004.

Findlen, Paula. "Jokes of Nature and Jokes of Knowledge: The Playfulness of Scientific Discourse in Early Modern Europe." *Renaissance Quarterly* 43, no. 2 (1990): 292–331.

Findlen, Paula. *Possessing Nature: Museums, Collecting, and Scientific Culture in Early Modern Italy*. Berkeley and Los Angeles: University of California Press, 1994.

Findlen, Paula. "Projecting Nature: Agostino Scilla's Seventeenth-Century Fossil Drawings." *Endeavour* 42, nos. 2–3 (2018): 99–132.

Finocchiaro, Maurice A., ed. and trans. *The Essential Galileo*. Indianapolis: Hackett, 2008.

Fiocca, Alessandra, and Andrea del Centina. "Borelli's Edition of Books V–VII of Apollonius's *Conics*, and Lemma 12 in Newton's *Principia*." *Archive for History of Exact Sciences* 74, no. 3 (2020): 255–279.

Fosi, Irene. *Inquisition, Conversion, and Foreigners in Baroque Rome*. Leiden: Brill, 2020.

Frank, Robert. *Harvey and the Oxford Physiologists: Scientific Ideas and Social Interaction; A Study of Scientific Ideas*. Berkeley and Los Angeles: University of California Press, 1980.

French, Roger. *William Harvey's Natural Philosophy*. Cambridge: Cambridge University Press, 1994.

Frost, Robert I. *The Northern Wars: War, State and Society in Northeastern Europe, 1558–1721*. New York: Pearson Education, 2000.

Furst, Branko. *The Heart and Circulation: An Integrative Model*. 2nd ed. Cham: Springer, 2020.

Gallois, Jean. *Conversations académiques, tirées de l'Académie de Monsieur l'Abbé Bourdelot*. Paris: Claude Barbin, 1674.

Galluzzi, Paolo. "L'Accademia del Cimento: 'Gusti' del principe, filosofia e ideologia dell'esperimento." *Quaderni storici* 16, no. 48(3) (1981): 788–844.

Garber, Daniel. "Steno, Leibniz, and the History of the World." In Andrault and Laerke, *Steno and the Philosophers*, 201–229.

Garboe, Axel. "Michael Lyser, a 17th Century Anatomist." *Acta Medica Scandinavica* 142, no. S266 (1952): 63–73.

Garboe, Axel. *Niels Stensens (Stenos) geologiske Arbejdes Skaebne: Et Fragtment af Dansk Geologis Historie*. Copenhagen: Danmarks Geologiske Undersøgelse, 1948.

Garboe, Axel. "Niels Stensen's Lost Geological Manuscript." *Meddelelser fra Dansk Geologisk Forening* 14, no. 3 (1960): 243–246.

García Valverde, José Manuel, and Peter Maxwell-Stuart. *Gómez Pereira's "Antoniana Margarita": A Work on Natural Philosophy, Medicine and Theology*. Leiden: Brill, 2019.

Giannini, Giulia. "An Indirect Convergence Between the Accademia del Cimento and the Montmor Academy: The 'Saturn Dispute.'" In *The Institutionalization of Science in Early Modern Europe*, edited by Mordechai Feingold and Giulia Giannini, 83–108. Leiden: Brill, 2020.

Gjellerup, S. M. "Eilersen, Jørgen." In *Dansk biografisk Lexikon, tillige omfattende Norge for tidsrummet 1537–1814*, edited by C. F. Bricka, 19 vols., 4:464–465. Copenhagen: Gyldendal, 1887–1905.

Gliozzi, Mario. "Angeli, Stefano degli." In DBI, 3:204–206.

Glisson, Francis. *Anatomia hepatis*. London: Du-Gard for Octavian Pulleyn, 1654.

Goldgar, Anne. *Impolite Learning: Conduct and Community in the Republic of Letters, 1680–1750*. New Haven, CT: Yale University Press, 1995.

Goldstein, Catherine. "Routine Controversies: Mathematical Challenges in Mersenne's Correspondence." In "La guerre en lettres: La controverse scientifique dans les correspondences des Lumières," special issue, no. 2, *Revue d'histoire des sciences* 66, no. 2 (2013): 249–273.

Gómez, Francisco, and Frank Sobiech. "Introduzione." In Vitoria and Gómez, *Scienza, filosofia e religione nell'opera di Niels Steensen*, 17–33.

Goodman, Dena. *The Republic of Letters: A Cultural History of the French Enlightenment*. Ithaca, NY: Cornell University Press, 1994.

Gould, Stephen Jay. *Hen's Teeth and Horse's Toes: Further Reflections in Natural History*. New York: W. W. Norton, 1983.

Gradi, Stefano. *Dissertationes physico-mathematicae quatuor*. Amsterdam: Daniel Elsevier, 1680.

Grafton, Anthony. *Defenders of the Text: The Traditions of Scholarship in an Age of Science, 1450–1800*. Cambridge, MA: Harvard University Press, 1991.

Grafton, Anthony. "The World of the Polyhistors: Humanism and Encyclopedism." In "The Culture of the Holy Roman Empire, 1540–1680" [special issue], *Central European History* 18, no. 1 (1985): 31–47.

Grafton, Anthony, and Joanna Weinberg. "Johann Buxtorf Makes a Notebook." In *Canonical Texts and Scholarly Practices: A Global Comparative Approach*, edited by Anthony Grafton and Glenn W. Most, 275–298. Cambridge: Cambridge University Press, 2016.

Graney, Christopher M. *Setting Aside All Authority: Giovanni Battista Riccioli and the Science Against Copernicus in the Age of Galileo*. Notre Dame, IN: University of Notre Dame Press, 2015.

Gray, Henry. *Anatomy of the Human Body*. 20th ed. Philadelphia: Lea & Febiger, 1918.

Gregory, Brad S. *The Unintended Reformation: How a Religious Revolution Secularized Society*. Cambridge, MA: Belknap Press of Harvard University Press, 2012.

Grell, Ole Peter. "Between Anatomy and Religion: The Conversions to Catholicism of the Two Danish Anatomists Nicolaus Steno and Jacob Winsløw." In *Medicine and Religion in Enlightenment Europe*, edited by Ole Peter Grell and Andrew Cunningham, 205–221. Burlington, VT: Ashgate, 2007.

Grell, Ole Peter. "Caspar Bartholin and the Education of the Pious Physician." In Grell and Cunningham, *Medicine and the Reformation*, 78–100.

Grell, Ole Peter. "In Search of True Knowledge: Ole Worm (1588–1654) and the New Philosophy." In *Making Knowledge in Early Modern Europe: Practices, Objects, and Texts, 1400–1800*, edited by Pamela H. Smith and Benjamin Schmidt, 214–232. Chicago: University of Chicago Press, 2007.

Grell, Ole Peter. "'Like the Bees, Who Neither Suck nor Generate Their Honey from One Flower': The Significance of the *peregrinatio academica* for Danish Medical Students in the Late Sixteenth and Early Seventeenth Centuries." In Grell, Cunningham, and Arrizabalaga, *Centers of Medical Excellence?*, 171–192.

Grell, Ole Peter, and Andrew Cunningham, eds. *Medicine and the Reformation*. New York: Routledge, 1993.

Grell, Ole Peter, Andrew Cunningham, and Jon Arrizabalaga, eds. *Centers of Medical Excellence? Medical Travel and Education in Europe, 1500–1789*. Burlington, VT: Ashgate, 2010.

Grosslight, Justin. "Small Skills, Big Networks: Marin Mersenne as Mathematical Intelligencer." *History of Science* 51, no. 3 (2013): 337–374.

Guerrini, Anita. *The Courtiers' Anatomists: Animals and Humans in Louis XIV's Paris*. Chicago: University of Chicago Press, 2015.

Guerrini, Anita. "Experiments, Causation, and the Uses of Vivisection in the First Half of the Seventeenth Century." *Journal of the History of Biology* 46, no. 2 (2013): 227–254.

Guerrini, Luigi. "Matematica ed erudizione: Giovanni Alfonso Borelli e l'edizione fiorentina dei libri V, VI e VII delle Coniche di Apollonio di Perga." *Nuncius* 14, no. 2 (1999): 505–658.

Hahn, Roger. *The Anatomy of a Scientific Institution: The Paris Academy of Sciences, 1666–1803*. Berkeley and Los Angeles: University of California Press, 1971.

Hall, Marie Boas. *Henry Oldenburg: Shaping the Royal Society*. Oxford and New York: Oxford University Press, 2002.

Hall, A. Rupert, and Marie Boas Hall, eds. *The Correspondence of Henry Oldenburg*. 13 vols. Madison: University of Wisconsin Press, 1965–1986.

Haller, Albrecht von. *Dissertationem inauguralem medicam et anatomicam de motu sanguinis per cor*. Gottingen, 1737.

Hallyn, Fernand, ed. *Metaphor and Analogy in the Sciences*. Origins 1. Dordrecht: Springer, 2000.

Hankinson, R. J. "Galen on the Limitations of Knowledge." In *Galen and the World of Knowledge*, edited by Christopher Gill, Tim Whitmarsh, and John Wilkins, 206–242. Cambridge: Cambridge University Press, 2009.

Hara, Kokithi. "Roberval, Gilles Personne de." In *Complete Dictionary of Scientific Biography*, 11:486–491. Charles Scribner's Sons, 2008.

Harsdörffer, Georg Philip. *Delitiae mathematicae et physicae*. 2 vols. Nuremberg: Jeremia Dümler, 1651–1653.

Harvey, William. *Exercitatio anatomica de motu cordis et sanguinis in animalibus*. Frankfurt: Gulielmus Fitzeri, 1628.

Hatch, Robert A. "The Republic of Letters: Boulliau, Leopoldo, and the Accademia del Cimento." In Beretta, Clericuzio, and Principe, *The Accademia del Cimento and Its European Context*, 165–181.

Heeffer, Albrecht. "*Récréations mathématiques* (1624): A Study on Its Authorship, Sources and Influence." *Gibecière* 1 (Summer 2006): 77–167.

Heilbron, J. L. *Electricity in the 17th and 18th Centuries: A Study of Early Modern Physics*. Berkeley and Los Angeles: University of California Press, 1979.

Heilbron, J. L. *Galileo*. Oxford and New York: Oxford University Press, 2010.

Henry, John. "Nicolas Steno (1638–1686): A Polymath Reassessed." *Isis* 111, no. 2 (2020): 365–367.

Hermansen, Viktor. "Niels Stensens Mutter." *Stenoniana Catholica* 6, no. 3 (1960): 71–81.

Heurnius, Johannes. *Praxis medicinæ nova ratio*. Rotterdam: Arnold Leers, 1650.

Hilarius, Georgius. *Progymnasmatum mathematicorum enchiridion*. Copenhagen: Pertrus Morsingi, 1656.

Hoboken, Nicolaus. *Novus ductus salivalis blasianus*. Utrecht: Johannis à Renswouw, 1662.

Hoch, Ella. "Diagnosing Fossilization in the Nordic Renaissance: An Investigation into the Correspondence of Ole Worm (1588–1654)." In Duffin, Moody, and Gardner-Thorpe, *A History of Geology and Medicine*, 307–327.

Hoffmann, C. G. *Scriptores rerum lusaticarum antiqui et recentiores*. Leipzig: David Richter, 1719.

Hollerbach, Teresa. *Sanctorius Sanctorius and the Origins of Health Measurement*. Cham: Springer, 2023.

Hom, Stephanie Malia. "Consuming the View: Tourism, Rome, and the Topos of the Eternal City." *Annali d'italianistica* 28 (2010): 91–116.

van Horne, Johannes. *Mikrokosmos seu brevis manuductio ad historiam corporis humani*. Leiden: Jacob Chouët, 1662.

Horstmanshoff, Manfred, Helen King, and Claus Zittel, eds. *Blood, Sweat and Tears: The Changing Concepts of Physiology from Antiquity into Early Modern Europe*. Intersections: Interdisciplinary Studies in Early Modern Culture 25. Leiden: Brill, 2012.

Hsu, Kuang-Tai. "The Path to Steno's Synthesis on the Animal Origin of *Glossopetrae*." In Rosenberg, *The Revolution in Geology from the Renaissance to the Enlightenment*, 93–106.

Huet, Pierre Daniel. *Censura philosophiae Cartesianae*. Paris: Daniel Horthemels, 1689.

Huet, Pierre Daniel. *Nouveaux mémoires pour servir à l'histoire du Cartesianisme*. Paris: Henry Desbordes, 1692.

Huisman, Tim. *The Finger of God: Anatomical Practice in 17th-Century Leiden*. Leiden: Primavera Press, 2009.

Hunter, Michael, Antonio Clericuzio, and Lawrence M. Principe, eds. *The Correspondence of Robert Boyle*. 6 vols. London: Pickering & Chatto, 2001.

Huygens, Christiaan. *Oeuvres complètes*. 22 vols. The Hague: M. Nijhoff, 1888–1950.

Huygens, Christiaan. *Systema saturnium*. The Hague: Adriaan Vlacq, 1659.

Iliffe, Rob. "Abstract Considerations: Disciplines and the Incoherence of Newton's Natural Philosophy." *Studies in History and Philosophy of Science* 35, no. 3 (2004): 427–454.

Iliffe, Rob. *Priest of Nature: The Religious Worlds of Isaac Newton*. Oxford and New York: Oxford University Press, 2017.

Iliffe, Robert. "Foreign Bodies: Travel, Empire and the Early Royal Society of London; Part 1, Englishmen on Tour." *Canadian Journal of History* 33, no. 3 (1998): 357–385.

Irwin, Joyce. "Anna Maria van Schurman and Antoinette Bourignon: Contrasting Examples of Seventeenth-Century Pietism." *Church History* 60, no. 3 (1991): 301–315.

Israel, Jonathan I. *Spinoza: Life and Legacy.* Oxford and New York: Oxford University Press, 2023.

John Paul II. "Homilia in Vaticana basilica habita ob decretos Dei Servo Nicolao Stensen." 23 October 1988. In *Acta Apostolicae Sedis*, 81:290–296. Vatican: Libreria Editrice Vaticana, 1989.

Johns, Adrian. "The Ambivalence of Authorship in Early Modern Natural Philosophy." In *Scientific Authorship: Credit and Intellectual Property in Science*, edited by Mario Biagioli and Peter Galison, 67–90. London: Routledge, 2003.

Jones, Colin. *Paris: The Biography of a City.* New York: Penguin Books, 2006.

Jordan, Hieronymus. *De eo quod divinum aut supernaturale est in morbis humani corporis.* Frankfurt: Johann Gottfried Schönwetter, 1651.

Jorink, Eric. "Cartesian Sex: Dutch Anatomists on Genitalia, Lust, and Intercourse (ca. 1660–1680)." In *Les libertains néerlandais/The Dutch Libertines*, edited by Nicole Gengoux, Pierre Girard, and Mogens Laerke, 247–288. Libertinage et philosophie à l'époque classique (xvie–xviiie siècle) 19. Paris: Classiques Garnier, 2022.

Jorink, Eric. "*Modus politicus vivendi*: Nicolaus Steno and the Dutch (Swammerdam, Spinoza and Other Friends), 1660–1664." In Andrault and Lærke, *Steno and the Philosophers*, 13–44.

Jorink, Eric. "The *Myologia* by Saeghemolen and van Horne in Context: Art, Science and Religion at Leiden University, ca. 1660." In *Quatre atlas de myologie de van Horne et Sagemolen: Approche pluridisciplinaire de dessins inédits du Siècle d'or neerlandais; Actes du colloque international des 18 et 19 juin 2021*, edited by Jean-François Vincent and Isabelle Bonnard, 39–64. Paris: Université Paris Cité, 2022.

Jorink, Eric. "'Outside God, There Is Nothing': Swammerdam, Spinoza, and the Janus-Face of the Early Dutch Enlightenment." In *The Early Dutch Enlightenment in the Dutch Republic, 1650–1750*, edited by Wiep van Bunge, 81–107. Leiden: Brill, 2003.

Kaplan, Benjamin J. *Divided by Faith: Religious Conflict and the Practice of Toleration in Early Modern Europe.* Cambridge, MA: Harvard University Press, 2007.

Kardel, Troels. "Nicolaus Steno's New Myology: Rather than Muscle, the Motor Fibre Should Be Called Animal's Organ of Movement." *Nuncius: Annali di storia di scienza* 23 (2008): 37–64.

Kardel, Troels. "Prompters of Steno's Geological Principles: Generation of Stones in Living Beings, Glossopetrae and Molding." In Rosenberg, *The Revolution in Geology from the Renaissance to the Enlightenment*, 127–134.

Kardel, Troels. "Steno on Hydrocephalus: Introduction to Niels Stensen's Letter 'On a Calf with Hydrocephalus,' with a Short Biography." *Journal of the History of the Neurosciences* 2, no. 3 (1993): 171–178.

Kardel, Troels. *Steno on Muscles: Introduction, Texts, Translations.* Transactions, American Philosophical Society 84, pt. 1. Philadelphia: American Philosophical Society, 1994.

Kardel, Troels. "Steno's Myology: The Right Theory at the Wrong Time." In Andrault and Laerke, *Steno and the Philosophers*, 138–173.
Kardel, Troels, and Paul Maquet, eds. *Nicolaus Steno: Biography and Original Papers of a 17th Century Scientist*. 2nd ed. Berlin: Springer, 2018.
Kenny, Neil. *The Uses of Curiosity in Early Modern France and Germany*. Oxford and New York: Oxford University Press, 2004.
Kermit, Hans. *Niels Stensen: The Scientist Who Was Beatified*. Leominster: Gracewing, 2003.
Kessel, Elisja Schulte van. "Gender and Spirit, *pietas et contemptus mundi*: Matron-Patrons in Early Modern Rome." In *Women and Men in Spiritual Culture, XIV–XVII Centuries: A Meeting of South and North*, edited by Elisja Schulte van Kessel, 47–68. The Hague: Netherlands Government Publishing Office, 1986.
Keyvanian, Carla. "Papal Urban Planning and Renewal: Real and Ideal, c. 1471–1667." In *A Companion to Early Modern Rome, 1492–1692*, edited by Pamela M. Jones, Barbara Wisch, and Simon Ditchfield, 305–323. Brill's Companions to European History 17. Leiden: Brill, 2019.
Kircher, Athanasius. *Magnes sive de arte magnetica*. Cologne: Iodocum Kalkoven, 1643.
Klestinec, Cynthia. *Theaters of Anatomy: Students, Teachers, and Traditions of Dissection in Renaissance Venice*. Baltimore, MD: Johns Hopkins University Press, 2011.
Klever, Wim. "Spinoza and van den Enden in Borch's Diary in 1661 and 1662." *Studia Spinozana* 5 (1989): 311–325.
Klever, Wim. "Steno's Statements on Spinoza and Spinozism." *Studia Spinozana* 6 (1990): 303–313.
Koch, Hans-Reinhard, and Konrad R. Koch. "Borri, the Prophet, on the 'Restitutio Humorum' and on Lens Aspiration in the 17th Century/Der Prophet Borri über die 'Restitutio Humorum' und die Linsen-Aspiration im 17. Jahrhundert." *Sudhoffs Archiv* 101, no. 2 (2017): 160–183.
Kuhn, Thomas. "Metaphor in Science." In *Metaphor and Thought*, edited by Andrew Ortony, 533–542. 2nd ed. Cambridge: Cambridge University Press, 1993.
Laerke, Mogens. "Leibniz and Steno, 1675–1680." In Andrault and Laerke, *Steno and the Philosophers*, 63–84.
Larroque, Philippe Tamizey de, ed. *Lettres de Jean Chapelain, de l'Académie Française*. 2 vols. Paris: Imprimerie Nationale, 1880–1883.
Laurenza, Domenico. "Images and Theories: The Study of Fossils in Leonardo, Scilla, and Hooke." *Nuncius: Annali di storia della scienza* 33, no. 3 (2018): 442–463.
Lazzareschi, Eugenio. *Lettere di Nicola Stenone a Lavinia Cenami Arnolfini*. Lucca: Scuola Tip. Artigianelli, 1936.
Lennon, Thomas M. "Pierre-Daniel Huet, Skeptic Critic of Cartesianism and Defender of Religion." In Nadler, Schmaltz, and Antoine-Mahut, *The Oxford Handbook of Descartes and Cartesianism*, 780–790.
Lennon, Thomas M. "Villebressieu, Étienne de." In *The Cambridge Descartes Lexicon*, edited by Lawrence Nolan, 747. Cambridge: Cambridge University Press, 2016.
Levitin, Dmitri. *The Kingdom of Darkness: Bayle, Newton, and the Emancipation of the European Mind from Philosophy*. Cambridge: Cambridge University Press, 2022.
Lewis, Charlton T. *An Elementary Latin Dictionary*. New York: American Book Company, 1890.
Lightman, Bernard, ed. *Rethinking History, Science, and Religion: An Exploration of Conflict and the Complexity Principle*. Pittsburgh: University of Pittsburgh Press, 2019.

Lilti, Antoine. *The World of the Salons: Sociability and Worldliness in Eighteenth-Century Paris*. Oxford and New York: Oxford University Press, 2015.
Lindeboom, G. A. *The Letters of Jan Swammerdam to Melchisedec Thévenot*. Amsterdam: Swets & Zeitlinger, 1975.
Lindeboom, G. A. *Dutch Medical Biography: A Biographical Dictionary of Dutch Physicians and Surgeons, 1475–1795*. Amsterdam: Rodopi, 1984.
Lipen, Martin. *Bibliotheca realis philosophica omnium materium*. Frankfurt: Johannes Frederici, 1682.
Lloyd, G. E. R. *The Revolutions of Wisdom: Studies in the Claims and Practices of Ancient Greek Science*. Berkeley and Los Angeles: University of California Press, 1989.
Long, Pamela O. *Artisan/Practitioners and the Rise of the New Sciences, 1400–1600*. Corvallis, OR: Oregon State University Press, 2011.
Long, Pamela O. *Openness, Secrecy, Authorship: Technical Arts and the Culture of Knowledge from Antiquity to the Renaissance*. Baltimore, MD: Johns Hopkins University Press, 2004.
Luria, Keith. *Sacred Boundaries: Religious Coexistence and Conflict in Early-Modern France*. Washington, DC: Catholic University of America Press, 2005.
Lüthy, Christoph H. "Atomism, Lynceus, and the Fate of Seventeenth-Century Microscopy." *Early Science and Medicine* 1, no. 1 (1996): 1–27.
Lux, David S. *Patronage and Royal Science in Seventeenth-Century France: The Académie de Physique in Caen*. Ithaca, NY: Cornell University Press, 1989.
Lux, David S., and Harold J. Cook. "Closed Circles or Open Networks? Communicating at a Distance During the Scientific Revolution." *History of Science* 36, no. 2 (1998): 179–211.
Luzzini, Francesco. *Il miracolo inutile: Antonio Vallisneri e le scienze della terra in Europa tra XVII e XVIII secolo*. Florence: Leo S. Olschki, 2013.
Lyby, Thorkild, and Ole Peter Grell. "The Consolidation of Lutheranism in Denmark and Norway." In *The Scandinavian Reformation: From Evangelical Movement to Institutionalisation of Reform*, edited by Ole Peter Grell, 114–143. Cambridge: Cambridge University Press, 1995.
Lyell, Charles. *Principles of Geology*. 3 vols. London: John Murray, 1830.
Lyser, Michael. *Culter anatomicus*. Copenhagen: Georg Lamprecht, 1653.
Machamer, Peter. "The Nature of Metaphor and Scientific Description." In Hallyn, *Metaphor and Analogy in the Sciences*, 35–52.
Maclean, Ian. "The Medical Republic of Letters Before the Thirty Years War." *Intellectual History Review* 18, no. 1 (2008): 15–30.
[Magalotti, Lorenzo]. *Saggi di naturali esperienze fatte nell'Accademia del Cimento*. Florence: Giuseppe Cocchini, 1667.
Maillard, Jean-François. "Descartes et l'alchimie: Une tentation conjurée?" In *Aspects de la tradition alchimique au XVIIe siècle: Actes du colloque international de l'Université Reims-Champagne-Ardenne, Reims, 28 et 29 novembre 1996*, edited by Frank Greiner, 95–109. Textes et travaux de Chrysopoeia 4. Paris: Arché, 1998.
Malloch, Archibald. *Finch and Baines: A Seventeenth Century Friendship*. Cambridge: Cambridge University Press, 1917.
Malpighi, Marcello. *The Correspondence of Marcello Malpighi*. Edited by Howard Adelmann. 5 vols. Ithaca, NY: Cornell University Press, 1975.
Malpighi, Marcello. *De pulmonibus observationes anatomicæ*. Bologna: Jo. Baptista Ferroni, 1661.

Malpighi, Marcello. *Opera posthuma*. Amsterdam: Georg Gaillet, 1700.

Maluf, N. S. R. "History of Blood Transfusion." *Journal of the History of Medicine and Allied Sciences* 9, no. 1 (1954): 59–107.

Manni, Domenico Maria. *Vita del letteratissimo Monsig. Niccolò Stenone di Danimarca*. Florence: Giuseppe Vanni, 1775.

Margócsy, Dániel. "Advertising Cadavers in the Republic of Letters: Anatomical Publications in the Early Modern Netherlands." *British Journal for the History of Science* 42, no. 2 (2009): 187–210.

Martin, Craig. "Conjecture, Probabilism, and Provisional Knowledge in Renaissance Meteorology." In "Evidence and Interpretation: Studies on Early Science and Medicine in Honor of John E. Murdoch," [special issue,] *Early Science and Medicine* 14, no. 1/3 (2009): 265–289.

Martin, Craig. *Renaissance Meteorology: Pomponazzi to Descartes*. Baltimore, MD: Johns Hopkins University Press, 2011.

Martinović, Ivica. "Stjepan Gradić On Galileo's Paradox of the Bowl." *Dubrovnik Annals* 1 (1997): 31–69.

Mattes, Johannes. "Mapping the Invisible: Knowledge, Credibility and Visions of Earth in Early Modern Cave Maps." *The British Journal for the History of Science* 55, no. 1 (2022): 53–80.

Mazur, Peter. *Conversion to Catholicism in Early Modern Italy*. New York: Routledge, 2016.

McLean, Paul D. *The Art of the Network: Strategic Interaction and Patronage in Renaissance Florence*. Durham, NC: Duke University Press, 2007.

Meijer, Cornelis. *Traité des moyens de rendre les rivières navigables avec plusieurs desseins de jetée*. Amsterdam: Pierre Mortier, 1696.

Mercati, Michele. *Metallotheca vaticana*. Rome: Johannes Maria Salvioni, 1719.

Meschini, Franco Aurelio. "Davisi, Urbano Giovan Francesco." In DBI, 33:171–173.

Middleton, W. E. Knowles. *The Experimenters: A Study of the Accademia del Cimento*. Baltimore, MD: Johns Hopkins University Press, 1971.

Middleton, William E. Knowles. "Science in Rome, 1675–1700, and the Accademia Fisicomatematica of Giovanni Giustino Ciampini." *British Journal for the History of Science* 8, no. 2 (1975): 138–154.

Miert, Dirk van. *Humanism in an Age of Science: The Amsterdam Athenaeum in the Golden Age, 1632–1704*. Translated by Michiel Welema, with Anthony Ossa-Richardson. Brill's Studies in Intellectual History 179. Leiden and Boston: Brill, 2009.

Miller, Peter N. *Peiresc's Europe: Learning and Virtue in the Seventeenth Century*. New Haven, CT: Yale University Press, 2000.

Miniati, Stefano. *Nicholas Steno's Challenge for Truth: Reconciling Science and Faith*. Milan: FrancoAngeli, 2009.

Mirto, Alfonso. *Alessandro Segni e gli Accademici della Crusca: Carteggio (1659–1696)*. Florence: Accademia della Crusca, 2016.

Mirto, Alfonso. *La biblioteca del Cardinal Leopoldo de' Medici: Catalogo*. Florence: Leo S. Olschki Editore, 1990.

Mirto, Alfonso. "Le lettere di Athanasius Kircher della Biblioteca Nazionale Centrale di Firenze." *Atti e memorie dell'Accademia toscana di scienze e lettere "La Colombaria"* 54 (1989): 129–165.

Mirto, Alfonso. "Lettere di Stefano Gradi ai Fiorentini: Viviani, Dati, Redi, Leopoldo e Cosimo III de' Medici." *Studi secenteschi* 49 (2008): 371–404.

Moe, Harald. "When Steno Brought New Esteem to Glands." In *Nicolaus Steno, 1638–16868: A Re-consideration by Danish Scientists*, edited by J. Poulsen and E. Snorrason, 51–96. Gentofte, Denmark: Nordisk Insulinlaboratorium, 1986.

Moe, Harald, and Finn Bojsen-Møller. "The Fine Structure of the Lateral Nasal Gland (Steno's Gland) of the Rat." *Journal of Ultra Structure Research* 36, no. 1 (1971): 127–148.

Mollerus, Johannes. *Bibliotheca septentrionis eruditi sive syntagma tractatuum de scriptoriibus illius*. Leipzig: Gothofr. Libezeitzzii, 1699.

Montanari, Tomaso. "Gradi, Stefano." In DBI, 58:361–363.

Moreni, Domenico. *Saggio di lettere d'Orazio Rucellai e di testimonianze autorevoli in lode e difesa dell'Accademia della Crusca*. Florence: Magheri, 1826.

Morello, Nicoletta. *La nascita della paleontologia nel Seicento: Colonna, Stenone e Scilla*. Milan: FrancoAngeli, 1979.

Moulin, Daniel de. "Paul Barbette, M.D.: A Seventeenth-Century Amsterdam Author of Best-Selling Textbooks." *Bulletin of the History of Medicine* 59, no. 4 (1985): 506–514.

Murphy, Kathryn. "Robert Burton and the Problems of Polymathy." *Renaissance Studies* 28, no. 2 (2014): 279–297.

Nadeau, Jean-Benoît, and Julie Barlow. *The Story of French*. New York: St. Martin's Press, 2006.

Nadler, Steven. "Spinoza, Descartes, and the 'Stupid Cartesians.'" In Nadler, Schmaltz, and Antoine-Mahut, *The Oxford Handbook of Descartes and Cartesianism*, 659–677.

Nadler, Steven, Tad M. Schmaltz, and Delphine Antoine-Mahut, eds. *The Oxford Handbook of Descartes and Cartesianism*. Oxford and New York: Oxford University Press, 2019.

Nagra, Gurjit, and Marc R. Del Bigio. "Pathology of Pediatric Hydrocephalus." In *Pediatric Hydrocephalus*, edited by Giuseppe Cinalli, M. Memet Ozek, and Christian Sainte-Rose, 1–25. Cham: Springer, 2018.

Newman, William R. *Atoms and Alchemy: Chymistry and the Experimental Origins of the Scientific Revolution*. Chicago: University of Chicago Press, 2006.

Newman, William, and Lawrence Principe. "Alchemy vs. Chymistry: The Etymological Origins of a Historiographic Mistake." *Early Science and Medicine* 3, no. 1 (1998): 32–65.

Nielsen, Henry, Helge Kragh, Kristian Hvidfeldt Nielsen, and Peter Kjaergaard, eds. *Science in Denmark: A Thousand-Year History*. Aarhus: Aarhus University Press, 2008.

Nirenberg, David. *Communities of Violence: Persecution of Minorities in the Middle Ages*. Princeton, NJ: Princeton University Press, 1996.

Noordergraaf, Gerrit J., Johnny T. Ottesen, Wil J. P. M. Kortsmit, Wil H. A. Schilders, Gert J. Scheffer, and A. Noordegraaf. "The Donders Model of the Circulation in Normo- and Pathophysiology." *Cardiovascular Engineering* 6, no. 2 (2006): 53–72.

Nordström, Johan. "Antonio Magliabechi och Nicolaus Steno: Ur Magliabechis Brev till Jacob Gronovius; with a Summary in English, 'Antonio Magliabecchi and Nicolaus Steno: From Magliabecchi's Letters to Jacob Gronovius.'" *Lychnos* 20 (1962): 1–42.

Nordström, Johan. "Swammerdamiana: Excerpts from the Travel Journal of Olaus Borrichius and Two Letters from Swammerdam to Thévenot." *Lychnos* 14 (1954–1955): 21–65.

Nowak, Gisela. *Maria Elisabeth von Rantzau: Ein Leben für Caritas und Einheit im Glauben*. Hildesheim: Bernward, 1984.

Nutton, Vivian. "*Physiologia* from Galen to Jacob Bording." In Horstmanshoff, King, and Zittel, *Blood, Sweat and Tears*, 27–40.

Nutton, Vivian. "The Rise of Medical Humanism: Ferrara, 1464–1555." *Renaissance Studies* 11, no. 1 (1997): 2–19.

Ogilvie, Brian W. *The Science of Describing: Natural History in Renaissance Europe*. Chicago: University of Chicago Press, 2006.

Olden-Jørgensen, Sebastian. "Jesuits, Women, Money or Natural Theology? Nicolas Steno's Conversion to Catholicism in 1667." In Andrault and Laerke, *Steno and the Philosophers*, 45–62.

Olden-Jørgensen, Sebastian. "Nicholas Steno and René Descartes: A Cartesian Perspective on Steno's Scientific Development." In Rosenberg, *The Revolution in Geology from the Renaissance to the Enlightenment*, 149–157.

Omodeo, Pietro. "Lodewijk de Bils' and Tobias Andreae's Cartesian Bodies: Embalmment Experiments, Medical Controversies and Mechanical Philosophy." *Early Science and Medicine* 22, no. 4 (2017): 301–332.

Osler, Margaret J. *Divine Will and the Mechanical Philosophy: Gassendi and Descartes on Contingency and Necessity in the Created World*. Cambridge: Cambridge University Press, 1994.

Park, Katharine. *Secrets of Women: Gender, Generation, and the Origins of Human Dissection*. New York: Zone Books, 2006.

Park, Katharine, and Lorraine Daston. "The Age of the New." In Park and Daston, *Early Modern Science*, 1–17.

Park, Katharine, and Lorraine Daston, eds. *Early Modern Science*. Vol. 3 of *The Cambridge History of Science*, edited by David C. Lindberg and Ronald L. Numbers. Cambridge: Cambridge University Press, 2008.

Parker, Charles H. *Faith on the Margins: Catholics and Catholicism in the Dutch Golden Age*. Cambridge, MA: Harvard University Press, 2008.

Pasnau, Robert. *After Certainty: A History of Our Epistemic Ideals and Illusions*. Oxford and New York: Oxford University Press, 2017.

Partington, J. R. *A History of Chemistry*. 4 vols. London: Macmillan, 1961–1964.

Patin, Guy. *Correspondance complète et autres écrits de Guy Patin*. Edited by Loïc Capron. Paris: Bibliothèque interuniversitaire de santé, 2018. www.biusante.parisdescartes.fr/patin/?do=pg&let=0152&cln=26.

Pecquet, Jean. *Experimenta nova anatomica... [et] dissertatio anatomica de circulatione sanguinis*. Paris: Sebastian & Gabriel Cramoisy, 1651.

Pecquet, Jean. *Experimenta nova anatomica... [et] dissertatio anatomica de circulatione sanguinis*. Paris: Sebastian & Gabriel Cramoisy, 1654.

Pecquet, Jean. " An Extract of a Letter of M. Pecquet to M. Carcavi, Concerning a New Discovery of the Communication of the Ductus Thoracicus with the Emulgent Vein." *Philosophical Transactions* 2, no. 25 (1667): 461–464.

Pender, Stephen. "Examples and Experience: On the Uncertainty of Medicine." *British Journal for the History of Science* 39, no. 1 (2006): 1–28.

Perrault, Charles. *Mémoires de ma vie*. Edited by Paul Bonnefon. Paris: Librairie Renouard, 1909.

Piso, Willem. *De Indiæ utriusque re naturali et medica libri quatuordecim*. Amsterdam: Ludovic et Daniel Elzevir, 1658.

Piso, Willem. *Historia naturalis brasiliæ*. Amsterdam: Franciscus Hack, 1648.
Placcius, Vincent. *De scriptis et scriptoribus anonymis atque pseudonymis*. Hamburg: Christianus Guthius, 1674.
Pomata, Gianna. "Observation Rising: Birth of an Epistemic Genre, ca. 1500–1650." In Daston and Lunbeck, *Histories of Scientific Observation*, 45–80.
Pomata, Gianna. "*Praxis Historialis*: The Uses of *Historia* in Early Modern Medicine." In Pomata and Siraisi, *Historia: Empiricism and Erudition in Early Modern Europe*, 105–146.
Pomata, Gianna. "Sharing Cases: The *Observationes* in Early Modern Medicine." *Early Science and Medicine* 15, no. 3 (2010): 193–236.
Pomata, Gianna, and Nancy G. Siraisi, eds. *Historia: Empiricism and Erudition in Early Modern Europe*. Boston: MIT Press, 2005.
Pomata, Gianna, and Nancy G. Siraisi. "Introduction." In Pomata and Siraisi, *Historia: Empiricism and Erudition in Early Modern Europe*, 1–38.
Poole, William. *The World Makers: Scientists of the Restoration and the Search for the Origins of the Earth*. Oxford: Peter Lang, 2010.
Popkin, Richard H. *Isaac La Peyrère: His Life, Work and Influence*. Leiden and New York: Brill, 1987.
Porter, Roy. "The Scientific Revolution and Universities." In *Universities in Early Modern Europe (1500–1800)*, edited by Hilde De Ridder-Symoens, 531–564. Vol. 2 of *A History of the University in Europe*, ed. Walter Rüegg. Cambridge: Cambridge University Press, 1996.
Poynter, F. N. L. "Nicolaus Steno and the Royal Society of London." In Scherz, *Steno and Brain Research in the Seventeenth Century*, 273–280.
Principe, Lawrence M. *The Secrets of Alchemy*. Chicago: University of Chicago Press, 2013.
Principe, Lawrence M. *The Transmutations of Chymistry: Wilhelm Homberg and the Académie Royale des Sciences*. Chicago: University of Chicago Press, 2020.
Ragland, Evan. "Between Certain Metaphysics and the Senses: Cataloging and Evaluating Cartesian Empiricisms." *Journal of Early Modern History* 3, no. 2 (2014): 119–139.
Ragland, Evan. "Chymistry and Taste in the Seventeenth Century: Franciscus Dele Boë Sylvius as a Chymical Physician Between Galenism and Cartesianism." *Ambix* 59, no. 1 (2012): 1–21.
Ragland, Evan R. "The Contested *Ingenia* of Early Modern Anatomy: Continuities and Conflicts in Medical Training at Leiden University, 1592–1678." In *Ingenuity in the Making: Matter and Technique in Early Modern Europe*, edited by Richard J. Oosterhoff, José Ramón Marcaida, and Alexander Marr, 112–130. Pittsburgh: University of Pittsburgh Press, 2021.
Ragland, Evan R. "Experimental Clinical Medicine and Drug Action in Mid-Seventeenth-Century Leiden." In "Testing Drugs and Trying Cures," special issue, *Bulletin of the History of Medicine* 91, no. 2 (2017): 331–361.
Ragland, Evan R. "Experimenting with Chymical Bodies: Reinier de Graaf's Investigations of the Pancreas." *Early Science and Medicine* 13, no. 6 (2008): 615–664.
Ragland, Evan R. *Making Physicians: Tradition, Teaching, and Trials at Leiden University, 1575–1639*. Clio Medica 106. Leiden: Brill, 2022.
Ragland, Evan R. "Mechanism, the Senses, and Reason: Franciscus Sylvius and Leiden Debates over Anatomical Knowledge After Harvey and Descartes."

In Distelzweig, Goldberg, and Ragland, *Early Modern Medicine and Natural Philosophy*, 173–206.
Räss, Andreas. *Die Convertiten seit der Reformation nach ihrem Leben und aus ihren Schriften dargestellt*. 13 vols. Freiburg: Herder'sche Verlagshandlung, 1866–1880.
Redi, Francesco. *Esperienze intorno alla generazione degl' insetti*. Florence: Stella, 1668.
Redi, Francesco. *Opere*. 9 vols. Milan, 1809–1811.
Redi, Francesco. *Osservazioni intorno agli animali viventi che si trovano negli animali viventi*. Florence, 1684.
Ricci, Michelangelo. *Geometrica exercitatio*. Rome: Nicolaus Angelus Tinassius, 1666.
Righini Bonelli, Maria Luisa. "The Accademia del Cimento and Niels Stensen." In Scherz, *Steno and Brain Research in the Seventeenth Century*, 253–260.
Riva, Michele Augusto, Marta Benedetti, Francesca Vaglienti, Chiara Torre, Gaspare Baggieri, and Giancarlo Cesana. "Guglielmo Riva and the End of Hepatocentrism: A 17th-Century Painting." *Vesalius: Acta Internationalia Historiae Medicinae* 20, no. 2 (2014): 69–72.
Roberts, Michael. "The Swedish Dilemma." In *The Thirty Years' War*, edited by Geoffrey Parker, 140–145. 2nd ed. London and New York: Routledge, 1997.
Rohault, Jacques. *Traité de physique, II. III. & IV. Partie*. Amsterdam: Jaques le Jeune, 1672.
Roos, Anne Marie. *Web of Nature: Martin Lister (1639–1712), the First Arachnologist*. History of Science and Medicine Library 22/Medieval and Early Modern Science 16. Leiden: Brill, 2011.
Rørdam, Holger. *De danske og norske Studenters Deltagelse i Kjøbenhavns Forsvar mod Karl Gustav*. Copenhagen: C. G. Iversen, 1855.
Rosenberg, Gary D. "Introduction: The Revolution in Geology from the Renaissance to the Enlightenment." In Rosenberg, *The Revolution in Geology from the Renaissance to the Enlightenment*, 1–11.
Rosenberg, Gary D. "Preface." In Rosenberg, *The Revolution in Geology from the Renaissance to the Enlightenment*, vii.
Rosenberg, Gary D., ed. *The Revolution in Geology from the Renaissance to the Enlightenment*. Boulder, CO: Geological Society of America, 2009.
Roux, Sophie. "An Empire Divided: French Natural Philosophy (1670–1690)." In *The Mechanization of Natural Philosophy*, edited by Daniel Garber and Sophie Roux, 55–98. Dordrecht: Springer, 2013.
Roux, Sophie. "The Condemnations of Cartesian Natural Philosophy Under Louis XIV (1661–91)." In Nadler, Schmaltz, and Antoine-Mahut, *The Oxford Handbook of Descartes and Cartesianism*, 755–779.
Rudwick, Martin. *The Meaning of Fossils: Episodes in the History of Palaeontology*. London: Macdonald, 1972.
Ruestow, Edward. *The Microscope in the Dutch Republic: The Shaping of Discovery*. Cambridge: Cambridge University Press, 1996.
Santorio, Santorio. *Commentaria in primam fen primi libri canonis Avicennae*. Venice: Jacob Sarcina, 1625.
Santorio, Santorio. *De statica medicina*. Leiden: David Lopes de Haro, 1642.
Scherz, Gustav. "Danmarks Stensen-Manuskript." *Fund og Forskning* (1959): 19–33.
Scherz, Gustav. "Introduction." In *Nicolaus Steno's Lecture on the Anatomy of the Brain*, edited by Gustav Scherz, 61–103. Copenhagen: Nyt Nordisk Forlag Arnold Busck, 1965.

Scherz, Gustav. "Niels Stensen's First Dissertation." *Journal of the History of Medicine and Allied Sciences* 15, no. 3 (1960): 247–264.
Scherz, Gustav. *Niels Stensen: Eine Biographie*. 2 vols. Leipzig: St. Benno-Verlag, 1987. [Scherz's first volume was translated into English in Kardel and Maquet, *Nicolaus Steno: Biography and Original Papers*, 1–410.]
Scherz, Gustav. "Niels Stensen und Galileo Galilei." In *Saggi su Galileo Galilei*, edited by Carlo Maccagni, 731–779. Pubblicazioni del Comitato nazionale per le manifestazioni celebrative del IV centenario della nascita di Galileo Galilei 3/2. Florence: Barbèra, 1972.
Scherz, Gustav. "Supplement: 2. Two Poems by Nicolaus Steno." In Scherz, *Nicolaus Steno and His Indice*, 290.
Scherz, Gustav, ed. *Nicolaus Steno and His Indice*. Acta Historica Scientiarum Naturalium et Medicinalium, Edit. Bibliotheca Universitatis Hauniensis 15. Copenhagen: Munksgaard, 1958.
Scherz, Gustav, ed. *Steno and Brain Research in the Seventeenth Century*. Analecta Medico-Historica 3. Oxford: Pergamon Press, 1968.
Schmaltz, Tad M. "The Early Dutch Reception of *L'Homme*." In *Descartes' "Treatise on Man" and Its Reception*, edited by Stephen Gaukroger and Delphine Antoine-Mahut, 71–90. Cham: Springer, 2016.
Schmaltz, Tad M. *Early Modern Cartesianisms: Dutch and French Constructions*. Oxford and New York: Oxford University Press, 2017.
Schmitt, Charles B. "Towards a Reassessment of Renaissance Aristotelianism." *History of Science* 11, no. 3 (1973): 159–193.
Schott, Gaspar. *Mechanica hydraulico-pneumatica*. Frankfurt: Henricus Pigrin, 1657.
Schylander, Cornelius. *Practica chirurgiae brevis et facilis*. Antwerp: Antonius Tilenius, 1577.
Serjeantson, Richard. "Proof and Persuasion." In Park and Daston, *Early Modern Science*, 132–178.
Seymour, Roger S., and Kjell Johansen. "Blood Flow Uphill and Downhill: Does a Siphon Facilitate Circulation Above the Heart?" *Comparative Biochemistry and Physiology Part A: Physiology* 88, no. 2 (1987): 167–170.
Shackelford, Jole. "Documenting the Factual and the Artifactual: Ole Worm and Public Knowledge." *Endeavour* 23, no. 2 (1999): 65–71.
Shackelford, Jole. "To Be or Not to Be a Paracelsian: Something Spagyric in the State of Denmark." In *Paracelsian Moments: Science, Medicine, and Astrology in Early Modern Europe*, ed. Gerhild Scholz Williams and Charles D. Gunnoe, Jr., 35–69. Sixteenth Century Essays and Studies 64. Kirksville, MO: Truman State University Press, 2002.
Shapin, Steven. *A Social History of Truth: Civility and Science in Seventeenth-Century England*. Chicago: University of Chicago Press, 1994.
Sibbald, Robert. *The Autobiography of Robert Sibbald, Knt., M.D., to Which Is Appended Some Account of His MSS*. Edited by James Maidment. Edinburgh: Thomas Stevenson and John Wilson, 1833.
Siegel, Rudolph E. *Galen's System of Physiology and Medicine: An Analysis of His Doctrines and Observations on Blood Flow, Respirations, Humors and Internal Diseases*. Basel: S. Karger, 1968.
Simon, Gérard. "Analogies and Metaphors in Kepler." In Hallyn, *Metaphor and Analogy in the Sciences*, 71–82.

Siraisi, Nancy G. "*Historia, actio, utilitas*: Fabrici e le scienze della vita nel Cinquecento." In *Il teatro dei corpi: Le pitture colorate d'anatomia di Girolamo Fabrici d'Acquapendente*, ed. Maurizio Rippa Bonati and José Pardo-Tomás, 63–73. Milan: Mediamed Edizioni Scientifiche Srl., 2004.

Skavlem, John H. "The Scientific Life of Thomas Bartholin." *Annals of Medical History* 3, no. 1 (1921): 67–81.

Sleigh, Charlotte. "Jan Swammerdam's Frogs." *Notes and Records: The Royal Society Journal of the History of Science* 66 (2012): 373–392.

Smith, C. U. M., Eugenio Frixione, Stanley Finger, and William Clower. *The Animal Spirit Doctrine and the Origins of Neurophysiology*. Oxford and New York: Oxford University Press, 2012.

Smith, Pamela H. *The Body of the Artisan: Art and Experience in the Scientific Revolution*. Chicago: University of Chicago Press, 2004.

Snorrason, Egill. *Niels Stensen: En students notater fra 1659*. Copenhagen: Mölnlycke, 1966.

Snorrason, Egill. "The Studies of Nicolaus Steno 1659 in Copenhagen Libraries." In Scherz, *Steno and Brain Research in the Seventeenth Century*, 69–93.

Sobiech, Frank. *Ethos, Bioethics, and Sexual Ethics in Work and Reception of the Anatomist Niels Stensen (1638–1686): Circulation of Love*. Dordrecht: Springer, 2016.

Sobiech, Franz. *Herz, Gott, Kreuz: Die Spiritualität des Anatomen, Geologen und Bischofs Dr. med. Niels Stensen (1638–86)*. Westfalia Sacra 13. Munster: Aschendorff Verlag, 2004.

Sorbière, Samuel. *Discours du sieur de Sorbiere, sur sa conversion à l'Église Catholique*. Paris: Antoine Vitré, 1654.

Sorbière, Samuel. *Relations, lettres et discours*. Paris: François Clousier, 1660.

Spinoza, Baruch. *Renati Descartes principiorum philosophiae . . . more geometrico demonstrate*. Amsterdam: Johannes Riewerts, 1663.

Spruit, Leen, and Pina Totaro. *The Vatican Manuscript of Spinoza's "Ethica."* Leiden: Brill, 2011.

Steno, Nicolaus. *Apologiæ prodromus, quo demonstratur, judicem Blasianum et rei anatomicae imperitum esse, et affectuum suorum servum*. Leiden: Jacob Chouet, 1663.

Steno, Nicolaus. *Chaos: Niels Stensen's Chaos-Manuscript, Copenhagen, 1659*. Complete ed. with introduction, notes and commentary by August Ziggelaar, SJ. Acta Historica Scientiarum Naturalium et Medicinalium 44. Copenhagen: Danish National Library of Science and Medicine, 1997.

Steno, Nicolaus. *Defensio et plenior elucidatio epistolae de propria conversione*. Hannover, 1680.

Steno, Nicolaus. *Defensio et plenior elucidatio scrutinii reformatorum*. Hannover: Wolfgang Schwendimann, 1679.

Steno, Nicolaus. *De musculis et glandulis observationum specimen*. Copenhagen: Matthia Godicchenius, 1664.

Steno, Nicolaus. *De solido intra solidum naturaliter contento dissertationis prodromus*. Florence: Stella, 1669.

Steno, Nicolaus. *Discours sur l'anatomie du cerveau*. Paris: Robert de Ninville, 1669.

Steno, Nicolaus. *Disputatio anatomica de glandulis oris*. Leiden: Johannes Elzevier, 1661.

Steno, Nicolaus. *Disputatio II de mediis nonnullis, genesin disputationis concernentibus . . . praeside Christiano Schioldborg, defendet Nicolaus Stenonis*. Copenhagen: Petrus Morsingi, 1657. Copenhagen, Royal Library of Denmark, 14,–143 4° 01108.

Steno, Nicolaus. *Disputatio physica de thermis.* Amsterdam: Joannem Revasteinium, 1660.
Steno, Nicolaus. *Elementorum myologiae specimen.* Florence: Stella, 1667.
Steno, Nicolaus. *Epistola de propria conversione.* Florence: Johannes Gugliantini, 1677.
Steno, Nicolaus. *Niels Stensen: A Danish Student in His Chaos-Manuscript 1659.* Edited by H. D. Schepelern. Copenhagen: University Library, 1987.
Steno, Nicolaus. *Observationes anatomicae quibus varia oris, oculorum, et narium vasa describuntur, novique salivae, lacrymarum et muci fontes deteguntur.* Leiden: Jacobus Chouët, 1662.
Steno, Nicolaus. "Observationes anatomicæ spectantes ova viviparorum." In Bartholin, *Acta medica et philosophica Hafniensia*, 2:210–218.
Steno, Nicolaus. *Occasio sermonum de religione.* Hannover: Wolfgang Schwendimann, 1678.
Steno, Nicolaus. "Ova viviparorum spectantes observationes aliæ." In Bartholin, *Acta medica et philosophica Hafniensia*, 2:219–232.
Steno, Nicolaus. "Prooemium demonstrationum anatomicarum in Theatro Hafniensi anni 1673." In Bartholin, *Acta medica et philosophica Hafniensia* 2:359–366.
Stensen, Niels. *Discours sur l'anatomie du cerveau.* Edited by Raphaële Andrault. Paris: Éditions Classiques Garnier, 2009.
Stevin, Simon. *De Beghinselen der Weeghconst.* Leiden: Christoffel Plantijn, 1586.
Stevin, Simon. *Les oeuvres mathématiques.* Leiden: Bonaventure and Abraham Elsevier, 1634.
Stolberg, Michael. "Empiricism in Sixteenth-Century Medical Practice: The Notebooks of Georg Handsch." *Early Science and Medicine* 18, no. 6 (2013): 487–516.
Stolberg, Michael. *Gabrielle Falloppia, 1522/23–1562: The Life and Work of a Renaissance Anatomist.* Abingdon: Routledge, 2023.
Stolberg, Michael. "Sweat: Learned Concepts and Popular Perceptions, 1500–1800." In Horstmanshoff, King, and Zittel, *Blood, Sweat and Tears*, 503–522.
Strocchia, Sharon. *Forgotten Healers: Women and the Pursuit of Health in Late Renaissance Italy.* Cambridge, MA: Harvard University Press, 2019.
Stroup, Alice. *A Company of Scientists: Botany, Patronage, and Community at the Seventeenth-Century Parisian Royal Academy of Sciences.* Berkeley and Los Angeles: University of California Press, 1990.
Struik, Dirk J. *The Land of Stevin and Huygens: A Sketch of Science and Technology in the Dutch Republic During the Golden Century.* Studies in the History of Modern Sciences 7. Dordrecht and Boston: D. Reidel, 1981.
Sturdy, David J. "The Accademia del Cimento and the Académie Royale des Sciences." In Beretta, Clericuzio, and Principe, *The Accademia del Cimento and Its European Context*, 181–194.
Suy, Raphael, Sarah Thomis, and Inge Fourneau. "The Discovery of the Lymphatics in the Seventeenth Century. Part III: The Dethroning of the Liver." *Acta Chirurgica Belgica* 116, no. 6 (2016): 360–397.
Suy, Raphael, Sarah Thomis, and Inge Fourneau. "The Discovery of the Lymphatic System in the Seventeenth Century. Part IV: The Controversy." *Acta Chirurgica Belgica* 117, no. 4 (2017): 270–278.
Swammerdam, Jan. *The Book of Nature, or The History of Insects.* London: C. G. Seyffert, 1758.

Swammerdam, Jan. *Bybel der Natuure, of Historie der Insecten/Biblia naturæ, sive Historia insectorum*. Leiden: Isaak Severinus, Boudewyn van der Aa, and Pieter van der Aa, 1738.

Swammerdam, Jan. *Miraculum naturæ, sive uteri muliebris fabrica*. Leiden: Severinus Matthæi, 1672.

Swammerdam, Jan. *Tractatus physico-anatomico-medicus de respiratione usuque pulmonum*. Leiden: Daniel, Abraham, and Adrian à Gaasbeeck, 1667.

Sylvius, Franciscus de Le Boe. *Disputationum medicarum pars prima*. Amsterdam: Johannes van den Bergh, 1663.

Sylvius, Franciscus de Le Boe. *Opera medica, hoc est, disputationum medicarum decas*. Geneva: Samuel de Tournes, 1681.

Taton, René. *Les origines de l'Académie royale des sciences*. Paris: Palais de la découverte, 1966.

Thévenot, Melchisedec. *Bibliotheca thevenotiana, sive catalogus impressorum et manuscriptorum librorum bibliothecae viri clarissimi D. Melchisedecis Thevenot*. Paris, 1694.

Thomson, Keith. *Fossils: A Very Short Introduction*. Oxford and New York: Oxford University Press, 2005.

Thoren, Victor E. *The Lord of Uraniborg: A Biography of Tycho Brahe*. Cambridge: Cambridge University Press, 1990.

Tolmer, Léon. "Vingt-deux lettres inédites d'André Graindorge à P.-D. Huet." *Mémoires de l'Académie nationale* 10 (1942): 245–337.

Totaro, Pina. "Steno in Italy: From Florence to Rome." In Andrault and Laerke, *Steno and the Philosophers*, 270–287.

Tutino, Stefania. *Uncertainty in Post-Reformation Catholicism: A History of Probabilism*. Oxford and New York: Oxford University Press, 2018.

Untzer, Matthia. *Anatomia mercurii spagyria*. Halle, 1620.

Vai, Gian Battista. "The Scientific Revolution and Nicholas Steno's Twofold Conversion." In Rosenberg, *The Revolution in Geology from the Renaissance to the Enlightenment*, 187–208.

van Horne, Johannes. *Mikrokosmos seu brevis manuductio ad historiam corporis humani*. Leiden: Jacob Chouët, 1662.

van Horne, Johannes. *Suarum circa partes generationis . . . observationum prodromus*. Leiden, 1668.

Vanpaemel, Geert. "Jesuit Science in the Spanish Netherlands." In *Jesuit Science and the Republic of Letters*, edited by Mordechai Feingold, 389–432. Cambridge, MA: MIT Press, 2003.

van Schooten, Frans. *Principia matheseos universalis seu introductio ad geometriæ methodum Renati des Cartes edita ab Er. Bartholino, Casp. Fil*. Leiden: Elsevier, 1651.

Varen, Bernhard. *Geographia generalis*. Amsterdam, 1650.

Veder, W. "Witsen (Mr. Nicolaas)." In *Nieuw Nederlandisch Biographisch Woordenboek*. 10 vols., 4:cols. 1473–1479. Leiden: A. W. Sijthoff, 1911–1937.

Vesling, Johan. *Syntagma anatomicum*. 2nd ed. Amsterdam, 1666.

Vigilante, Magda. "Dati, Carlo Roberto." In DBI, 33:24–28 .

Visser, R. P. W. "Theorie en Praktijk van Swammerdams Wetenschappelijke Methode in Zijn Entomologie." *Tijdschrift voor de Geschiedenis der Geneeskunde, Natuurwetenschappen, Wiskunde en Techniek* 4 (1981): 63–73.

Vitoria, María Ángeles, and Francisco Javier Insa Gómez, eds. *Scienza, filosofia e religione nell'opera di Niels Steensen: Atti della giornata di studio su Niccolò Stenone, Pontificia Università della Santa Croce, Roma, 7 maggio 2019*. Florence: Pagnini Editore, 2020.

Waddell, Mark A. *Jesuit Science and the End of Nature's Secrets*. Burlington, VT: Ashgate, 2015.

Waddell, Mark A. "The World, as It Might Be: Iconography and Probabilism in the *Mundus subterraneus* of Athanasius Kircher." *Centaurus: An International Journal of the History of Science and Its Cultural Aspects* 48, no. 1 (2006): 3–22.

Wahl, Charlotte. "Between Cosmopolitanism and Nationalism: The Role of Expatriates in the Dissemination of Leibniz's Differential Calculus." *Almagest: International Journal for the History of Scientific Ideas* 5, no. 2 (2014): 40–68.

Wallace, William A. *Causality and Scientific Explanation*. 2 vols. Ann Arbor: University of Michigan Press, 1972–1974.

Wallace, William A. *The Modeling of Nature: The Philosophy of Science and the Philosophy of Nature in Synthesis*. Washington, DC: Catholic University of America Press, 1996.

Warntz, William. "Newton, the Newtonians, and the *Geographia Generalis Varenii*." *Annals of the Association of American Geographers* 79, no. 2 (1989): 165–191.

Wear, Andrew. "William Harvey and the 'Way of the Anatomists.'" *History of Science* 21, no. 3 (1983): 223–249.

Westfall, Richard S. *The Construction of Modern Science: Mechanisms and Mechanics*. Cambridge: Cambridge University Press, 1971.

Wharton, Thomas. *Adenographia, sive glandularum totius corporis descriptio*. London: J. G., 1656.

Wharton, Thomas. *Thomas Wharton's Adenographia*. Translated by Stephen Freer. With a historical introduction by Andrew Cunningham. Oxford and New York: Oxford University Press, 1996.

Wilson, Leonard G. "The Transformation of Ancient Concepts of Respiration in the Seventeenth Century." *Isis* 51, no. 2 (1960): 161–172.

Witsen, Nicolaus. *Aeloude en Hedendaegsche Scheeps-Bouw en Bestier*. Amsterdam: Commelijn & Broer and Jan Appelaer, 1671.

Worm, Ole. *Museum wormianum, seu historia rerum rarioum*. Leiden: Johann Elsevier, 1655.

Wower, Johann von. *De polymathia tractatio: Integri operis de studiis veterum*. Hamburg: Froben, 1603.

Yamada, Toshihiro. "Kircher and Steno on the 'Geocosm,' with a Reassessment of the Role of Gassendi's Works." In *The Origins of Geology in Italy*, ed. Gian Battista Vai, W. Glen Caldwell, and E. Caldwell, 65–80. Boulder, CO: Geological Society of America, 2006.

Yeo, Richard. *Notebooks, English Virtuosi, and Early Modern Science*. Chicago: University of Chicago Press, 2014.

Yeo, Richard. "Thinking with Excerpts: John Locke (1632–1704) and His Notebook." *Berichte zur Wissenschafts-Geschichte* 43, no. 2 (2020): 180–202.

Zarri, Gabriella. *Uomini e donne nella direzione spirituale (secc. XIII–XVI)*. Spoleto: Fondazione CISAM, 2016.

Index

Page numbers followed by "f" and "t" refer to figures and tables, respectively.

Académie de Physique de Caen, 135, 162; influence of Steno on, 136, 143–46. *See also* Graindorge, André
Académie Française: Chapelain's membership in, 138, 160
Académie Royale des Sciences: and blood transfusions, 49; foundation and significance of, 2, 8n19, 16, 135, 138, 162, 269; influence of Accademia del Cimento on, 174; members, 160, 162–63; prehistory, 135, 138; Steno's potential membership, 162–63
Academy of Queen Christina, 276
Accademia del Cimento: dissolution of, 204; experiments at, 209, 231; foundation of, 166; goals of, 167; influence on Steno's work, 209–10, 231, 239; and Medici court, 208–10, 231; publication of *Saggi*, 167, 202–3; scientific activities, 167–68, 203–5; Steno's relationship with, 167–68, 202–5
Accademia della Crusca, 200
accuracy. *See* certainty
Acquapendente, Hieronymus Fabricius ab. *See* Fabricius ab Acquapendente, Hieronymus
Albert the Great, 227
Aldrovandi, Ulisse, 228
Alexander VII, Pope, 184, 250

All Souls, Feast of, 240, 261, 267, 276
Amsterdam: Blasius's work in, 63, 65; disputes, 68–70; intellectual environment, 95, 108, 124, 127; shipbuilding culture, 127; Spinoza's activities in, 8, 95, 108, 127, 245, 247–48; Steno and Swammerdam's meeting in, 147; visits to chymical workshops, 128–29.
Amsterdam Athenæum, 65, 70–71, 73, 91–92
analogies: between body and Earth, 231–34; chymical, 222–23, 232; in Earth history, 216–19, 231–34; macrocosm and microcosm, 225, 232; mechanical, 12, 15, 99–100, 106, 119, 135, 139, 150–52, 154–55, 212, 214, 222, 225, 232; reasoning with, 12, 99, 106, 135, 151; in scientific explanation, 225–26, 232–33; use in anatomy, 212–16, 222–26, 232–33
anatomical theaters, 46, 48, 84, 143, 280–81
anatomists: authorship and academic practices, 69–70, 73; controversy and disputes, 63, 64, 84, 87; interdisciplinary approaches, 15–16, 23, 46, 49, 56, 62, 95–96, 119, 135, 212, 214, 216, 218, 228; and mathematicians, 23, 26, 31, 56–60, 134, 137, 151, 160; observations and methods, 67, 75, 76, 145, 147, 154; Republic of Letters, 68, 91

anatomy: Aristotelian influence on, 77–78, 212; brain, 135, 142, 144, 156, 159; comparative method in, 212–16, 218; disciplinary breadth, 12–16, 76, 94–96, 134, 166–68; Galenic, 44–45, 47, 78, 105, 119–21, 123, 150, 155, 170, 175; *historia anatomica*, 77–82, 92; mathematization of, 12–16, 96–98, 104–5, 123–25, 165, 192–93, 271, 273; modern, 4, 63, 94, 152n100, 153n103, 175–78, 193–97, 205; post-Harveian, summary of, 44–47, 147, 150, 152; Steno's studies, 48–51, 56–60, 134–65. *See also* animals; dissections; glands; observations; vivisections

animal spirits: skepticism by Steno, 156, 192, 223–24, 257, 260; Swammerdam's views on, 154–57; theory of, 223–24; traditional theories of, 155. *See also* vital spirits

animals: calves, 76–77, 97, 100, 144, 279; chickens, 155; cows, 77n80, 97; dogs, 48–49, 67, 111, 114, 140, 142, 144, 147, 152, 154, 156; humans (cadavers), 48n151, 67, 77n80, 97, 152; lambs, 77n80; lions, 48n151; lobsters, 208; mammals, 214; owls, 48n151; rabbits, 77n80; rays, 106–8, 214; sharks, 207–9, 210–12, 214, 216, 218, 227–28, 217f, 220f; sheep, 65; soul of, 95, 112, 119–22; swans, 111; torment of, 120

Anna de' Medici, 205
Annalena convent, 265
Antwerp, 131
Apollonius, 125, 172
Archimedes, 198
Arezzo, 206
Aristotelianism: challenges to, 168, 178
Aristotle, 39, 41, 122; comparative method of, 212; *Meteorology*, 227, 229; use of analogies, 225–26
arithmetic, 24, 25n29, 30, 96
Arnolfini, Lavinia Cenami, 18n55, 276; influence on Steno's conversion, 253–54, 257, 261–62; spiritual guidance of Steno, 262, 266
artisanal epistemology, 271

artisanal skills: and dissections, 16, 48, 62, 145; and intellectual studies, 51. *See also* artisanal workshops

artisanal workshops, 16, 20, 26, 271, 272; chymical, 128; of glassmaking, 33; of goldsmithing, 20–21, 26, 31, 48; influence on Steno's mechanics, 126–27; and intellectual studies, 27; and lenses, 26, 61, 126; and measurements, 62; shipyards, 127; in Steno's family, 26, 48. *See also* goldsmiths

Aselli, Gaspare, 16, 45–46, 80; lacteal vessels, 78, 82

atomism, 12, 35, 61, 109. *See also* corpuscularianism

atoms, 12, 40, 50, 61, 102, 231–33. *See also* corpuscularianism

Augustine, Saint, 258
Auzout, Adrien, 58, 60, 141n40, 199
Avianus, Wilhelm, 26
Avicenna, 227

Bacon, Francis, 37n90, 61
Bandini, Angelo Maria, 271
Bang, Thomas, 37n90
Barbette, Paul, 69, 74, 84, 91
Bartholin, Caspar the Elder, 22, 24, 32, 44, 46
Bartholin, Erasmus, 24, 124n162, 172
Bartholin, Thomas, 31–32, 37n91; *Acta medica et philosophica Hafniensia*, 205, 274, 280; anatomical contributions, 44–49, 78n86, 82, 94–95, 103–4, 169, 170, 175, 205; and Blasius, 69–70, 83; *Cista medica Hafniensia*, 172; collaboration with Lyser, 72–74; *De libris legendis*, 34; *De peregrinatione medica*, 32; influence on Steno, 169, 170, 175, 205; letters to Steno, 64–65, 94–95; praises Steno, 75, 84, 91; rejection of Cartesian anatomy, 110; role in Republic of Letters, 68, 74, 84, 86–87; son of Caspar, 22, 32
Bartholin family, 32, 170
Basel, 74
Basilica di San Lorenzo, 3

Bek-Thomsen, Jakob, 209–10
Bellini, Lorenzo, 171, 175–76, 186, 276
Bernier, François, 158–59
Bernini, Gian Lorenzo, 182
Bertoloni Meli, Domenico, 13
Bils, Lodewijk de, 69, 76n72, 132; and lymphatics, 114–17
Blair, Ann, 33
Blasius, Gerard, 95, 122, 147, 171, 216, 275; claim of discovery, 68–70, 73; early interactions with Steno, 63–65; *Medicina generalis*, 68–69, 72–73, 83, 87; meetings in Amsterdam, 147; responses and controversy, 75, 83–84, 87–91; Steno's refutation of, 159
Blasius, Johannes, 69–70
Blasius controversy, 63–65, 68–70, 73–75, 83–84, 87–91; impact on Steno's career, 64, 92
blood, 45–51, 56–58, 92, 116–17, 147; bloodletting, 60; experiments on, 147; flow in vena cava, 150, 152; and gland function, 97, 100; and heart function, 103–4; mechanical explanation of, 152; and muscle studies, 102–5; quantification of flow, 102, 123; and tear production, 102
blood circulation: chymical analogies for, 222–23; discovery and reception, 45–46, 49, 57n190, 102; and glands, 102–3; Harvey's discovery, 102; *historia*, 78, 82; mechanical explanations, 56–57, 60, 150–52; and observations, 76; research by Steno, 135, 150–52, 175, 177, 203, 219; Steno's views, 102–3; theories of Harvey, 147, 150. *See also* respiration
blood transfusions, 49–50
Bodenhausen, Rudolf Christian von, 270
Boerhaave, Herman, 158
Bologna, 169, 171, 182
Bona, Giovanni, 198, 251
Borch, Ole, 37–39, 62, 245–46, 226, 231; friendship with Steno, 63, 67–68, 95, 111, 117, 120, 127–28, 169–71; intellectual exchanges, 111, 120, 127–28, 137–40, 154; journal of, 140; microscope of, 28, 29f; in Paris with Steno, 137, 139–40, 142; Steno's friend, 63, 67–68, 137–40, 154; travel diary, 111–12, 127–28; travels in Italy, 169–71; visits to workshops, 127–28
Borel, Pierre, 21, 76; and Descartes, 61; influence on Steno, 16, 134; in Montpellier, 165; *observationes*, 39, 51–54, 55–56. *See also observationes*
Borelli, Giovanni Alfonso, 153; *De motu animalium*, 14, 166; departure from Accademia del Cimento, 204; influence on Steno, 125; initial skepticism of Steno, 171; later acknowledgment, 198; and Malpighi, 126; mathematician, 171, 175, 184–86, 188–89, 198; *Theorica mediceorum planetarum*, 196
Borelly, Jacques, 232
Borri, Giuseppe, 245
Boulliau, Ismael, 173
Bourdelot, Pierre, 134, 143, 155, 170; academy of, 137–40, 158, 161; Steno's attendance at, 139
Boyle, Robert, 199, 269
Brahe, Tycho, 31–32, 39
brain: anatomy of, 135, 142, 144, 156, 159; Descartes's work on, 219; Steno's dissection of, 219–21, 221f; Steno's research on, 135, 156, 159; Willis's work on, 219. *See also* anatomy
Briggs, Henry, 26
broadsheets, 87, 88f, 91
Bruce, Robert, 165
Brunswick court, 6

cabinets of curiosities, 7, 65, 226
Caen. *See* Académie de Physique de Caen
calf, large-headed, 279
Calvin, John, 244
Calvinism: in Denmark, 258; in Netherlands, 245–49
Cambridge, 14
Cartesianism, 24, 51; criticism by Steno, 61–62, 159–60, 247–48; influence in Paris, 160–61; influence on Steno, 246–47; Leiden anatomists' rejection

Cartesianism (*Cont.*)
of, 110–12; materialism, 95, 109–10, 120–21; opposition to, 95, 109–12, 120–21, 159–61; rejection by Roberval, 160; Steno's rejection of, 109–12, 120–21, 159–60, 168; Swammerdam's views on, 157, 113. *See also* Descartes, René

Casserius, Julius, 76

Cassini, Gian Domenico, 162

Cathedral School of Copenhagen, 21, 31, 34, 37, 41, 64–65, 140, 243; and languages, 22, 256, 276; and mathematics, 23, 41, 94, 124, 273, 275. *See also* humanist tradition; liberal arts

Catholic Church: censorship by, 228–29, 239; conversion efforts, 249–50, 262; doctrine of papal infallibility, 241, 258; Eucharistic beliefs, 255–56; full communion with, 18; magisterial authority, 258–59; minority communities, 2, 130; patronage of science, 250–51; role in Steno's conversion, 241–42, 257–60; Steno's conversion to, 240. *See also* papal infallibility

Catholicism, 109, 129–31, 270, 276–77; in Denmark, 244–45; in the Netherlands, 245–48. *See also* Catholic Church

Centuriatores Magdeburgenses, 256. *See also* Lutheranism

certainty, search for: in anatomy, 64, 92, 95, 97, 117–19, 144, 159, 212, 218–22, 225, 226, 227, 228, 230, 231; definition of, 11; in Earth history, 226–31, 239; and focus, 38–44; and interdisciplinary methods, 136, 222, 230–31, 239; and mathematics, 24–25, 28, 30, 41–43, 104–5, 118–19, 222, 224, 225; and modesty, 92; and observations, 21, 64, 83, 117–19, 144, 210; and philosophical detachment, 231, 239; in religion, 18, 240, 242–43, 251, 255–61, 267–68; as research motivation, 9, 17, 64, 67, 76, 210; Steno's probabilistic approach, 42, 118–19, 186

Chapelain, Jean, 134, 136, 138, 140, 158–64; awareness of Steno's book, 199; role in Académie Française, 136, 138; role in Académie des Sciences, 138, 162, 173, 199, 201

"Chaos" manuscript, 19–62, 66, 85, 212, 269, 278; name of, 35; physical description, 19, 38

Christina, Queen of Sweden, 182, 249–50; academy of, 276

chyle, 45, 47, 56–58, 115, 117, 153, 156, 233

chymistry, 35, 37, 39, 44, 46, 82, 93, 271, 275; in "Chaos" manuscript, 20–21, 51; chemical reactions, 12; and colors, 27, 116–17; definition of, 1n2; epistemic role, 27, 177, 202; and glands, 98–99, 116–17; and measurements, 28; and medicine, 56; in scientific culture, 147, 154; and Steno on hot springs, 70–71

Cicero, 92

circulation of blood. *See* blood circulation

civility, 9, 90–91. *See also* virtues

Clerselier, Claude, 160

Colbert, Jean-Baptiste, 138, 158, 162–63; patronage of *Journal des sçavans*, 140

collaboration: in academic disputations, 70; with Borelli on anatomical research, 171, 175, 184–86, 188–89, 198; with Dati on scientific exchanges, 172, 176, 200, 204; in dissections, 67; and friendships, 8; impact on Steno's research, 168, 170, 199–200; intellectual, 95, 124–25; and interdisciplinarity, 12; with Magalotti on dissections and observations, 168, 176–78, 200; with mathematicians, 124–25; with Redi on anatomical observations, 169–70, 176, 200, 205; in research, 72; role in scientific networks, 172–74; scientific, 7–9, 16, 18, 135, 147–54; with Viviani on manuscript editing, 194–96

Collège de Clermont, 160

Collège Royal, 160

Cologne, 249

colors, 40, 116–17; epistemic role of, 27, 116–17; Steno's experiments on, 177–78
Commenius, Jan, 130
commonplace books, 20, 33–34, 39, 52f, 53f, 55
comparative method: in anatomy, 212–16, 218; in Earth history, 216–19, 234
Copenhagen, 24, 32–33, 35, 208, 212, 226, 273, 276, 280; anatomical research in, 205–6; departure from, 63, 65; dissections in, 47–49, 106–8; and goldsmithing, 21, 27; libraries in, 37; Lutheranism in, 241, 245; and religion, 21; siege of, 19, 27, 31, 48; Steno's birthplace, 2; Steno's education in, 256; Steno's origins in, 143, 169–70, 172, 205–6; Steno's return to, 5f, 7, 16, 106, 132; Steno's studies in, 2, 4, 5, 16, 134, 140; Steno's upbringing in, 241, 245. *See also* Cathedral School of Copenhagen; Denmark; University of Copenhagen
Copenhagen Cathedral School. *See* Cathedral School of Copenhagen
Corpus Christi, Feast of, 255, 258
Cornelio, Tommaso, 182
corpuscularianism, 35, 42, 51, 231. *See also* atomism
Cosimo III de' Medici, 3, 174–75
Council of Trent, 255
craftsmanship skills. *See* artisanal skills
creativity, 4
Croese, Gerard, 87n127
Croone, William, 164n170, 165, 177–78, 208
cross-sections: in anatomy, 190n125, 219, 221f, 223–25; in geology, 219, 220f, 235–38
curiosity, 18n55. *See also* cabinets of curiosities

d'Aguillon, François, 131
Dal Prete, Ivano, 227
Darwin, Charles, 239
Dati, Carlo, 208, 216, 227; contributions to Accademia del Cimento, 204; friendship with Steno, 200; and *Metallotheca Vaticana*, 216–18; scholar, 172, 176, 200, 204; and scientific exchanges, 172, 176, 200, 204; and Steno's work, 208, 216–18, 227
de Bils, Lodewijk, 69, 76, 114–17, 132
de Graaf, Regnier, 130–31, 156, 164, 214, 246, 276
de Keyser, Ludovicus, 87n127
de Raey, Johannes, 112–13, 120
de Saint-Vincent, Grégoire, 131
de Sallo, Denis, 140–42
degli Angeli, Stefano, 198
Denmark, 65, 84, 130; artisanal workshops, 8, 33; censorship in, 245; humanist tradition, 22; Lutheranism in, 44, 241, 245, 258; science golden age in, 31; and Steno, 2, 109, 132, 143, 170, 180, 203, 266; Steno's family in, 263–64. *See also* Copenhagen
Descartes, René, 24, 219; anatomical ideas, 109–12, 159; *De homine*, 112, 155, 223, 269; influence on Steno, 24, 50, 56, 61, 109–10, 134–35, 246–47; *L'Homme*, 156, 159; mechanical philosophy, 50, 61, 135, 151; opposition to, 109–12, 120–21, 159–61; *Principia philosophiæ*, 51, 120, 235–36, 237; rejected by Steno, 43, 61–62, 164, 247–48; and soul, 120–21; theory of Earth's formation, 236; theory of muscle contraction, 223; on vacuum, 11, 41, 61, 137; work on brain, 219; work on respiration, 148. *See also* Cartesianism
Deusing, Anton, 86n122, 115n101, 171, 189, 275
diagrams, 125; in anatomy, 103n34, 151f, 190, 224–25; in geology, 234–39
digestion, 45, 47, 50, 117, 233
disciplinary boundaries: of anatomy, 12–13; anatomy and mathematics, 94–95, 104–5; anatomy and mechanics, 99–100; experimental programs across disciplines, 161; mathematics and nature, 1, 14; modern, 3–4, 12; and salons, 154; Steno's interdisciplinary methods, 134, 137

disputes, 8, 16, 63, 64, 84, 87; with Blasius, 94, 122
dissection techniques: epistemic value, 60; multiple dissections for certainty, 76–77, 82; narratives of, 48–49, 56, 64, 78–80; as novel practice, 46; semi-public, 47–48; by young Steno, 48–49
dissections, 270, 279, 280; in Copenhagen, 106–8; epistemic value, 145; influence on Graindorge, 144–46; narratives of, 145; in Paris, 140–42; semi-public, 142; of shark, 207–9, 210–12, 214, 216, 218, 227, 228; by Steno, 140, 142, 144, 147, 154, 156, 205; Steno's skills in, 3, 62, 94–96, 100–102, 106–7, 145, 207, 212, 214, 216, 218, 227–28; techniques, 96–98, 100–102, 106–7, 145, 75–80; textbook of, 46. See also observations; vivisection
Dominici, Stefano, 209
Drexel, Jeremias, 34–35, 37n90, 39n99, 52n172
ducts: salivary, 16, 94–98, 100–101, 106–7, 144; thoracic, 47, 56–59, 137. See also salivary duct; thoracic duct
ductus blasianus, 87, 89f. See also salivary duct
ductus stenonianus, 63, 67–68, 87, 90n134. See also salivary duct
ductus whartonianus, 66, 81

Earth, strata, 3, 218, 234, 236, 255, 261, 284. See Steno's laws
Earth history. See history of the Earth
elasticity of the air, 12, 58, 150
Elba, island of, 206
Elementorum myologiæ specimen: on fossils, 211–12, 218–19, 226–28, 234; imprimatur of, 210–11, 228; manuscript of, 211–12; on muscle contraction, 210–11, 223–25, 234; on ovaries, 214–15, 215, 239; publication of, 211–12, 216, 239; on shark dissection, 207–8, 210–12, 218
Elsevier, 74
empiricism, 65, 77, 87, 259. See also epistemology; observations; reasoning

encyclopedism, 9, 11
Enden, Franciscus van den, 246
Epicureanism, 38, 61, 122–23
epistemic genre, 13, 16, 54, 55, 56, 169, 194–96; definition of, 54; *observationes*, 96, 102–3, 117–19. See also *observationes*
epistemology, 4, 42, 108–19, 122, 271. See also observations; probabilism; reasoning
Erasistratus, 119
Erculiani, Camilla, 229n101
Eucharist, 255–56. See also Corpus Christi, Feast of
Euclid, 23, 39, 190, 192, 225, 234
experimental methods, 135, 137, 147–54. See also physico-mathematics
experimental philosophy, 167, 174
Eyssonius, 70

Faber, Johannes, 250
Fabri, Honoré, 183, 185–88, 199, 213n34, 250, 275
Fabricius ab Acquapendente, Hieronymus, 77–78, 147–48, 270
Falloppio, Gabriele, 45n133, 85n119, 228
Favereau, Jacques, 39n99
Ferdinand II de' Medici, 2, 167, 174, 199–201, 205, 207, 230
fibers: of heart, 212; and history of Earth, 233; and mathematics, 23, 104–6, 190–94, 224, 257, 273; and mechanics, 127, 131; of muscles, 144–45, 156, 175, 178, 183, 277
final causes, 119–22
Finch, John, 251
Findlen, Paula, 228
Fiorentini, Francesco Maria, 253
Flanders, 6, 130–31, 246
Flavia del Nero, Maria, 276, 280; account of Steno's conversion, 242, 262n132, 265; spiritual influence on Steno, 265–66
Florence: Accademia del Cimento in, 208–10, 231, 239; conversion in, 22, 44; court culture in, 166–206; mathematical legacy, 94; Medici court in, 208–10, 231, 239, 251–52; National

Library, 19; scientific life in, 167–68; Steno's arrival at, 2, 5f, 162; Steno's attachment to, 2–3; Steno's conversion in, 241, 253–54, 261–62; Steno's fossil research, 17; Steno's later work in, 94, 125; Steno's life in, 251–52; Steno's move to, 142, 162–63; Steno's stay in, 170–72, 176–78, 199–201

fluids: in anatomy, 232–33; in Earth history, 231–33

focused interdisciplinarity, 15–16, 18, 135, 210, 240, 271, 273; in Borel and Pecquet, 16; definition of, 10–12; and religion, 243; in Steno's work, 95–96, 123–25

fossils, origins of, 273; comparative method for, 216–19; and Earth history, 216–19, 226–30, 234–35; in *Metallotheca Vaticana*, 216–18; and shark's teeth, 207–9, 211, 212, 216–19, 217, 220, 227–29; Steno's conjectures on, 228–30; Steno's early ideas on, 166, 196, 198, 206; Steno's *historia* of, 218–19, 226–28

Fracassati, Carlo, 176, 278

Frank, Robert, 13

Frénicle de Bessy, Bernard, 139

friendships: with Borch, 63, 67–68, 95, 111, 117, 120, 127–28, 137–40, 154, 169–71; with Dati, 172, 176, 200, 204; with Gradi, 174, 182, 198; and itinerary, 7, 9; with Lavinia Cenami Arnolfini, 253–54, 255, 257, 261–62, 263, 266; with Leibniz, 2, 6; with Magalotti, 168, 176–78, 200; with Maria Flavia del Nero, 265–66, 267; with Redi, 169–70, 176, 200, 205, 243, 251–52, 253, 267; and scientific collaboration, 8, 16, 95, 109, 111, 120, 124–25, 127–28, 135, 147, 154, 168, 170, 199–200; with Spinoza, 109, 127; and Steno's career, 168, 170, 199–200; and Steno's conversion, 242, 243, 251–52, 253–54, 255, 256, 259, 261–62, 263, 264, 265–66, 267; with Swammerdam, 135, 147–54, 157–58; with Viviani, 168, 194–96, 200–201

Fuiren, Henrik, 37n91

Galen, 15, 105, 119–21, 123, 150, 155, 170, 175, 213–14; anatomy, 44–45, 47, 55; medicine, 55; perspiration, 30; theories of animal spirits, 155; theories of respiration, 150

Galilei, Galileo. See Galileo

Galileo, 19, 21, 24, 38, 250; *Dialogue on the Two Chief World Systems*, 251; historical significance, 1, 167, 199; influence on Steno, 2, 18, 94, 167, 199; *Two New Sciences*, 2

Gassendi, Pierre, 60, 232; influence on Steno, 15, 95, 119, 122–23, 134; mechanical philosophy, 41–42, 60–61, 122–23; at Mersenne's meetings, 137; notes on, 38–42, 54, 60; and vacuum, 11

Genoa, 248

geological research: Steno's work in, 166, 198, 206

geology, 273, 278; and anatomy, 13, 208, 210, 212, 216, 218, 225, 227, 230, 234, 239; discipline of, 4n11; founding, 1; modern, 3; Steno's contribution to, 1, 3, 207, 208, 209, 210, 212, 218, 219, 225, 227, 228, 230, 231, 232, 234, 235, 239. *See also* history of the Earth; Steno's laws

geometrical models, 166, 192–94

geometry, 14, 22–24; in anatomy, 57, 104–5, 123–25; certitude of, 11, 24; Galileo's use of, 18; geometrical model, 16; in medicine, 59–60; Steno's work in, 4

Germany, 2, 172, 250, 266

Gersdorff, Joachim, 37

Gesner, Conrad, 226

Ghent, 131

glands: and chymical experiments, 116–17; *historia* of, 16; lacrimal, 99–102; maxillary, 66, 75, 81–82, 96–100, 124; mechanical understanding, 4, 99–100; of the mouth, 75–82; parotid, 63, 66, 75–77, 81–82, 96–98, 100–101; salivary, 96–98, 100–101, 144, 146; skin, 213–14; Steno's research on, 212–14, 219; studies of, 75–82, 96–102, 116–17, 156

Glauber, Johannes, 128

Glisson, Francis, 49
glossopetrae. *See* tongue stones
God, 3, 32n61, 133, 183, 245, 247, 254, 280; and anatomy, 132, 268, 277; grace of, 258–62; prayers to, 40, 43–44; servants of, 264; in Steno's mechanical philosophy, 119–22
goldsmiths, 2, 8, 16, 20–21, 26–278, 234
Golius, Jacob, 97, 124–25, 131, 172
Gómez Pereira, Antoniana Margarita, 112
Gradi, Stefano, 174n45, 180, 184–85, 187, 198, 276
Grafton, Anthony, 22, 55
Graindorge, André, 136, 143–46, 249
Guerrini, Anita, 12

Habsburg court, 270
Hahn, Roger, 135
Halle, 206
Hamburg, 1n3, 2, 277, 280
handicraft skills. *See* artisanal skills
Hannover, 2, 6, 268, 277, 280
Harvey, William, 45–47, 50n166, 56–57, 82, 86, 213–14; broad reading, 15; and circulation, 102, 147, 150, 152; *De motu cordis,* 103, 175; and *historia,* 78; influence on Steno, 102–3, 105, 147, 175, 177; and life in blood, 49n157; and observations, 76, 147; on respiration, 147, 150; as traveling scholar, 6
heart, 4, 103–4, 139, 156, 212–13
Hedwig, Dorothea, 266
Hilarius, Georgius, 23, 41, 124, 275
Hildesheim, 266
Hippocrates, 155
historia: in anatomy, 212, 218–19; in Earth history, 218–19, 226–28; and Medici court, 209–10. See also *historia anatomica*
historia anatomica: epistemic value, 77–82, 92
history of the Earth, 4, 17, 118; analogies in, 216–19, 231–34; and anatomy, 208, 210, 212, 216–19, 225, 227, 230, 234, 239; and chymistry, 231–32, 233, 239; comparative method in, 216–19, 234; diagrams in, 235, 236, 237, 238, 239; and fossils, 216–19, 226–30, 234, 235; and interdisciplinary methods, 212, 222, 225, 230–31, 239; learned from strata, 3; and mathematics, 234, 239; and search for certainty, 218, 219, 226–31, 239; Steno's contributions to, 207–10, 216–39; Steno's interest, 9. *See also* geology
Hoboken, Nicolaus, 86–87, 89–90
Holy Office, 248, 252, 262
Holy Trinity, 259
Homberg, Wilhelm, 141
Hooke, Robert, 148
Horace, 85
hospitals, 77n80, 81, 201, 272
hot springs, 70, 72, 227
humanist tradition, 12, 15, 39, 64, 85, 92, 119; at Cathedral School, 21–22, 34; and religion, 243, 256; and Steno's education, 37, 64, 95, 119. *See also* liberal arts; reading; textual techniques; writing
Huygens, Christiaan, 6, 24, 125, 160, 162, 164, 173, 269; at Montmor's meetings, 138–39
Huygens, Constantin, 125
hydrocephalus, 279
hydrostatics, 4, 11, 14, 56, 124, 222, 271

iatromechanics, 273
Immaculate Conception, Feast of, 241
imprimatur, 183n80, 190n121, 210–11, 230
Innsbruck, 205–6, 279
Inquisition, 109, 251
interdisciplinary methods: in anatomy, 12, 51, 57–58, 94–96, 98–100, 102–6, 118–19, 135, 212, 222–26, 232–34, 239; in Earth history, 212, 222, 225, 230–31, 239; and epistemology, 15; focused, 10–12, 18, 95–96, 123–25, 135; and mobility, 7; research, 271, 273, 275; and search for certainty, 222, 230–31, 239; significance of, 4, 135. *See also* disciplinary boundaries; focused interdisciplinarity; polymathy
intersection of disciplines, 4, 5, 9–12, 94, 95, 99, 116. *See also* interdisciplinary methods

Isidore of Seville, 227
Issy, 145

Jesuits: colleges, 8, 130–31, 250; conversion efforts, 245, 249, 256, 262; former, 128, 246; mathematicians, 11, 14, 24–26, 126, 130–31, 183, 186; scholars, 34
John Paul II, Pope, 3, 243
Jonston, Johannes, 226
Journal des sçavans, 141, 269
journals, scientific. *See* scientific journals

Kallenbach, Regina, 22n11, 263
Kaplan, Benjamin, 246
Kardel, Troels, 273, 279
Kerckring, Theodor, 3n5, 128–29, 275
Kepler, Johannes, 225
Kircher, Athanasius, 9–10, 212, 227, 228, 229; on fossils, 228; influence on Steno, 124, 186, 250; *Magnes sive de arte magnetica*, 123–24; and mathematics, 24–26, 39, 42; *Mundus subterraneus*, 229; Steno studies, 38–40, 51, 56, 61; and Steno's work, 228–29
knife, anatomical, 46, 48, 85, 94, 135, 154, 208, 272
knowledge: anatomical, 16; breadth, 4, 9; cross-disciplinary, 4, 11–12; rigorous, 15; universal, 10. *See also* intersection of disciplines
Krag, Otto, 65

La Barre, Jean-Baptiste de, 249, 265
La Peyrère, Isaac, 229
laboratories, 20, 26, 37–38, 128, 162, 202, 275. *See also* artisanal workshops
lacteals, 45, 47, 56–57, 59, 82, 115–16
Lærke, Mogens, 274
Langelott, Joel, 74
Laurentian Library, 256, 271
laws of nature, 3, 4, 13, 18, 277. *See also* Steno's laws
Le Blanc, Vincent, 33, 43n128
Leibniz, Gottfried Wilhelm, 2, 6, 9–10, 109, 268, 274, 277
Leiden, 24, 46, 57, 272, 275; anatomical research in, 94–95, 97, 100, 103, 106, 109–11, 115, 124, 128, 130–31; Blasius controversy in, 73–74, 84, 91–92; disputations, 73–74; intellectual community in, 109–10, 112; intellectual environment, 168, 172, 203; mathematical studies in, 124–25, 131; patrons in, 9; Steno's education in, 63–64, 66–67; Steno's studies in, 2, 7, 63–64, 66–67, 73–74, 84, 91–92, 94–95, 96, 124, 135, 140, 154, 168, 172, 203, 222, 239; Steno's time in, 94–95, 97, 109–10; Steno's works in, 7; Swammerdam's studies in, 154; University of, 246–47. *See also* Low Countries; Netherlands; University of Leiden
Leonarts, Joannes, 87n127
Leopoldo de' Medici, 208, 269; correspondence with scholars, 170, 172, 174–75, 199–201, 203–4; distribution of Steno's book, 198; founder of Accademia del Cimento, 166–68; patronage of Steno, 174, 199–201
Leurechon, Jean, 25–26
Libavius, Andreas, 49–50
liberal arts, 10, 16, 20, 22. *See also* humanist tradition
ligatures, 45, 115–16, 272. *See also* dissections
Lister, Martin, 165
liver, 45, 47, 49, 57, 86, 115, 117
Livorno, 176, 206–7, 251, 255
London, 68
Louis XIV, 1, 2, 135, 138, 160, 162
Louvain, 248
Low Countries: intellectual environment, 95–96, 108, 124, 127; Steno's experiences in, 95–96, 108, 124, 127–29; Steno's visit, 16. *See also* Netherlands
Lower, Richard, 148
Lucca, 196, 229–30, 253–54
Luther, Martin, 244, 258
Lutheranism: in Denmark, 241, 245, 258; Steno's abandonment of, 2, 8, 258; Steno's upbringing in, 241, 245
Lyell, Charles, 239

lymphatic system, 47, 144, 175
Lyon, 170
Lyser, Michael, 46, 72–73

macrocosm and microcosm, analogy of, 225, 232
Magalotti, Lorenzo, 208, 231, 267, 270; contributions to Accademia del Cimento, 204; on dissolution of Accademia del Cimento, 204; friendship with Steno, 168, 176–78, 200; *Saggi di naturali esperienze*, 167, 202–3, 209, 231; secretary of Accademia del Cimento, 167
Magliabechi, Antonio: correspondence with Steno, 175–76; librarian, 171, 175–76
Malpighi, Marcello, 15, 105, 110, 121, 126, 171, 182, 213; *De pulmonibus*, 126; meeting with Steno in Rome, 182; on respiration, 147–48; and Steno, 16, 147
Malta, 226, 230
manners, 9, 134, 144, 163–65, 168, 189, 252–53. *See also* civility; modesty; virtues
manuscripts, 196, 211–12, 216–18
Marcgrave, Georg, 107–8
Marguerite of Lorraine, 142
mathematicians: and anatomists, 56–60, 105–8, 108, 126, 153, 275, 276; Jesuit, 11, 14, 24–26, 131, 186; and mechanics, 125–26; and observations, 123–25; and philosophers, 136–38, 159; Steno as, 192, 198, 252, 271
mathematics, 1, 271, 272, 275; in anatomy, 56–59, 96–98, 104–5, 123–25, 134, 137, 153, 166, 190, 192–94; and certainty, 104–5, 118–19; in Earth history, 234, 239; mixed, 14, 26, 99, 123–24; pure, 14, 24, 96, 123; and rigor, 28, 31; in science, 6, 13–14, 24–25, 134, 137, 153. *See also* mathematicians; physico-mathematics
mathematization, 6, 12–15, 249, 271–72, 275
maxillary gland, 66, 75, 81–83, 96–100, 124. *See also* glands
measurements, 31; in anatomy, 96–98, 102; and chymistry, 28–30, 62;
gland weights, 97t; and Harvey, 45; quantification of blood flow, 102; Steno's interest in, 20, 26–31, 50, 57–58; use in anatomical arguments, 96–98, 102; of weights, 27–28, 30, 58, 96–97. *See also* quantification; Santorio, Santorio
mechanical analogies: blood flow, 150–52; in muscle studies, 154–55; since Galen, 15; in Steno's research, 135, 150–52, 154
mechanical devices, 25, 42, 51, 123–24, 212, 272
mechanical philosophy, 1, 6, 60–61, 122–23, 269, 273, 283
mechanics, 222, 225; in anatomy, 16, 99–100, 106, 119, 135, 151; and certainty, 11, 125; in Earth history, 232; and Galileo, 9; and machines, 99–100, 106, 125; as physico-mathematics, 14; Steno's interest in, 4, 8, 135, 151
mechanisms, 95, 119, 122, 153, 155, 177, 234, 261; definition of, 11–12
Medici court, 16, 208, 209, 210, 228, 239; censorship at, 228–29; influence on Steno's work, 209–10, 239; intellectual environment, 166–206; patronage of Steno, 251–52; religious tolerance at, 251; Steno's activities at, 166–206
medicine, 20, 22, 56, 59, 76, 96, 112, 124, 128, 172, 256; Galenic, 44; quantitative, 31, 59; Steno's practice of, 15; Steno's studies of, 2, 21, 23, 31–34
Mentel, Jacques, 33
Mercati, Michele, 216, 218, 227, 228; on fossils, 216–18; *Metallotheca vaticana*, 216–18, 227
Mercenne, Pierre de, 33, 57
Mersenne, Marin, 68, 136–37, 160, 183n79
Meteorology (Aristotle), 227, 229
microscopes, 27–29, 38, 56, 61, 126–27, 147
Milan, 169, 229
milky vessels. *See* lacteals
mobility: and interdisciplinarity, 7; and itineracy, 18; modern, 6. *See also* pilgrimages
models, scientific, 12, 17; in anatomy, 192–94, 210, 212, 222–26, 232–34;

in Earth history, 231–34; influence of Euclid and Galileo, 192, 199; mathematical, 4, 94, 104–5, 123–25, 223–25, 234; mechanical, 99–100, 106, 119, 178, 192–97
modesty, 9, 75, 90–93, 185, 189, 267
Montmor, Henri Louis de, 138, 158; academy of, 143, 173
Montpellier, 142, 165, 209
Morel, Claude, 142–43, 161
Moulin, Pierre du, 130
muscles: anatomy of, 135, 139, 141, 144; contraction of, 16, 106, 154–55, 210–11, 223–25, 234; intercostal, 104–5, 139; mathematical model of contraction, 104, 210, 212, 223–25, 234; model of motion, 4, 14, 16. *See also* fibers
myology. *See* muscles

Naples, 182, 201
National Library of Florence, 19, 172, 278
natural history, 6, 218, 226, 239; and anatomy, 226; and fossils, 226–28; Steno's familiarity with, 226–28
natural philosophers, 2, 11, 134, 136, 159
natural philosophy, 14, 39–40, 44, 51, 60–61, 178, 183, 277; Cartesian, 161; and de Raey, 112–13; experimental, 167–68
Netherlands, 6, 8, 157, 164; cultural context, 127–29; intellectual environment, 95, 108, 124, 127; religious diversity in, 245–46; Steno in, 16; Steno's experiences in, 95, 108, 124, 127–29, 245–47, 258; Steno's studies in, 143. *See also* Low Countries
nerves, 51, 96–97, 100, 121, 154–56, 208, 212
nervous system, 156n119, 219
new sciences, 2, 13, 21, 39, 142
Newton, Isaac, 9, 11–13, 18
notebooks, 16, 20, 31, 33, 43, 50, 85; and excerpts, 51–54, 60–61; laboratory, 252, 275; management of, 34–37
notes: kinds of, 19; marginalia, 25, 35, 36f, 53f, 54, 54n172, 58, 60–61; personal, of Steno, 35, 36f, 41, 42, 50, 58, 61. *See also* "Chaos" manuscript; textual techniques

note-taking, 20, 85, 87, 92; excerpting, 25–27, 30, 38, 51; intensive, 56; methods of, 34–36, 38, 42, 54–55; and observations, 55; and polymaths, 10; reasons for, 33–34; and Steno's overall work, 13. *See also* commonplace books; notebooks; *observationes*

observationes, 30; *anatomicæ*, 13, 16, 56, 84–86, 91, 169, 194–96, 274–75; *medicæ*, 52–54, 55–56, 85; and note-taking, 55
observations: in dissections, 141, 142, 154; in Earth history, 278; emphasis on, 141, 142, 144, 154; methods, 75–80; significance, 67, 75–83, 84–86, 91–92, 277; techniques, 75–80. *See also* certainty, search for; reasoning
Oldenburg, Henry, 6, 68, 199
ovaries, 4, 214–16, 239

Padua, 77, 169. *See also* University of Padua
Palazzo Vecchio, 201–2
papal infallibility, 241, 258. *See also* Catholic Church
parallelepiped model, 223–24
Paris: Catholicism in, 248–49; intellectual environment, 168, 170, 172–73, 199, 202; School of Medicine, 140–42; scientific academies in, 134, 136–39; Steno's arrival, 2, 7, 9, 16, 134; Steno's dissections at, 141–42; Steno's time in, 168, 170, 172–73, 199, 202; Steno's travels to, 249, 264
Park, Katharine, 12
Parker, Charles, 246
particles in motion. *See* atoms
Pascal, Blaise, 58, 137, 264
patronage in science: impact on Steno's career, 166, 170, 174–75, 199–201; by Louis XIV, 135, 162; Medici family's support, 166–67, 172–73, 199; Steno's views on, 162
patrons, 2, 7, 9, 15, 208, 210, 228, 230
Paulli, Jakob, 66n18, 69n35
Paulli, Simon, 38, 48n151, 124n162, 132n216, 226
Peace of Westphalia, 2

Pecquet, Jean, 25, 46, 82, 134, 137, 150, 152, 163; *Experimenta nova anatomica*, 47, 53f, 54, 56–60, 86, 137, 175; influence on Steno, 76, 134; member of Académie Royale des Sciences, 163
Peiresc, Nicolas-Claude Fabri de, 6
peregrinatio medica: Steno's travels as, 169, 180, 276
Perriquet, Marie, 264
perspiration, 30–31; anatomy of, 213–14
Petit, Pierre, 139, 173
Phillipeaux, 161
Philosophical Transactions, 49, 148, 205, 269, 278
physico-mathematics: in anatomy, 222; in Earth history, 232; in Paris, 137; Steno and Swammerdam's use of, 135, 147–54. *See also* experimental methods
physiology: definition of, 4n9. *See also* anatomy
pilgrimages, 18; medical, 7–8, 32–33, 46, 62; religious, 8
Pisa, 174–76, 196, 251, 272, 275
Piso, Willem, 85n117, 106–8, 132
Pistoia, 253
Plater, Felix, 111
poems, 69–70
polyhistors, 9
polymaths: definition of, 9–10; golden age of, 4; and note-taking, 10. *See also* humanist tradition
polymathy, 9–10; modern views, 4. *See also* intersection of disciplines
Pomata, Gianna, 13, 54, 274
priority disputes. *See* disputes
probabilism, 41–42, 118, 122, 229, 261. *See also* epistemology; reasoning
probes, 66, 78–80, 116–18, 272
Protestantism, 243

quantification, 26, 28–31, 50, 81, 96–102, 132, 272. *See also* mathematics
Questiers, Catharina, 129–30, 133, 246, 264

Rantzau, Maria Elisabeth von, 249, 264–66
Ray, John, 165

reading, 4, 10, 12, 14, 15, 66, 72, 73f, 78, 85, 172; and note-taking, 87, 194–96; Steno's practices, 94, 110, 112. *See also* humanist tradition
reasoning, 12, 113, 118, 155, 259, 261; analogical, 139, 154, 277; quantitative, 26, 31, 96; and the senses, 82–83, 86, 113–15, 119, 219. *See also* epistemology; probabilism
Redi, Francesco, 18, 143, 169–70, 176, 200, 204–5, 252–53, 276
Regius, Henricus, 34n76, 51, 61, 120
religious conversion: in early modern Europe, 244–45; role of women in, 253–54, 261–66; Steno's reasons for, 257–60
reproductive system, female: Steno's research on, 214–16, 239; and van Horne's work, 216. *See also* ovaries
Republic of Letters: Bartholin's influence, 68, 74, 84, 91; correspondence between Florence and Paris, 172–74, 199; role in controversy, 64, 68, 74, 84, 86, 91; and Steno, 9, 15, 64, 74, 84, 92, 95, 108, 141
respiration: research by Steno and Swammerdam, 135, 147–54; theories of, 147–50. *See also* blood circulation
Ricci, Michelangelo, 182, 199, 267, 275
Riolan, Jean, 56–57
Roberval, Gilles Personne de, 25n28, 58, 59f, 60, 137, 160
Rohault, Jacques, 160–61
Rome, 128, 132, 275, 276; intellectual community in, 182–84, 198; papal patronage in, 250; Steno's visit to, 169, 176, 182–84, 198, 250, 254–55
Rondelet, Guillaume, 226
Rotterdam, 130
Royal Society of London, 148, 165, 199, 208, 269
Rubens, Peter Paul, 131
Rucellai, Orazio, 267
Rudbeck, Olaf, 72n54
Rudwick, Martin, 207, 208

Saint-Côme, confraternity, 142
salivary duct: anatomy, 75–77, 81–82; discovery of, 63–68, 75–77, 83–84,

87–90, 94–96; *ductus stenonianus*, 63, 67–68; studies of, 96–98, 100–101, 144. *See also* ducts
San Gaetano, Church, 280
Santa Maria del Fiore (cathedral), 268
Santa Maria Nuova (hospital), 201–2
Santissima Annunziata (church), 266, 280
Santorio, Santorio, 25, 30–31, 50, 132, 213–14
Savignani, Emilio, 256, 262
Schneider, Conrad, 111
Schott, Gaspar, 25–26, 126
Schuyl, Florentius, 112
Schwaz, 206
Schwenter, Daniel, 25–26
science: definition of, 1n2; modern, 4, 12
scientific academies, 2, 6, 8–9, 16; in Florence, 167, 174, 203–5; in Paris, 174. *See also* Académie de Physique de Caen; Académie Royale des Sciences
scientific controversies: Blasius controversy, 63–65, 68–70, 73–75; responses and impact, 83–84, 87–91. *See also* Blasius controversy
scientific journals, 8, 140–41, 269, 274
scientific research: and religious conversion, 243, 257–60
Scilla, Agostino, 235
Seger, Georg, 74
Segni, Alessandro, 170, 202
Sennert, Daniel, 34–35, 40
Serrarius, Petrus, 245–46
Settala, Manfredo, 229
shark: dissection of, 207–9, 210–12, 214, 216, 218, 227–28; and fossils, 207–9, 211–12, 216–19, 227–29; head of, 216, 217; significance in stratigraphy, 218–19, 228–29; and Steno's geological theories, 207–9, 218–19, 227–29; teeth of, 207–9, 211–12, 216–19, 217f, 220f, 227–29; tongue stones comparison, 211–12, 216–19, 227–29
Sibbald, Robert, 67
Siena, 208
skin glands, 213–14. *See also* glands
Skippon, Philipp, 165
social contexts, 4, 6–9

sola scriptura, 258
Sonnenthal, Baron, 128
Sorbière, Samuel, 57, 59, 60, 130, 137–38, 158, 161, 249
souls, 40, 91, 109, 130, 175, 263; animals, 95, 112; and body, 268, 280; rational, 119–22, 160; tranquility of, 253, 255, 258–59, 261. *See also* All Souls, Feast of; animals
Spinoza, Baruch, 8, 109, 127, 245, 247–48
Steno, Nicolaus: *Apologia prodromus*, 87–93; artisanal background, 26–27, 271, 272; beatification, 3, 243; biography, 2–3, 6, 94–95, 106, 132; as bishop, 2; *De musculis et glandulis*, 102–6, 132, 139, 141, 154, 156, 171, 274; *De propria conversione*, 242, 259–60; *De solido*, 18, 196, 269, 273, 277, 278; *Defensio epistolae*, 248; devotion to, 3; *Discours sur l'Anatomie du Cerveau*, 135, 142, 156, 159, 164; *Disputatio anatomica*, 73–74; *Disputatio physica de thermis*, 70–71; education of, 65, 71–72, 256; family of, 263–64; founder of geology, 1, 166, 207; influence on Académie de Physique de Caen, 143–46; interactions with Medici family, 174–75, 199–201; interactions with women, 253–54, 261–66; itinerant career, 7, 9, 18; knowledge breadth, 4, 9, 10, 16; *Liber de solido*, 230, 279; prayers of, 19, 21, 40, 43, 44; as priest, 2, 6, 242, 257, 266, 268–69, 278, 280; relationship with Accademia del Cimento, 203–5; religious life, 269, 270, 274, 276, 277, 280; as researcher only, 15; tomb, 3; travel preparation, 32–33, 39, 43, 62; travels of, 5f, 106, 127–31, 162, 208–10, 212–35, 245–47, 249–51, 254–55; writing style, 78–80, 85–86, 91–92. *See also Elementorum myologiæ specimen*
Steno's laws, 3, 207; law of lateral continuity, 235; law of original horizontality, 218; law of superposition, 218. *See also* geology; history of Earth
Stevin, Simon, 99, 125, 127
stratigraphy, principles of, 207, 218
Stuart, David, 130

Svane, Hans, 65
Swammerdam, Jan, 16, 28, 214, 216, 247–48, 275, 277; anatomical research, 113, 129, 133; Cartesian influences on, 113, 157; collaboration with Steno, 113, 129, 133, 135, 147–54; microscopic techniques, 113, 133, 147; research on respiration, 147–50, 155, 157
Sweden: military alliance, 1; throne of, 249
Sylvius, Franciscus, 46, 75, 81, 91; influence on Steno, 96, 98–99, 103, 111, 128; Steno's professor, 66–67, 81

teleology, 120–21. *See also* final causes
telescopes, 27–28, 31, 38, 56, 127
testicles, female. *See* ovaries
textual techniques, 12–13; manuscript editing, 196; note-taking, 194–96; in science, 87–90, 96, 102–3, 117–19; Steno's use of, 64, 87–88. *See also* note-taking; reading; writing
theology, 269, 274, 280; controversial, 128, 167, 229n101; in Denmark, 22–23, 39; in Italy, 183; in the Netherlands, 130; Steno's works, 89, 243–45, 256; and women, 265, 276
Thévenot, Melchisédech, 213, 264, 270, 277; academy of, 138–40, 143–48, 158, 170, 204, 270; correspondence with Steno, 161–64, 213; and Florence, 173, 186, 188; patron of Steno and Swammerdam, 157–58, 173, 248, 264, 277
Thirty Years War, 1, 2, 264
thoracic duct, 47, 56–59, 67, 82, 115–16, 137, 175, 233. *See also* lymphatic system
tongue stones, 209, 211–12, 217f, 218, 220f, 226–27, 230, 256. *See also* fossils, origins of
Trutwin, Tilmann, 251
Turin, 171
Tuscany, 167, 204, 207, 241; dissections in, 175, 178, 214; and Earth history, 209, 255, 278

University of Copenhagen, 245; anatomical research at, 205;
anatomical theater, 46; Bartholin's influence, 31–32, 47; curriculum, 23–24, 31; faculty, 31–32; mathematics at, 23–24; medical training, 21, 31–32; Steno's studies, 2, 16, 19, 21, 23, 24, 31, 32, 47, 170, 205
University of Leiden, 272; chymistry at, 223; intellectual environment, 110, 124–25; Steno's medical degree, 2; Steno's studies at, 63, 66–67, 73–74, 84, 91–92, 94–95, 96, 124
University of Padua, 6, 30, 270
University of Paris, 33, 56, 272
University of Pisa, 174–75, 272
Untzer, Matthias, 27, 39, 52f
Utrecht, 86

vacuum: and Aristotelians, 168; and Cartesians, 161; experiments on, 56, 58, 137, 148, 203, 269; and Gassendi, 11; and Steno, 40–41, 61, 150
Vallisneri, Antonio, 229n101
valves: in lymphatics, 57, 114, 116, 118; in veins, 57n190
van den Broecke, Petrus, 174
van Helmont, Johannes, 225
van Horne, Johannes, 91, 111–12, 114n99, 115n101, 216; Steno's professor, 66–68, 73–74, 124n162
van Schooten, Frans, 125
Varen, Bernhard, 26, 33, 41
Vatican, 216, 250. *See also* Vatican Library
Vatican Library, 180, 184, 187, 276
Vesalius, Andreas, 44, 105, 270
Vienna, 270
Villebressieu, Étienne de, 160–61
Vincent de Paul, 264
Virgin Mary, 265–66
virtues, 9, 90–91, 93, 129, 185, 259, 263–64, 268. *See also* civility; manners; modesty
vital spirits, 50, 156. *See also* animal spirits
Viviani, Vincenzo: collaboration with Steno, 14, 128, 169, 194–96, 199–203, 225; and Erasmus Bartholin, 24, 172; friendship with Steno,

16, 168, 200–201, 254, 267, 275; mathematician, 188–89, 192, 200–201
vivisection, 45–46, 48–49, 60, 67, 128, 147n80, 150, 154
Volterra, 206
Vorstius, Adolphus, 91

Walaeus, Johannes, 46, 104
Weinberg, Joanna, 55
Westfall, Richard, 273
Westphalia, Peace of. *See* Peace of Westphalia
Wharton, Thomas, 65–66, 81, 96–97, 100–101, 119

Willis, Thomas, 135, 159, 219
Witsen, Nicolaus, 127–28, 133
women: influence on Steno's conversion, 253–54, 261–66; spiritual leadership of, 263–66
women in science, 276
workshops. *See* artisanal workshops
Worm, Ole, 7, 31–32, 226
writing, 7, 14; editorial choices, 85–86, 169, 194–96; narrative style, 78–80; publications by Steno, 87–90

Zaandam, 127

www.ingramcontent.com/pod-product-compliance
Lightning Source LLC
Chambersburg PA
CBHW022030290426
44109CB00014B/814